Black Beaches and Bayous

The BP Deepwater Horizon Oil Spill Disaster

Edited by

Lisa A. Eargle and Ashraf Esmail

University Press of America,® Inc.
Lanham · Boulder · New York · Toronto · Plymouth, UK

4501 Forbes Boulevard
Suite 200
Lanham, Maryland 20706
UPA Acquisitions Department (301) 459-3366

10 Thornbury Road
Plymouth PL6 7PP
United Kingdom

Printed in the United States of America
British Library Cataloging in Publication Information Available

Library of Congress Control Number: 2012944508
ISBN: 978-0-7618-5978-9 (paperback : alk. paper)
eISBN: 978-0-7618-5979-6

™ The paper used in this publication meets the minimum requirements of American National Standard for Information Sciences—Permanence of Paper for Printed Library Materials, ANSI Z39.48-1992

We dedicate this book to the individuals, families, and communities along the Gulf Coast of the US who have experienced this tragedy and to those who are tirelessly working towards recovery.

CONTENTS

Figures vii
Tables viii
Foreword by Dr. George L. Amedee ix
Preface by Dr. Lisa A. Eargle and Dr. Ashraf M. Esmail xi
Acknowledgments xiii

Chapter One—The U.S. Oil Industry's Safety Record and the Need For More Domestic Oil Production 1
Jude Clemente

Chapter Two—Applying Technological Disaster Process Models to the BP Deepwater Horizon Oil Spill Disaster 21
Lisa A. Eargle, Ashraf M. Esmail, Jas M. Sullivan, and Shyamal K. Das

Chapter Three—Beyond Petroleum in the Gulf of Mexico: The Characteristics and Consequences of Catastrophe 33
John Barnshaw and Lynn Letukas

Chapter Four—Corporate Catastrophes from UC Bhopal to BP Deepwater Horizon: Continuities in Causation, Corporate Negligence, and Crisis Management 53
Tomás Mac Sheoin and Stephen Zavestoski

Chapter Five—The Effects of Oil Spills on Ecological Systems 95
Jeffrey R. Wozniak

Chapter Six—The Gulf Oil Spill, Ecological Debt, and Environmental Justice in Louisiana: Lessons From Sociology 105
Timothy J. Haney

Chapter Seven—A New Geography of Trouble 119
Daina Cheyenne Harvey

Chapter Eight—Ecological Identity and Disaster Recovery in an Oil-Stained Landscape: Current and Future Policy Implications 135
DeMond Shondell Miller, Jason David Rivera, and Brandon Eric Fleming

Chapter Nine—The Crude Awakening: Gulf Coast Residents Reflect on the BP Oil Spill and the 2010 Hurricane Season 159
Michelle Meyer Lueck and Lori Peek

Chapter Ten—The Ninety-Day Storm: Mississippi Community Response to the Deepwater Horizon Oil Spill 181
Jason S. Gordon and A. E. Luloff

Chapter Eleven—Disaster Vulnerability: The Differential Impact of the Deepwater Horizon Disaster Among Alabama's Gulf Coast Residents 205
James Hawdon and John Ryan

Chapter Twelve—Disaster Phases, Structural Vulnerability and Crime 231
Kelly Frailing and Dee Wood Harper

Chapter Thirteen—Hazard, Outrage and Locality: An Analysis of Two Oil Spills 249
Amanda K. Goddard, Kenneth A. Lachlan and Patric R. Spence

Chapter Fourteen—Resisting Corporatism: Citizens Fight Back Against the BP Deepwater Horizon Oil Spill Disaster 263
Stan C. Weeber

Chapter Fifteen—Disaster Distrust: Risk Assessment, Citizen Science and Technolegal Debates in the BP Oil Spill 281
Sabrina McCormick

Chapter Sixteen—The President, the News, and the Oil Spill: An Examination of National and State Newspapers' Framing of Obama and His Administration's Response to the Deepwater Horizon Gulf Oil Spill 293
Jas M. Sullivan and Meghan S. Sanders

Editors 309
Contributors 311

FIGURES

Figure 1.1: If Offshore U.S. Oil is Constrained from Delivering as Projected by 2025 3
Figure 1.2: The Decline in the Volumes and Incidents of U.S. Offshore Oil Spills Above 100 Gallons (1974-2004) 4
Figure 1.3: Rising CAFE Standards: Americans Drive More and Use More Fuel 7
Figure 1.4: Constant Increases in U.S. Oil Demand, GDP, and Population (1990-2030) 9
Figure 1.5: Canada's Projected Expanding Oil Export Capacity (2007-2035) 12
Figure 1.6: Mexico's Dramatic Shift from Oil Exporter to Oil Importer (2007-2035) 13
Figure 3.1: Continuum of Exception Occasions 38
Figure 3.2: Conceptual Characteristics of Catastrophe 39
Figure 9.1: Location of Respondents to the Open-ended Survey Question 165
Figure 11.1: Hypothesized Relative Vulnerability of Affluent and Disenfranchised for Natural Disasters 220
Figure 11.2: Hypothesized Relative Vulnerability of Affluent and Disenfranchised for Technological Disasters 221
Figure 16.1: Praise and Blame Frames of Obama and Obama Administration by Newspaper Ideology 305

TABLES

Table 4.1: Fatal Offshore Incidents 64
Table 4.2: Safety Incidents Involving BP in England, Scotland and the USA–1999-2010 67
Table 9.1: Background Characteristics of Respondents 166
Table 10.1: Key Informant Characteristics 188
Table 11.1: Crosstabulation of Spill Negatively Affected Work Situation by Community 213
Table 11.2: Logistic Regression of if Spill Negatively Affected Work Situation 214
Table 11.3: Crosstabulation of Knowing Someone Who Moved Because of Spill by Community 214
Table 11.4: Logistic Regression of Knowing Someone Who Moved from Community due to Oil Spill 215
Table 11.5: Crosstabulation of Effect of Spill on Community Solidarity by Community 216
Table 11.6: Logistic Regression of Oil Spill Led to Corrosive Community 217
Table 11.7: Descriptive Statistics 224
Table 13.1: Appeals and Mentions of Hazard and Outrage by Location 258
Table 13.2: Number of appeals for Image Restoration by Location 258
Table 16.1: Summary of Oil Spill News Content 301
Table 16.2: Summary of Variables of Interest by Newspaper 302

FOREWORD

For residents living along the Gulf Coast Region, the BP Deep Water Horizon Gulf Oil Spill Disaster represented the third in a series major disasters of such enormous proportions to devastate the region within a five year period. The Katrina and Rita disasters were natural disasters aggravated by man-made blunders. The BP Deep Water Horizon Gulf Oil Spill was a man-made disaster that assaulted a naturally developed and ecologically balanced productive coast.

The examination of Katrina and Rita within the academic community focused its vast amount of analysis on the multiplicity of problems, issues, and challenges associated with this disaster viewed from a multi-disciplinary perspective while ensuring the inclusion of the wide range perspectives and voices of the full range of actors and stakeholders. The concerns of government, industry, small businesses, residents, environmentalist, the media, activists on all sides, and the wide spectrum of scientific and nonscientific advocates and experts are reflected in much of the scholarly work resulting from these first two natural disasters. Similarly, any in depth study of the BP Deep Water Horizon Gulf Oil Spill requires the same degree of scrutiny with contributions coming from a full range of scholars within the academic community while ensuring that the perspectives and voices covering the many actors and stakeholders impacted by this major disaster are also heard.

This book offers the kind of expanded treatment necessary for a proper examination of one of the most significant man- made disasters in recent times. In addition to providing a full account of the BP Deep Water Horizon Gulf Oil Spill disaster and an analysis of its full context within the science of disasters, this book provides a very interesting and rigorous examination of this disaster inclusive of social, psychological, economic, political, environmental, and energy policy impacts and ramifications. It meets the criteria for the kind of multi-disciplinary analysis that is essential to understanding the full meaning and implications of such a horrible disaster, assuming that fully understanding the meaning and implications of a disaster of this magnitude is possible. Just as important, the perspectives of nearly all of the key stakeholders are also presented in the book.

In offering this collection of essays, this book provides its readers the information and analyses essential to understanding the lessons to be learned from the past and for the present, and, God forbid if another disaster of this type should occur, a path for addressing the spills of the future.

George L. Amedee, Ph.D.
Professor of Political Science
Southern University at New Orleans

PREFACE

When the BP Deepwater Horizon oil well exploded on April 20, 2010, killing eleven workers and spilling millions of gallons of crude oil into the Gulf of Mexico, it was considered one of the worst technological disasters in American history. However, technological accidents—in particular oil spills—are relatively common event in America and other parts of the world (National Wildlife Federation, 2010). What makes this disaster different is the type of environment in which it took place—over a mile below the ocean's surface, where humans could not physically go, but relied strictly on technology to access the resources and to subsequently stop the spill (*Times-Picayune* Staff, 2010). As societies become increasingly dependent on technologies to access previously inaccessible resources, the risks involved in a potential disaster also increase (Vago, 1999). When these disasters do occur, the costs associated with clean-up and recovery operations are high and the impacts on people, communities, social institutions, and the natural environment are long-term and not fully known (Bell, 2004).

The objectives of this book, *Black Beaches and Bayous: The BP Deepwater Horizon Oil Spill Disaster*, are multifold. First, we want to further our understanding of how technological disasters are created and unfold. This includes examining the features common to this oil spill disaster with other types of technological disasters. For example, in the chapter, "From Union Carbide's 'Bhopal' to British Petroleum's Deepwater Horizon'," Sheoin and Zavestoki examine the similarities between two different process industry disasters.

Secondly, we want to document the impacts that these disasters have on the well-being of people, communities, social institutions, and the environment. Several authors in multiple chapters address the mental and physical health of residents after the spill. Haney in "The Gulf Oil Spill, Ecological Debt, and Environmental Justice in Louisiana," and Harvey in the "New Geography of Trouble," examine how the Oil Spill is another disaster in a series of negative events that have affected the residents of the Gulf Coast region. Miller, Rivera and Fleming examine how the Oil Spill has impacted Gulf Coast residents' sense of identity and connected to place. Sullivan and Sanders in "The President, the News, and the Oil Spill," examine the media's framing of the Obama Administration's response to the oil spill. Frailing and Harper examine how disasters affect crime rates. Wozniak in "Understanding The Effects of Oil Spills on Ecological Systems" examines the impact of the natural environment.

Third, we hope to stimulate further analysis and discussion on ways to prevent and/or mitigate similar disasters and disaster outcomes in the future. Weeber in "Resisting Corporatism," examines how community residents decided what were the issues surrounding the oil spill and how *they* would create recovery. Goddard, Lachlan and Spence in "Hazard, Outrage and Locality" discuss the issues involved in effective risk communication. Barnshaw and Letukas address differences between the concepts of disaster and catastrophe and how these determine what types of responses are most effective.

Fourth, we wanted to examine the Oil Spill using a variety of scholars from different academic discipline and sector backgrounds. Although several of the book's contributors have backgrounds in sociology, other contributors approach the Oil Spill topic from the lenses of energy and technology, communications, political science, and criminal justice, and are employed in settings ranging from independent researcher to academia to federal government agencies. By examining this topic from many different angles, we hope to develop a more holistic understanding of the disaster event and other issues surrounding it.

While no one book can address all of the issues surrounding a disaster or by itself prevent future events from occurring, we do believe it will facilitate further discussion of the event, which will lead to new understandings that can be applied in the fields of disaster mitigation and emergency management.

Lisa A. Eargle, Ph.D. and Ashraf M. Esmail, Ph.D.

REFERENCES

Bell, Michael. 2004. *An Invitation to Environmental Sociology. Second Edition.* Thousand Oaks., CA: Pine Forge.

National Wildlife Federation. 2010. *Assault on America: A Decade of Petroleum Company Disaster, Pollution, and Profit.* Reston, VA: National Wildlife Federation.

Times-Picayune Staff. 2010a. "How the Gulf of Mexico oil spill happened: a graphic presentation." *The Times-Picayune*, May 7.

Vago, Steven. 1999. *Social Change. Fourth Edition.* Upper Saddle River, NJ: Prentice Hall.

ACKNOWLEDGMENTS

The editors would like to thank all of the contributors to this volume, for their interest and hard work on such an important topic as the BP Deepwater Horizon Oil Spill disaster.

CHAPTER ONE
THE U.S. OIL INDUSTRY'S SAFETY RECORD AND THE NEED FOR MORE DOMESTIC OIL PRODUCTION

JUDE CLEMENTE

BACKGROUND

> *In terms of non-OPEC. . . we are expecting that in three, four years' time the production of conventional oil will come to a plateau, and start to decline. In terms of the global picture, assuming that OPEC will invest in a timely manner, global conventional oil can still continue, but we still expect that it will come around 2020 to a plateau as well, which is, of course, not good news from a global-oil-supply point of view,*
> Fatih Birol, Chief Economist at the International Energy Agency (IEA), 2009 (Monbiot, 2009)

In its latest forecast, *International Energy Outlook 2010*, the U.S. Energy Information Administration (EIA) projects that domestic oil (all liquids) consumption will increase by 8 percent by 2030. (U.S. Department of Energy, July 27, 2010) U.S. crude oil production has basically been sliced in half since peaking in 1970 at 9.7 million barrels per day (b/d), and foreign sources now constitute a growing 60 percent of the 20 to 21 million b/d that we consume. Meanwhile, global demand is rapidly outpacing supply and is expected to increase by nearly 25 percent by 2030. The recession, resource nationalism, and the maturation of many of the world's giant oil fields have put global production after 2015 in serious jeopardy. In fact, there is a strong probability that we are now at peak oil production. In the *Joint Operating Environment (JOE) 2010* report, the U.S. Joint Forces Command, the Military's transformation leader, delineated the widening gap between global oil production and consumption: (U.S. Joint Forces Command, February 18, 2010) To generate the oil required worldwide by the 2030s, we will need to find an additional 1.4 million b/d every year until that time (the equivalent of adding a Saudi Arabia's worth of production every seven years).

- World surplus oil production capacity could entirely disappear by 2012, and as early as 2015, the shortfall in output could reach almost 10 million b/d.
- Global oil production and distribution infrastructure must receive extraordinary new investment if demand is to be satisfied at a cost comparable with economic growth and prosperity. The International Energy Agency (IEA) indicates this investment level is $6.3 trillion from 2007 to 2030. (International Energy Agency, November 12, 2008)

Nevertheless, oil is entrenched in the U.S. energy system and is expected to hold its approximately 40 percent grip on total energy consumption until at least 2030. Without increased domestic production, we face the prospect of extended supply shortages, rising prices, worsening trade deficits, and even greater national security vulnerabilities. The April 2010 BP Deepwater Horizon spill in the Gulf of Mexico, however, has hastily put the U.S. oil industry on the defensive. Just weeks before the 2010 mid-term elections, the Obama administration lifted the nearly six-month ban on offshore energy exploration but the Institute for Energy Research says a "permitorium" remains a guiding policy.(Institute for Energy Research, October 12, 2010)

A study (October 2010) by Southern Methodist University concluded that the drastic slowdown in the issuing of new drilling permits in the Gulf's shallow waters alone is jeopardizing nearly 40,000 jobs. (Snow, October 7, 2010) According to a study (July 2010) by the Louisiana State University (LSU), a drilling moratorium in the region may have actually been worse than the spill itself. The report found that a six-month moratorium in the region could trigger a loss of more than 8,000 jobs, almost $500 million in wages, and over $2.1 billion in economic activity in the Gulf region alone.(LSU News, July 26, 2010) By "stifling one of the area's primary economic engines, the administration is crippling the local economy and risking long-term consequences," the LSU study concluded.

Perhaps even more worrisome, a moratorium or a permit slowdown poses the very real question: "Where will we get our incremental oil?" The EIA specifically credits deep offshore fields in the Gulf of Mexico—Great White, Norman, Tahiti, Gomez, Cascade, and Chinook—for a projected 36 percent jump in domestic oil production from 2008 to 2035. (U.S. Department of Energy, May 11, 2010) The EIA projects U.S. imports, "which grew steadily from the mid-1980s to 2005, decline significantly from 2008 to 2035." As demonstrated in Figure 1.1, if policy constrains this offshore stockpile from development, our rank as the second highest source of non-OPEC incremental supply (after Brazil) will fade, and risky outside suppliers will gain leverage.

Figure 1.1: If Offshore U.S. Oil is Constrained from Delivering as Projected by 2025

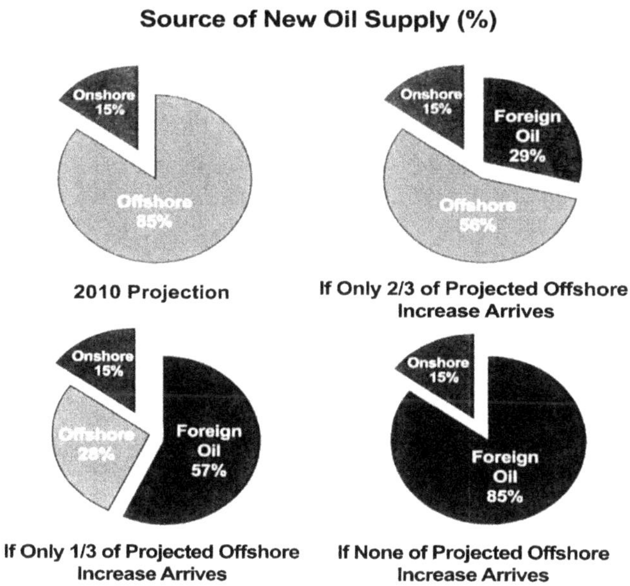

Source: developed from U.S. Energy Information Administration (*International Energy Outlook 2010*)

The present analysis outlines the three principal reasons why any lost opportunity for the U.S. to reverse the known costs and risks of dependence on foreign oil would be regrettable: 1) the Deepwater Horizon spill is an aberration, 2) the necessity to produce more oil will remain a hallmark of U.S. energy security until at least 2030, and 3) U.S. imports from both Canada and Mexico, North American oil that has kept our dependence on less friendly suppliers in check, now confront their own distinct challenges in meeting our growing needs.

THE U.S. OIL INDUSTRY'S SAFETY RECORD

Although the human and mechanical errors that led to the Deepwater Horizon spill are both identifiable and preventable, some opponents of offshore drilling have taken advantage of the incident to call for continuing moratoria or even cessation of deepwater drilling. This group is apparently beginning its mission by seeking a permanent ban on drilling in all federal waters off the Pacific Coast. Indeed, the Deepwater Horizon spill has morphed into an all-out assault on the U.S. oil industry's established safety record:

- "President Obama is right to impose the moratorium on deepwater drilling . . . But the moratorium should cover shallow water drilling as well. . . the President should go further," Frances Beinecke, president of the National Resources Defense Council (and member of the National Commission on the BP Deepwater Horizon Oil Spill) (Beinecke, May 27, 2010)
- "This spill reveals a serious lack of safety surrounding the oil industry," Amy Klobuchar, U.S. Senator, (D-Minnesota) (Klobuchar, May 18, 2010)
- "It is an unacceptable risk associated with offshore gas or oil drilling at this time," Louie Miller, state director of the Mississippi chapter of the Sierra Club (Kirgan, May 2, 2010)

The President of the Louisiana Mid-Continent Oil and Gas Association, Chris John disagrees: "Like any industry, many of the energy industries come with risks. But it does not stop our need for energy to fuel our economy and secure America. And the track record has been pretty good." (Alpert, April 26, 2010) The U.S. Minerals Management Service (MSS) states the case more bluntly: "The Nation's record for safe and clean offshore natural gas and oil operations is excellent." (U.S. Department of Interior, January 13, 2000) Despite heavy increases in transport, spill volumes and incidents have significantly diminished in recent decades. Offshore, this progress is particularly obvious since 1989, when the Exxon Valdez spill prompted majors to replace single-hull tankers with double-hull ships. By high-lighting the need for stronger legislation, inflaming public sentiment, and spurring U.S. Congress to enact comprehensive oil spill legislation, the incident helped lead to the decreases seen in Figure 1.2. The Oil Pollution Act of 1990: 1) expanded and clarified the authority of the federal government, 2) created new spill prevention

Figure 1.2: The Decline in the Volumes and Incidents of U.S. Offshore Oil Spills Above 100 Gallons (1974-2004)

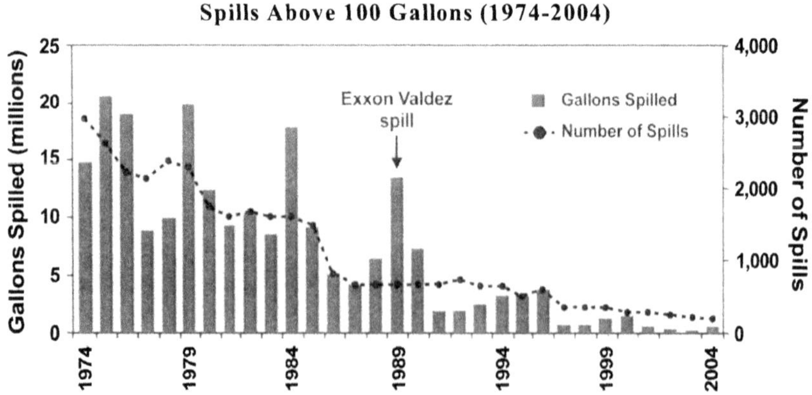

Source: developed from U.S. Coast Guard Oil Spill Compendium data.
Note: The U.S. Coast Guard spill database (as of Fall 2010) included data only through 2004 (via website: http://www.uscg.mil)

prevention and preparedness requirements, and 3) strengthened existing liability provisions.

Over the thirty-year time period depicted, while total imports nearly tripled, average annual volumes and incidents of offshore spills in the U.S. dropped by approximately 93 percent and 87 percent respectively. (Ramseur, April 30, 2010) Environmental Research Consulting (ERC) notes that spills above 100 gallons represent 99 percent of the total offshore spill volume but only 8 percent of the total offshore spill number. (Etkin, 2001) To put this amount in better perspective, one million gallons of oil is 23,810 barrels of oil. Therefore, offshore operations in the U.S., in an entire year, spill less than 0.1 percent of what the country consumes in a single day. According to the MMS, roughly 19,150 wells were drilled offshore in the 2000s, and only 0.001 percent of the total offshore yield has spilled in the past thirty years. (Cassidy, June 28, 2010) As for the Pacific Coast specifically, in the 40 years since the Santa Barbara spill, California's offshore wells have produced over one billion barrels of oil without a significant incident.

The claim that a "lack of safety" surrounds the U.S. oil industry is also called into question by sharp declines in onshore spill volumes and incidents. From 1987 to 2005, for instance, data gathered from the U.S. Environmental Protection Agency (EPA, 2009) and ERC (2001) show that the average annual volumes and incidents of onshore spills each plummeted by roughly 95 percent (see Ramseur, 2010). Spill statistics verify the tremendous strides that have been made in the industry. And we can expect more vigilance in the years ahead because companies realize aging infrastructure will see higher throughput and transport. As reasoned by Lee Hunt, President of the International Association of Drilling Contractors, companies have a "vested interest in being as safe as possible." (Gold and Power, May 7, 2010) In addition to lost product, the EPA says it costs about $220 to clean up a single gallon of oil. (U.S. Environmental Protection Agency, April 29, 2009) The Deepwater Horizon spill has caused a major shakeup at BP and is projected to ultimately cost the company as much as $63 billion. (Corkery, June 16, 2010)

Even worse, BP's reputation appears irreparably damaged. The company has been relatively quick to spread blame rather than accepting responsibility as operator. BP's own investigation, released in September 2010, did not "identify any single action or inaction" that caused the incident but rather a "complex and interlinked series of mechanical failures, human judgments, engineering design, operational implementation . . . came together to allow the initiation and escalation of the accident. Multiple companies, work teams and circumstances were involved over time." (Bergin and Rascoe, September 8, 2010) BP has publicly blamed both Transocean, the owner of the rig and the company hired to drill the well, and Halliburton, the oilfield service company that had cemented the well but apparently did not complete the seal when the explosion occurred. The exact causes of the Deepwater Horizon spill could remain a mystery for some time but we do know that there were key failures on at least three fronts:

- The blowout preventer failed to close off the well automatically in response to the sudden emergency.

- The failure of the cement, the casting, or both is the "one thing we know with certainty," says Steve Newman, Chief Executive Officer of Transocean. (MSNBC News, May 11, 2010)
- Oversight and response failures led to the Deepwater Horizon having a "dead man switch" to activate the blowout preventer but lacking the separate, remote-control switch that is commonly used by deepwater experts Brazil and Norway.

Similar to the ExxonValdez spill in 1989, the Deepwater Horizon incident represents a possible turning point for the U.S. oil industry. As part of their commitment to protect the environment and ensure workplace safety, oil companies continually look to deploy advancing technologies in all areas of operation. For example, the use of dispersants, a new spill mitigation technique that was recently reviewed and approved by the EPA, has proven effective in the Gulf. Four of the majors, Chevron, ConocoPhillips, ExxonMobil, and Shell, have committed $1 billion in start-up costs to build and deploy a swift response system that will be available to capture and contain oil in the event of a future deepwater well blowout. (Chevron, July 21, 2010) The transition toward deeper water and frontier regions signifies the offshore industry in particular must be fully prepared to prevent and respond to a similar incident in the future, especially considering the U.S. oil requirement will maintain its ascent.

RISING U.S. OIL NEEDS

"We have set in motion a national policy aimed at both increasing gas mileage and decreasing [greenhouse gas] pollution for new trucks and cars sold in the United States of America," U.S. President Barack Obama, announcing the proposal of new federal laws to increase Corporate Average Fuel Economy (CAFE) requirements, May 19, 2009 (PBS, May 19, 2009)

The transportation sector accounts for nearly 70 percent of total domestic oil demand, and passenger vehicles are the single largest driver of U.S. oil consumption. Passenger vehicles and light-duty trucks in the U.S. are currently required to get an average of 27.5 miles per gallon (mpg). In April 2010, with the goal of reducing "our dependence on oil while helping folks spend a little less at the pump," the Obama administration mandated that these vehicles reach an average fuel efficiency of 35.5 mpg by 2016. (Obama, March 31, 2010) The Natural Resources Defense Council claims the raised standard will save Americans $65 billion in fuel costs by cutting oil consumption by 1.3 million b/d by 2020. (Eilperin, April 2, 2010)

The common assumption that increased fuel efficiency standards for vehicles will assuredly cut into U.S. oil consumption highlights the ongoing triumph of wishful thinking over rational, evidence-based understanding. At the time of the 1973-1974 Arab oil embargo the U.S. imported about 33 percent of the 17 million b/d of oil that it consumed, 15 percent of which came from the Persian Gulf. (U.S. Department of Energy, July 29, 2010) With the security, political, and economic vulnerabilities of this dependence abruptly exposed, the federal government enacted CAFE standards in 1975. Automakers were required to meet certain mpg efficiency targets for new vehicles starting in model year 1978. This mandated standard has been elevated for passenger vehicles and light-duty trucks from 18 to 27.5 mpg. (U.S. Department of Transportation, November 2004) Figure 1.3 illustrates an increased CAFE has not had its desired and anticipated results. Americans continue to drive more and use more fuel. The U.S. now imports 60 percent of the 20 to 21 million b/d of oil that it consumes.

Figure 1.3: Rising CAFE Standards: Americans Drive More and Use More Fuel

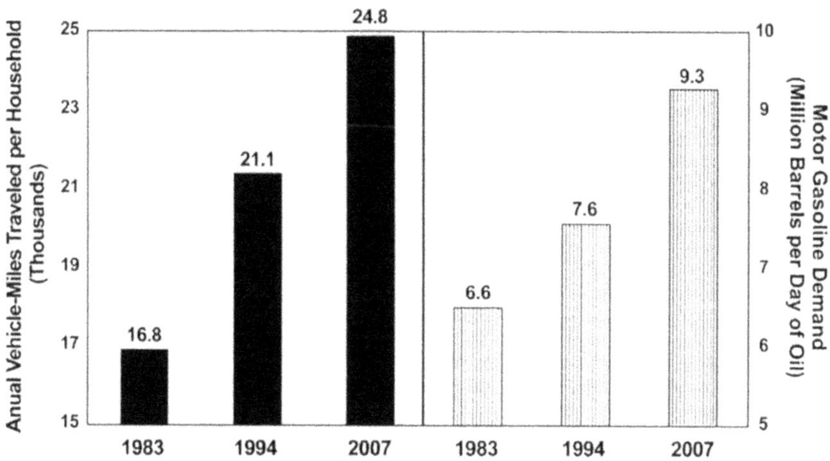

Source: developed from U.S. Energy Information Administration (*Household Vehicles Energy Use: Latest Data & Trends 2009*)

Efficiency gains in the U.S. vehicle sector have fallen victim to *The Jevons Paradox*, named after the father of energy economics, William Stanley Jevons, in 1865. In stark contrast to the virtual panacea CAFE enthusiasts claim that they are, efficiency improvements tend to diminish the need for a fuel, thereby reducing its costs and counter-intuitively increasing its demand. Indeed, in 2004, the National Commission on Energy Policy found that if U.S. Congress mandated the vehicle fleet raise its average fuel economy for passenger vehicles and light-duty trucks from 27.5 to 44 mpg—a staggering 60 percent increase—domestic oil consumption could still swell by 3.7 million b/d by 2025. (Puplava and Puplava, July 22, 2005) Jim Kliesch, a vehicle efficiency expert with the Union of Concerned Scientists, calculates that doubling the current efficiency standards would increase driving for the average American by more than 7 percent. (O'Carroll, May 22, 2009) "CAFE standards are regressive," says Stephen DeCanio, an economics professor at the University of California at Santa Barbara, "CAFE is not the way to go." (Roth-Furchtgott, May 28, 2009) CAFE has been "one really bad mistake. It didn't meet any of the goals," adds Jim Johnston, former President of General Motors. (Rutledge, 2006, 125)

One main problem has been that consumers often lack a real incentive to purchase a new vehicle. CAFE standards make new vehicles more expensive because they make manufacturing costs higher. CAFE standards also force automakers to build dangerously lighter and smaller vehicles that burn less fuel; the National Academy of Sciences has linked mileage standards to an average of 2,000 deaths per year. (The National Academies, 2002, 28) Due to better quality and luxury, more Americans are keeping their vehicles longer, driving them more, and frequently buying a second or even a third vehicle. Going forward, the constant need for more domestic oil production is one of the only certainties that our murky energy future holds:

- Today, there are over 265 million oil-only vehicles on U.S. roads; Dargay et al. (2007) project there will be 314 million such vehicles in 2030. (Dargay, Gately and Sommer, January 2007)
- The U.S. puts between 12 and 16 million new oil-only vehicles on its roads each year; in 2009, domestic sales of hybrids were less than 300,000 units. (Hybrid Cars, March 4, 2010)
- As seen with the now infamous "Cash for Clunkers" program, U.S. vehicle fleet penetration rates are exceeding slow; Hirsch et al. (2005) found that half of all vehicle models bought in a given year are still on the road seventeen years later. (Hirsch, Bezdek and Wendling, February 2005)
- Oil (and energy) consumption is more responsive to income than it is to price; the EIA notes "Since 1983. . .the growth in vehicle-miles traveled has basically mirrored the increases in real disposable income." (U.S. Department of Energy, November 2005)

Walter McManus, Director of the Office for the Study of Automotive Transportation at the University of Michigan, has documented the failure of American consumers to respond to higher gasoline prices by shifting to more fuel-efficient vehicles. For example, from 2001 to 2004, when the (inflation adjusted) price of gasoline increased 18 percent, there was no observable impact on new vehicle purchases. (Rutledge, 2006, xii) In fact, large pick-ups and luxury Sport Utility Vehicles, having the lowest fuel economy, actually gained market share during the period. Americans have consistently chosen to forgo efficiency improvements by driving safer, larger, and more powerful vehicles. Many insist this discretion is a consumer's fundamental right because the higher profit margins on these gas-guzzlers allow retailers to sell them at a discounted price.

As demonstrated in Figure 1.4, U.S. (a) oil demand is expected to preserve its lockstep increase with the two leading indicators of future energy demand: incremental growth in (b) gross domestic product (GDP) and (c) population. The correlation between these three variables is a steady drumbeat. Over the next 20 years, the EIA expects the U.S. to add more than two Japans to its GDP ($8.7 trillion) and expand its population by the size of France (65 million people). (U.S. Department of Energy, July 27, 2010) Thus, the Administration's projection that

U.S. oil demand will increase by 8 percent by 2030 is markedly conservative. Unfortunately, imports from our two most secure suppliers will struggle to meet our growing needs.

Figure 1.4: Constant Increases in U.S. Oil Demand, GDP, and Population (1990-2030)

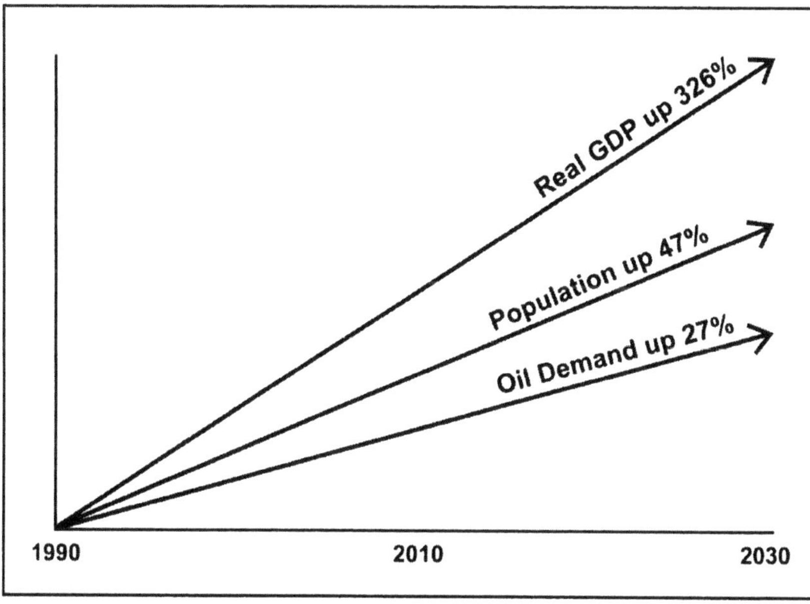

Source: developed from U.S. Energy Information Administration (*International Energy Outlook 2010*)

CANADA'S OIL COULD BE FORCED ABROAD

In the event the United States seeks to eliminate its reliance upon the oil that comes from oil sands, then it will be sold somewhere else in the world, Gary Mar, Alberta, Canada's diplomatic representative to the U.S., 2009 (India Environmental Portal, June 2009)

As a dependable supplier and long-time ally, Canada is an indispensable safeguard of U.S. energy security. In fact, Canada has been the largest U.S. oil supplier every year since 2003 (Saudi Arabia) and meets about 10 percent of our total consumption. In 2009, supplies from Canada surpassed the combined imports from Saudi Arabia and Nigeria, respectively our second and fourth highest foreign sources of oil. The EIA reports that Canada exports nearly 60 percent of the 3.4

million b/d that it produces to the U.S. (U.S. Department of Energy, Energy Administration, 2010) BP's *Statistical Review of World Energy June 2010* estimates Canada's proven oil reserves at 33.2 billion barrels. (BP, June 2010, 6) BP does not include, but separately records, an additional 143 billion barrels of recoverable oil sands in the western province of Alberta, second globally to Saudi Arabia and more oil than there is in either Iran or the Former Soviet Union.

The EIA reports that Alberta's massive oil sands resource currently yields about 1.4 million b/d and accounts for an escalating 42 percent of Canada's total production. Output in the area is expected to more than double by 2017 and soar to five million b/d by 2030. A relatively flat population growth rate will have Canada's organic demand hovering around 2.3 million b/d. Much of this surplus oil is anticipated to be headed south, where U.S. refineries, especially in the Midwest and Gulf Coast, are continually being reconfigured to handle these heavier supplies. TransCanada's $12 billion Keystone pipeline system currently brings oil non-stop from Canada to market hubs in the U.S. Midwest, and the proposed 1,661-mile Keystone Gulf Coast Expansion Project would serve the vital Texas market. (TransCanada, October 5, 2010)

The Canadian-American oil alliance is a prime example of the firm link between stable and affordable exporters and U.S. energy security. Without access to Canada's oil, we would be forced to shop overseas, where questionable supplies depend upon risky and/or hostile producers. Alberta has an oil production potential of over 100 years, and the province is prized as the world's most predictable region still capable of expanded production. (Jamasmie, September-October 2010) Both China and India are well aware of the unparalleled potential this "free market" resource holds, as they strive to meet the surging energy needs of almost 40 percent of humanity. But, as an unconventional resource, oil sands present more challenges than conventional development and are typically only economically viable with higher prices. The rebound in prices to the $65 to $75 range has investors, producers, and the Alberta and Canadian governments primed to meet the energy security needs of the U.S. but U.S. policy could make it easier for other countries to gain access to this critical resource.

With the oil sands' bitumen requiring energy-intensive, emissions-heavy processing, U.S. climate change legislation could effectively deny them access to the U.S. market. The Low Carbon Fuel Standard (LCFS) is an example of the dual edged sword that characterizes our attempt to combine energy and environmental legislation. In January 2010, California became the first U.S. state to formally implement a LCFS, requiring state suppliers to track and report the "fuel pathway," or life-cycle, greenhouse gas (GHG) emissions of their transportation fuels. California's LCFS mandates that the "carbon intensity" of the state's transportation fuels in 2020 is 10 percent less than it was in 2006. (Crane and Prusnek, January 8, 2007) Carbon intensity is measured in two parts: 1) the direct emissions associated with producing, transporting, and using the fuel and 2) the indirect effects, including those caused by changes in land use. Lost in the debate, DNV, a leading international provider of services for managing risk, notes that "measuring the direct full life-cycle GHG emissions of transportation fuels is not technologically feasible." (DNV, 2009)

Other climate change legislation will also likely limit the availability of Canadian oil to the U.S. market. For example, the American Clean Energy and Security Act of 2009 (ACES), colloquially known as the Waxman-Markey Bill, seeks to "reduce carbon emissions from major U.S. sources by 17 percent by 2020 and over 80 percent by 2050 compared to 2005 levels." (U.S. House of Representatives, June 23, 2009) ACES narrowly passed in the House of Representatives in June of 2009, and the bill is still under consideration in the Senate (as of Fall 2010). Under ACES, life-cycle GHG emissions include "direct emissions and significant indirect emissions such as significant emissions from land use changes" as determined by the Administrator of the EPA, Lisa Jackson.(U.S. Environmental Protection Agency, May 2009)

The energy security concern stems from a provision in ACES that would grant President Obama the power to slap tariffs on imports that have a carbon footprint larger than U.S.-made goods. Although Kroft et al. (2009) suggest ACES does not directly threaten U.S. imports from oil sands because it does not include an LCFS, President Obama has made clear the EPA could issue waivers for other states to adopt their own standard. (Kroft, Elnekave, and Lees, September 30, 2009) The governors of at least eleven Northeast and Mid-Atlantic states have signed a memorandum of understanding, committing to develop a California-like regional LCFS by 2011.

The environmental benefits of laws that, even inadvertently, target oil sands in the transition to a low-carbon economy appear minimal. IHS Cambridge Energy Research Associates, one of the most referenced energy consulting firms in the world, has determined that the "wells-to-wheels" GHG emissions from oil sands—from extraction and processing through combustion of its refined product—are only 5 percent to 15 percent higher than the oil processed in the U.S. (IHS Cambridge Energy Research Associates, May 19, 2009)

Further, from an energy security perspective, the tariffs proposed in ACES, which would come into effect in 2020, would be implemented at the worst possible time. The oil sands will triple Canada's export capacity in the next decade (see Figure 1.5). This opening has prompted the arrival of China's global equity oil quest to North America. In 2009, PetroChina purchased a $1.7 billion (60 percent) stake in two of Canada's oil sands projects. (Fekete, September 1, 2009) Energy-hungry India has also expressed a keen interest in developing the resource. These two ready markets will surely negate any GHG emissions saved by curtailed U.S. involvement in the oil sands.

In addition, the Harper administration has its own climate change goals: to decrease GHG emissions 20 percent below 2006 levels by 2020 and 60-70 percent by 2050. (Natural Resources Canada, October 12-14, 2009) These policies could eventually collide with the "proportionality rule" of the 1994 North American Free Trade Agreement, which basically declares that Canada itself cannot try to limit is exports of oil sands production to the U.S.

For the U.S., the importance of Alberta's oil sands goes beyond energy security. A 2009 study by the Canadian Energy Research Institute found that, as production and investment in the oil sands rises, demand for U.S. goods and services would increase the U.S. GDP by $34 billion in 2015 and $42.2 billion in 2025. (Snow, October 26, 2009) The report, *Canada's Oil Sands and Economic Impact on the USA*, found that oil sands development would add 342,000 new U.S. jobs from 2011 to 2015. Mexico, our next most important oil supplier, also faces policy issues that could limit the country's exports to the U.S. but production will be constrained by below-ground realities.

Figure 1.5: Canada's Projected Expanding Oil Export Capacity (2007-2035)

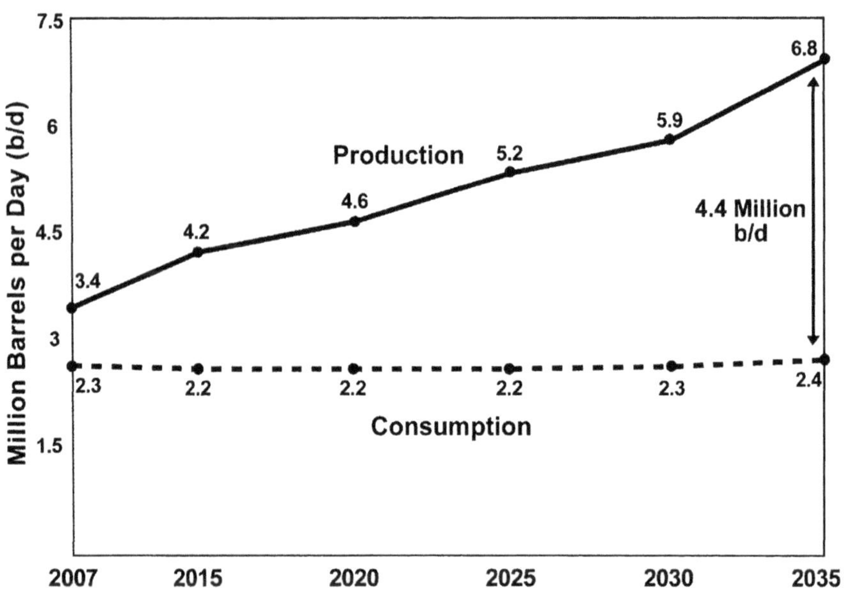

Source: developed from U.S. Energy Information Administration (*International Energy Outlook 2010*)

MEXICO'S EXPORTS COULD BE EVEN LESS THAN ADVERTISED

BP's *Statistical Review of World Energy June 2010* estimates Mexico's proven oil reserves at 11.9 billion barrels. Since its discovery in 1976, Mexico's giant oil field, Cantarell, located offshore in the Gulf of Mexico, forty-seven miles northeast of Ciudad del Carmen, Campeche, has solidified the country as a consistent top-three U.S. oil supplier. Blessed with unique geology and location, Cantarell was soon yielding over one million b/d from just forty wells. (Lynch, April 24, 2009)

By the mid-1990s, however, declining natural gas pressure forced Mexico's state-owned oil monopoly, Petróleos Mexicanos (PEMEX), to install a nitrogen injection plan to reverse a production slump at its most essential resource. The program lifted production to more than two million b/d by 2001 and made Cantarell the second largest oil field in the world, behind Saudi Arabia's legendary Ghawar. Nitrogen enhancements are simply short-term solutions, however, and Cantarell's share of Mexico's total output plummeted from 60 percent in 2005 to less than 20 percent in 2010. (Campbell, May 3, 2010) Peak oil production in Mexico has been reached.

In response, President Felipe Calderón's historic energy reform bill finally passed through Mexico's Congress in October of 2008. With the purported purpose of tapping the deepwater reserves of its Gulf, the PEMEX Law enacts a new legal framework for the company by amending the laws and regulations that govern its activities. PEMEX is finally scheduled to offer performance-based contracts at shallow water and mature field projects as a possible prelude to more attractive deepwater contracts. As the final region unexplored enough to offer hope of an industrial rebound, some estimate Mexico's side of the Gulf of Mexico could contain as many as 50 billion barrels of oil. (*Daily News & Economic Review*, August 17, 2010) But, the EIA concludes these resources will not be enough (see Figure 1.6).

Figure 1.6: Mexico's Dramatic Shift from Oil Exporter to Oil Importer (2007-2035)

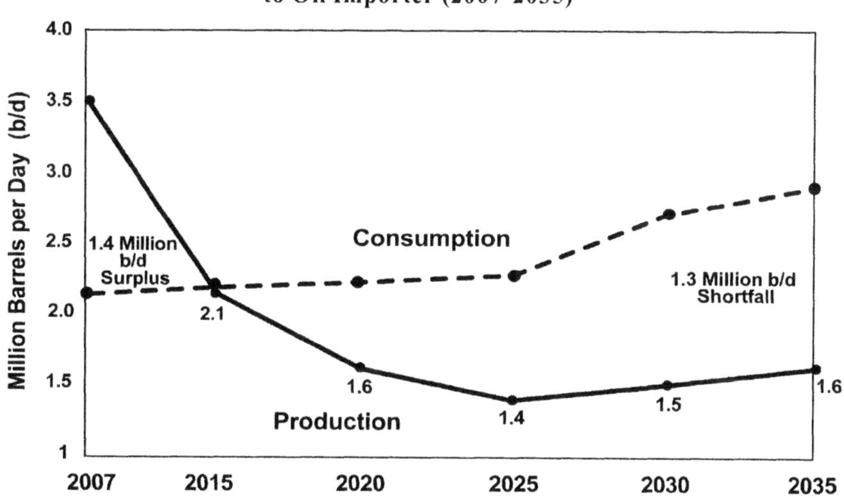

Source: developed from U.S. Energy Information Administration (*International Energy Outlook 2010*)

And other independent analysts are even more skeptical. Simply put, Mexico has still not laid out specific ways to improve deepwater exploration and production in its Gulf. The performance-based service contracts permitted are typically reserved for shallow water operations—areas that Mexico has already exploited. The PEMEX Law *does not change* the constitutional principles that reserve oil production solely for the Mexican government, and outside oil companies are unlikely to be willing to accept cash in lieu of discovered oil. Implementation of the reforms has progressed far too slowly. PEMEX continues to operate like a politicized government agency instead of the globally competitive, highly dynamic oil company that it must now become.

Reduced oil production from Mexico will have far reaching implications. In the U.S., a stable source of supply will be eroded to the detriment of both reliability and energy security. After a record high of 1.7 million b/d in 2004, imports from Mexico have decreased 30 percent, and the fact is we should be preparing for even less oil from Mexico than is being projected. (U.S. Department of Energy Information Administration, September 29, 2010) PEMEX officials themselves have acknowledged that the two domestic oil regions holding nearly 75 percent of Mexico's non-Cantarell reserves, the Chicontepec Basin and Ku-Maloob-Zaap, will peak in 2016 and 2010 respectively, making the EIA's projected late 2020s uptick rather optimistic. (Clemente, August 2008) In addition, Mexico's rising domestic oil needs, stemming from a growing population, economy, and vehicle consumption rate, suggest the decline in U.S. imports from Mexico will accelerate. Mexico will not only be producing less oil, the country will be needing more.

As important as these consequences are for the U.S. oil outlook, however, they pale in significance compared to the impact reduced oil production will have on Mexico itself—a nation which has literally transformed its socioeconomic profile with oil exports. Oil sales constitute 40 percent of the federal budget, and record revenues pay for schools, roads, hospitals, and other imperative societal infrastructure. The government will be hard pressed to maintain social programs, economic development, and infrastructure expansion. The effect may very well take a heavy toll on the hopes and dreams of millions of Mexican families.

The last time Mexico experienced turmoil of this potential magnitude was during the revolution era from 1910 to 1940, when a quarter of the population left the country, most arriving in the U.S. A similar mass migration today would obviously have far greater ramifications because Mexico's population has increased five-fold in the last half century. The U.S. could be faced with an unprecedented humanitarian crisis along its southern border. Indeed, the development of Canada's massive oil sands resource is as much related to North American social stability as it is to the continent's energy security. Mexico is now expected to become a net importer of oil in 2015 (see Figure 1.6), sending the country into territory that was unimaginable just a few years ago.

CONCLUSION

The dangers that even well-intended climate change legislation could quickly bring to bear cannot be discounted. Oil and oil imports will remain a critical component of the U.S. energy portfolio for decades to come. Historical evidence suggests that efficiency gains tend to increase fuel consumption over time because they lower prices. Importantly, the push for more wind and solar power will not alter our fundamental need for more oil because oil generates less than 2 percent of all U.S. electricity. As supplies from our two most secure suppliers become less predictable, rising domestic needs could easily oblige a global supply chain dominated by unfriendly and/or unstable producers—over 80 percent of the world's reserves are controlled by nine national oil companies within a single cartel, the Organization of Petroleum Exporting Countries. Consequently, we should pursue policies designed to enhance, not impede, the domestic production of oil, particularly our offshore treasure trove.

As stated by Louisiana Representative Bill Cassidy, the call to end or downsize offshore oil production "is an emotional overreaction" and the repercussions of "such short-sighted policy are immediate and severe." (Cassidy, June 28, 2010) The Gulf of Mexico, for instance, yields one of every four barrels that we produce, and a drilling ban off California's coast would squander at least 10 billion barrels. (Warth, June 18, 2008) The MMS estimates the U.S. Outer Continental Shelf holds between 67 and 115 billion barrels of technically recoverable oil, or potentially enough oil to displace all of our imports for the next twenty-five years. (U.S. Department of the Interior Minerals Management Service, September 17, 2010) The improving safety record of the U.S. oil industry justifies a dismantling of barriers to the leasing of federal lands by offering royalty reductions, tax breaks, and other economic incentives to boost production. Instead of picking winners and losers with policy, all energy technologies should be allowed to compete in meeting our goals to enhance energy security, cut GHG emissions, and enhance sustainability.

REFERENCES

Alpert, B. (April 26, 2010). "Oil rig explosion used by opponents of Obama plan to expand offshore drilling," *The Times-Picayune*. Retrieved September 20, 2010 from http://www.nola.com/politics/index.ssf/2010/04/oil_rig_explosion_used_by_oppo.html.

Beinecke, F. (May 27, 2010). "Obama Starts to Address Offshore Hazards and Calls for Clean Energy Action." Huffington Post. Retrieved October 1, 2010, from http://www.huffingtonpost.com/frances-beinecke/obama-starts-to-address-o_b_592251.html.

Bergin, T. and Rascoe, A. (September 8, 2010). "BP and partners trade blame for oil spill." Retrieved September 19, 2010, from http://www.reuters.com/article/idUSTRE6871GC20100908.

BP. (June 2010). "Statistical Review of World Energy June 2010." London, UK.

Campbell, R. (May 3, 2010). "Mexico eyes up to 14 oil deals by year-end." Reuters. Retrieved September 29, 2010, from http://uk.reuters.com/article/idUKTRE64244R20100503.

Cassidy, B. (June 28, 2010). "Cassidy: Gulf Spill Shouldn't End Offshore Drilling." Roll Call. Retrieved September 23, 2010, from http://www.rollcall.com/issues/55_154/ma_congressional_relations/47787-1.html.

Cassidy, B. (June 28, 2010). "Cassidy: Gulf Spill Shouldn't End Offshore Drilling." Roll Call. Retrieved September 25, 2010, from http://www.rollcall.com/issues/55_154/ma_congressional_relations/47787-1.html.

Chevron. (July 21, 2010). "New Oil Spill Containment System to Protect Gulf of Mexico Planned by Major Oil Companies." Press Release. Retrieved September 18, 2010, from http://www.chevron.com/chevron/pressreleases/article/07212010_newoilspillcontainmentsystemtoprotectgulfofmexicoplannedbymajoroilcompanies.news.

Clemente, J. (August 2008). "Cantarell is Not Mexico's Only Production Problem." *Pipeline & Gas Journal.* Retrieved October 1, 2010, from http://www.peakoil.net/files/Cantarell %20Is %20Not %20Mexico %E2 %80 %99s %20Only %20Oil %20Production %20Problem.pdf.

Corkery, M. (June 16, 2010). "BP Oil Spill Costs: $20 Billion? Try $63 Billion." *The Wall Street Journal.* Retrieved October 7, 2010, from http://www.blogs.wsj.com/deals/2010/06/16/bp-oil-spill-costs-20-billion-try-63-billion/.

Crane, D. and Prusnek, B. (January 8, 2007). "The Role of a Low Carbon Fuel Standard in Reducing Greenhouse Gas Emissions and Protecting Our Economy." Office of the Governor. Retrieved September 22, 2010, from http://gov.ca.gov/fact-sheet/5155.

Daily News & Economic Review. (August 17, 2010). "Mexico eyes deeper oil drilling in shadow of Gulf disaster." Retrieved September 26, 2010, from http://www.hurriyetdailynews.com/n.php?n=mexico-eyes-deeper-oil-drilling-in-shadow-of-gulf-disaster-2010-08-17.

Dargay, J., Gately, D., and Sommer, M. (January 2007). "Vehicle Ownership and Income Growth, Worldwide 1960-2030." New York University. Retrieved September 11, 2010, from http://www.econ.nyu.edu/dept/courses/gately//DGS_Vehicle %20Ownership_2007.pdf.

DNV. (2009). "A Primer on the California Low Carbon Fuel Standard (LCFS) and Midwestern Corn Ethanol." Retrieved October 6, 2010, from http://www.dnv.us/focus/climate_change/hot_topics/A_Primer_on_the_California_Low_Carbon_Fuel_Standard_and_Midwestern_Corn_Ethanol.asp.

Eilperin, J. (April 2, 2010). "Emissions limits, greater fuel efficiency for cars, light trucks made official." *The Washington Post.* Retrieved October 7, 2010, from http://www.washingtonpost.com/wp-dyn/content/article/2010/04/01/AR2010040101412.html.

Etkin, D. (2001). "Analysis of Oil Spill Trends in the United States and Worldwide." Environmental Research Consulting. Retrieved September 27,

2010, from http://www.environmental-research.com/publications/pdf/spill_statistics/paper4.pdf.

Fekete, J. (September 1, 2009). "China's $1.9B bid in oilsands sparks energy security debate." *Calgary Herald*. Retrieved September 28, 2010, from http://www.calgaryherald.com/business/China+oilsands+sparks+energy+security+debate/1949947/story.html.

Gold, R. and Power, S. (May 7, 2010). "Oil Regulator Ceded Oversight to Drillers." *The Wall Street Journal*. Retrieved October 11, 2010, from http://www.online.wsj.com/article/SB10001424052748704370704575228512237747070.html.

Hirsch, R., Bezdek, R., and Wendling, R. (February 2005). "Peaking of World Oil Production: Impacts, Mitigation, & Risk Management." U.S. Department of Energy. Retrieved October 1, 2010, from http://www.netl.doe.gov/ publications/others/pdf/oil_peaking_netl.pdf.

Hybrid Cars. (March 4, 2010). "Best Selling Hybrid Cars in 2009." Retrieved October 2, 2010, from http://www.hybrid-cars.org/announcements/top-hybrid-car-2009.

IHS Cambridge Energy Research Associates. (May 19, 2009). "Oil Sands Move from the 'Fringe to Center' of Energy Supply." Press Release. Retrieved October 11, 2010, from http://www.cera.com/aspx/cda/public1/news/press Releases/press ReleaseDetails.aspx?CID=10329.

India Environmental Portal. (June 2009). "Oil Sands Production to Increase, Despite Move Toward Low-Carbon Fuels." Retrieved September 23, 2010, from www.indiaenvironmentportal.org.in/files/Oil Sands Production to Increase.doc.

Institute for Energy Research. (October 12, 2010). "Good Bye Moratorium, Hello Permitorium." Retrieved October 17, 2010, from http://www.institute forenergy research.org/2010/10/12/good-bye-moratorium-hello-permitorium/

International Energy Agency. (November 12, 2008). *World Energy Outlook 2008*. Presentation to the Press. London. Retrieved October 3, 2010, from http://www.iea.org/speech/2008/Birol_WEO2008_PressConf.pdf.

Jamasmie, C. (September-October 2010). "Oil Sands." *Mining Magazine*. Retrieved October 14, 2010, from http://magazine.mining.com/issues/1009/Vol03-06-TheChallengesAndPotentialOfCanadasOilSands-06-09.pdf.

Kirgan, H. (May 2, 2010). "Mississippi Sierra Club calls Deepwater Horizon oil spill America's Chernobyl." *Mississippi Press*. Retrieved September 24, 2010, from http://blog.gulflive.com/mississippi-press-news/2010/05/spill_is_americas_chernobyl.html.

Klobuchar, A. (May 18, 2010). "Klobuchar Presses Administration Officials, Oil Executives to Swiftly Resolve the Gulf Oil Disaster and Implement Safeguards to Prevent Future Oil Spills." Senator Amy Klobuchar. Retrieved October 2, 2010, from http://klobuchar.senate.gov/newsreleases_detail.cfm?id=325099&.

Kroft, J., Elnekave, R., and Lees, M. (September 30, 2009). "Canadian implications of U.S. climate change regulation." Stikeman Elliot. Retrieved October 7, 2010, from http://www.stikeman.com/cps/rde/xchg/se-en/hs.xsl/13062.html.

LSU News. (July 26, 2010). "LSU's Joseph Mason Testifies Before U.S. Senate on July 27." Louisiana State University. Retrieved September 28, 2010, from http://appl003.lsu.edu/UNV002.nsf/%28NoteID%29/766ADD35F6249BB48625776C007399BA?OpenDocument.

Lynch, M. (April 24, 2009). "Cantarell field in Mexico falling fast as gas cap expands down structure." Gerson Lehrman Group. Retrieved September 27, 2010, from http://www.glgroup.com/News/Cantarell-field-in-Mexico-falling-fast-as-gas-cap-expands-down-structure-37962.html.

Monbiot, G. (April 14, 2009). "We spend millions on smallpox, but nothing on this far greater threat." *UK Guardian*. Retrieved September 22, 2010, from http://www.guardian.co.uk/commentisfree/2009/apr/14/george-monbiot-smallpox-oil-supply.

MSNBC News. (May 11, 2010). "Oil executives shift spill blame in Senate." Retrieved October 3, 2010, from http://www.msnbc.msn.com/id/37079761/.

The National Academies. (2002). *Effectiveness and Impact of Corporate Average Fuel Economy (CAFE) Standards*. Washington, D.C.: The National Academies Press.

Natural Resources Canada. (October 12-14, 2009). "Overview of Canadian Policy and Project Activities in Carbon Capture and Storage (CCS)." Retrieved October 3, 2010, from http://www.egcfe.ewg.apec.org/publications/proceedings/CFE/Korea_2009/ %28Session %202-B %29Weyburn-Midale %20CCS %20Project %20Update.pdf.

Obama, B. (March 31, 2010). "Remarks by the President on Energy Security at Andrews Air Force Base." The White House, Office of the Press Secretary. Retrieved September 22, 2010, from http://www.whitehouse.gov/the-press-office/remarks-president-energy-security-andrews-air-force-base-3312010.

O'Carroll, E. (May 22, 2009). "Does increasing fuel efficiency make people drive more." *The Christian Science Monitor*. Retrieved October 12, 2010, from http://www.csmonitor.com/Environment/Bright-Green/2009/0522/does-increasing-fuel-efficiency-make-people-drive-more.

PBS. (May 19, 2009). "New Environmental Standards for Cars." Retrieved October 15, 2010, from http://www.pbs.org/newshour/extra/video/blog/2009/05/new_environmental_standards_fo.html.

Puplava, J. and Puplava, C. (July 22, 2005). "Unloved and Underowned." PFS Group. Retrieved September 25, 2010, from http://www.financial sense. com/Market/puplava/2005/0722.html.

Ramseur, J. (April 30, 2010). "Oil Spills in U.S. Coastal Waters: Background, Governance, and Issues for Congress." Congressional Research Service. Retrieved September 20, 2010, from http://assets.opencrs.com/rpts/RL33705_20100430.pdf

Roth-Furchtgott, D. (May, 28, 2009). "Obama Should Ditch Deadly CAFÉ Standards." Real Clear Markets. Retrieved September 30, 2010, from http://www.realclearmarkets.com/articles/2009/05/obama_should_ditch_deadly_cafe.html.

Rutledge, Ian. (2006). *Addicted to Oil: America's Relentless Drive for Energy Security.* I. B. Tauris & Co. Ltd, Publishers. London, UK.

Snow, N. (October 26, 2009). "CERI sees U.S. economic gains from Canadian oil sands." *Oil & Gas Journal.* Volume 107. Issue 40. p. 32.

Snow, N. (October 7, 2010). "SMU: Shallow-water permit delays threaten 40,000 jobs." *Oil & Gas Journal.* Retrieved October, 9, 2010, from http://www.ogj.com/index/article-display/0861882623/articles/oil-gas-journal/general-interest-2/2010/10/smu_-shallow-water.html.

TransCanada. (October 5, 2010). "Keystone Pipeline Project." Retrieved October 11, 2010, from http://www.transcanada.com/keystone.html.

U.S. Department of Energy, Energy Information Administration. (November 2005). "Household Vehicles Energy Use: Latest Trends & Data." Retrieved September 20, 2010, from http://www.eia.doe.gov/emeu/rtecs/nhts_survey/ 2001/tablefiles/es0464(2005).pdf.

U.S. Department of Energy, Energy Information Administration. (May 11, 2010). *Annual Energy Outlook 2010.* Retrieved September 30, 2010, from http://www.eia.doe.gov/oiaf/aeo/, p. 77.

U.S. Department of Energy, Energy Information Administration. (July 27, 2010). *International Energy Outlook 2010.* Retrieved October 6, 2010, from http://www.eia.doe.gov/oiaf/ieo/index.html.

U.S. Department of Energy, Energy Information Administration. (July 29, 2010). "U.S. Imports by Country of Origin." Retrieved October 15, 2010, from http://tonto.eia.doe.gov/dnav/pet/pet_move_impcus_a2_nus_ep00_im0_mbblpd_a.html.

U.S. Department of Energy, Energy Information Administration. (September 29, 2010). "Crude Oil and Total Petroleum Imports Top 15 Countries." Retrieved October 6, 2010, from http://www.eia.doe.gov/pub/oil_gas/petroleum/data_publications/ company_level_imports/current/import.html.

U.S. Department of Energy, Energy Information Administration. (2010). "Country Analysis Briefs: Canada." Retrieved October 7, 2010, from available at: http://www.epa.gov/oem/docs/oil/fss/fss09/denning.pdf. http://www.eia.doe.gov/cabs/Canada/Background.html.

U.S. Department of the Interior, Minerals Management Service. (January 13, 2000). "Overview of Safety Program." Retrieved October 6, 2010, from http://www.gomr.mms.gov/homepg/offshore/safety/overview.html

U.S. Department of the Interior, Minerals Management Service. (September 17, 2010). "Outer Continental Shelf Oil & Gas Assessment 2006." Retrieved October 9, 2010, from http://www.boemre.gov/revaldiv/RedNat Assessment.html.

U.S. Department of Transportation, National Highway Traffic Safety Administration. (November 2004). "Automotive Fuel Economy Program Annual Update Calendar Year 2003." Retrieved September 23, 2010, from http://www.nhtsa.dot.gov/cars/rules/cafe/FuelEconUpdates/2003/index.html.

U.S. Environmental Protection Agency. (April 29, 2009). "Measuring Benefits of Oil Spill Prevention: Methods and Approaches." Retrieved October 8, 2010, from available at: http://www.epa.gov/oem/docs/oil/fss/fss09/denning.pdf.

U.S. Environmental Protection Agency. (May 2009). "EPA Lifecycle Analysis of Greenhouse Gas Emissions from Renewable Fuels." Retrieved September 30, 2010, from http://www.epa.gov/oms/renewablefuels/420f09024.html.

U.S. House of Representatives. (June 23, 2009). "The American Clean Energy And Security Act (H.R. 2454)." Committee on Energy and Commerce. Retrieved September 15, 2010, from http://www.energycommerce.house.gov/ Press_111/20090623/hr2454_rulessummary.pdf.

U.S. Joint Forces Command. (February 18, 2010). *Joint Operating Environment (JOE) 2010*. Retrieved October 1, 2010, from http://www.peakoil.net/files/JOE2010.pdf.

Warth, G. (June 18, 2008). "Region: President Bush Proposes Offshore Oil Drilling." *North County Times*. Retrieved September 19, 2010, from http://www.nctimes.com/news/local/sdcounty/article_1f5beecf-d0c8-5524-877e-5268f60c3676.html.

CHAPTER TWO
APPLYING TECHNOLOGICAL DISASTER PROCESS MODELS TO THE BP DEEPWATER HORIZON OIL SPILL DISASTER

LISA A. EARGLE, ASHRAF M. ESMAIL, JAS M. SULLIVAN, AND SHYAMAL K. DAS

INTRODUCTION

This chapter examines the BP Oil Spill within the context of several Technological Disaster Process Models. The BP Deepwater Horizon Oil Spill disaster is one of the costliest disasters in US history, surpassing that of the Exxon Valdez wreck and many natural disasters. Its effects on society and the environment are likely to be felt for years, if not decades (National Wildlife Federation, 2010).

In the sections below, we first provide a brief summary and timeline for the BP Deepwater Horizon disaster's events, from April to September 2010, as presented by the BBC News (published September 19, 2010). Next we discuss several models of the technological disaster process, including the Corrosive Community model (Picou and Marshall, 2007), Socio-Technical Disaster Model (Aini and Fakhrul-Razi, 2010), a model of organizational factors and errors that contribute to disasters (Shaluf, Ahmadun, and Shariff, 2003), and Disaster Resilience of Place Model (Cutter, Barnes, Berry, Burton, Evans, Tate, and Webb, 2008). We examine these models to see what features or phases they have in common. Using these common features of the models, we highlight important characteristics of and stages in the unfolding of the Deepwater Horizon Oil Spill disaster. Information regarding what events took place before and after the disaster come from a content analysis of reports published on-line by the New Orleans' Times-Picayune newspaper (NOLA.Com) during the April to September 2010 time period.

A BRIEF SUMMARY AND TIMELINE

On April 20, 2010, an explosion occurs on the BP Deepwater Horizon oil drilling rig in the Gulf of Mexico, located approximately 50 miles from Venice, LA. Search and rescue and fire response efforts begin. Two days later, the rig sinks to the ocean floor, and 11 missing workers are presumed dead. On April 26, officials report oil is leaking from the damaged well. BP sends robots underwater to make an unsuccessful attempt to activate the blow-out preventer valves. Oil leak estimates are 1000 barrels per day. On April 29, the US Coast Guard sets fire to a growing oil slick and sprays dispersants in an unsuccessful attempt to stop the oil from reaching shore. Oil washes ashore onto Louisiana wetlands on April 30. President Obama enacts a drilling moratorium.

On May 2, President Obama travels to the Gulf Coast and vows to hold BP accountable for the clean-up costs associated with the oil spill. On May 8, BP tries the "Top Hot" strategy, a siphon attached to the damaged oil well pipe. This strategy is shown unsuccessful, as ice crystals form in and clog the siphon. Oil leak estimates are revised to 5000 plus barrels per day. On May 10, BP considers the "Junk Shot" strategy, to fill the damaged well with all sorts of rubber junk and mud, but concludes the strategy is too risky and calls it off. Meanwhile, Congressional hearings into the disaster begin. Testimony by oil executives is lacking, which offends members of Congress.

On May 14, BP siphons oil from the water into collecting ships. Oil leak estimates continue to rise, with some estimates as high as 70,000 barrels per day. On May 26, the "Top Kill" strategy (pump mud into well) fails. On June 2, criminal inquiries by the US Justice Department into the disaster begins. BP establishes a $2 billion compensation fund for those impacted by the disaster. By July 6, oil reaches all Gulf Coast states' shores. A new well cap and additional skimming ships are brought to the oil well site. The new oil well cap is installed on July 15 and the leak is stopped. Drilling to complete two relief wells continues.

After many controversial remarks, Tony Hayward, CEO of BP resigns and is reassigned elsewhere within the corporation. On August 4, BP successfully employs the "Static Kill" strategy, where mud followed by cement is pumped into the damaged well via the relief well. On August 16, new regulations in the drilling permit application process are implemented. Environmental assessments are required for all future drilling projects.

On September 3, the failed blow-out preventer device is brought ashore for study. At that time, the costs associated with the oil spill are estimated to be $6 billion. BP also publishes its own study, based upon an internal investigation of the incident on September 8. The report blames many other entities and is accused of understating its own responsibility for the disaster. After several tests by the government and others, the Deepwater Horizon oil well is declared "dead." National Commission hearings continue around the Gulf Coast states, revealing many mistakes were made, as this chapter is being written (BBC News, 2010).

TECHNOLOGICAL DISASTER PROCESS MODELS

While much of the disaster theory literature addresses the stages through which natural disasters unfold, there is a growing segment of disaster scholarship that examines how technological disasters are created and unfold. One contribution to this literature is the work of Picou and his associates in the development of the Corrosive model (e.g. Picou, Marshall, and Gill, 2004). According to Picou and Marshall (2007), there *may* be as many as eight phases in a technological disaster are similar to those of a natural disaster. The first phase, *Warning*, occurs when a potential technological hazard presents the possibility of a turning into a disaster event exists in the near future. The second phase, *Threat*, occurs when it is evident that a disaster event is likely to occur. The third phase, *Impact*, occurs when the actual disaster event happens.

The remaining phases of technological disaster, however, are different from those of a natural disaster. After the event, the *Rescue* phase (fourth phase) occurs as people and property are removed from the impact area. Since technological disasters are created by the actions or negligence of humans, the fifth phase involves assigning *Blame* to some person(s), organizations, or institution for the disaster. Additional disaster events may evolve from the initial disaster event, in which people and communities recycle back through the phases of Warning, Threat, Impact, Rescue and Blame. If a society continues to cycle through these five phases repeatedly without moving forward into *Restoration, Reconstruction, and Recovery* phases (where damages are repaired and losses are replaced), a loss of faith in societal institutions (such as government, businesses, and justice system) will occur (Picou and Marshall, 2007).

Another technological disaster process model, Socio-Technological Disaster, is proposed by Aini and Fakhrul-Razi (2010). This model proposes that there are ten phases through which a disaster can evolve. The first phase, *Operation*, occurs when a factory, mine or distribution facility is built. The second phase, *Incubation*, occurs when mistakes are made in the construction or operation of a facility. The third phase, *Forewarning*, occurs when it appears that an accident event is forthcoming. Often, clues that an accident is going to occur are ignored, misinterpreted, or not reported to the proper responsible authorities. Preventive measures to prevent the accident event do not occur. The *Activation* phase occurs when a disaster catalyst is in place to trigger a disaster. A catalyst could be something such as a spark to ignite flammable materials or increasing pressure on a weakened pipe.

When the catalyst triggers the accident, the *Onset* phase occurs. An explosion, fire, or collapse of a facility happens. After some amount of time has elapsed, detection of the accident happens. Actions are taken (*Response* phase) to limit the impact of the event. The effectiveness of response actions are influenced by the presence (or absence) of trained personnel, proper equipment, and coordination of efforts among responding organizations.

After the Response phase, the *Inquiry and Reporting* phase takes place. Investigations into the causes of the disaster occur. After some length of time, *Feedback* occurs, where investigation findings are presented to society and recommendations are made by investigators to compensate victims and prevent similar disasters in the future. Based upon the Feedback received from investigators, victims and government agencies take action (e.g. lawsuits, criminal trials) against those found responsible for the disaster (*Social Justice* phase). The last phase, *Reform*, occurs when new regulations and sometimes new agencies are created to prevent future disasters from occurring (Aini and Fakhrul-Razi, 2010).

A third contribution to the technological disaster literature comes from Shaluf, Ahmadun, and Shariff (2003). They discuss the different types of organizational factors that can contribute to a technological disaster. They argue that the presence of *Factors* such as unqualified or poorly trained workers, poor communication, high work load, management attitudes, and poor project management can create different types of *Errors* within an industrial facility. These errors include Technical, Operator, Management, Organizational, and Operational Errors, and Ignoring of Warning Signs. *Technical Errors* include inappropriate design and installation, poor maintenance, and failure of equipment. *Operator Errors* include poor inspection and maintenance, improper use of equipment, and failure to report problems to managers. *Management Errors* include inadequate planning, budgeting, and policy making.

Organizational Errors include inadequate resource allocations, neglect of safety issues, and communication failures. *Operational Errors* include operators pushing the wrong button, incorrect use of hazardous substances, incorrect repair, and disconnected safety systems. *Ignoring of Warning Signs* include failure to investigate near-accidents, accidents and violations; failure to implement audit recommendations; and failure to follow inspector recommendations. *Defense Errors* include inadequate alarm systems, insufficient personnel to handle an emergency situation, and shortage of safety equipment.

These errors, if left unchecked, eventually create a *Triggering Event*. These events include explosions and toxic releases. The Triggering Event can escalate into a *Large-Scale Disaster* if *External Factors* (such as infrastructure access and emergency preparedness) are inadequate. A Large-Scale Disaster is an event that impacts large human populations and their surrounding environments (Shaluf, Ahmadun, and Shariff 2003).

A fourth model that can be applied to understanding technological disasters is the Disaster Resilience of Place Model (DROP Model), developed by Cutter and her associates (2008). Some useful features of the DROP model, in addition to its application to multiple types of disasters, are that (1) it focuses on the resilience of places, not on populations or social institutions per se; (2) resilience to disasters is viewed as a situation that exists prior to and continues to evolve during and after a disaster event occurs; (3) therefore, the model can be used to examine a disaster at one point in time and/or over a period of time; (4) the model can be applied to both rapidly and slowly unfolding disasters; and (5) the model considers the influences of political policies, locally and beyond, on disaster resilience.

There are several components to the DROP model. The DROP model argues that there are *Antecedent Conditions* (precursors) existing in a society's natural, built, and social environments that contribute to the potential for a disaster to occur. Within these conditions, a *Disaster Event* occurs. The characteristics of the disaster event interact with the *Coping Responses* of society (efforts made to limit the disaster event's impacts) to the event. This interaction determines the extent and nature of the disaster event's Impact on society. If the event does not exceed the *Absorptive Capacity* of society (society's ability to adequately respond to the event), then the *Degree of Recovery* (return to pre-disaster societal conditions) that the society experiences will be relatively high. If the impact exceeds the absorptive capacity of society, then the degree of recovery will be determined by the ability of society to *Improvise* (develop new strategies to respond to the event) and engage in *Social Learning* (apply the new strategies to the event). If improvising and learning do not occur, then recovery will be hampered and relatively low. Depending on the degree of recovery, disaster *Mitigation* (ability to prevent future disasters) and *Preparedness* (ability to respond to potential events of a similar nature) for future events will be affected. These conditions in turn will be part of the antecedent conditions for a potential future disaster event (Cutter et. al, 2008).

COMMON FEATURES OF THE MODELS

One feature common to all four of these models are that there are precursor conditions that exist before the disaster event that make the disaster event possible. These precursor conditions include human error, ineffective response to errors, warning signs, and vulnerabilities in a society's social, natural and built environment. Second, there is the disaster event itself, which is triggered by these precursor conditions. Third, for an event to be deemed a "disaster," there must be some sizeable impact on human populations and their environment. After the disaster event is detected, rescue and recovery operations take place.

The models, with the exception of the Shaluf et. al (2003) model, also agree that investigations into the cause and impacts of the disaster are instigated. During these investigations, blame for the disaster is assigned and discussions of the prevention of future disasters emerge. Efforts into recovery and reconstruction and social justice are also initiated. These models agree that the success of these efforts can vary greatly.

DATA AND METHODS

Our information for the analysis portion of this chapter comes from articles published on-line by the New Orleans' Times-Picayune newspaper from April 20, 2010 until September 30, 2010. This content analysis is guided by the factors we found common to the technological disaster process models discussed above. We are specifically looking in the newspaper articles for discussions of precursor conditions, disaster event triggers, disaster impacts, assigning blame, disaster prevention, recovery, and seeking social justice. While we recognize that most technological disasters can take years or decades to completely unfold and that we are only examining what has occurred within a five month period, we believe our study will provide insight and understanding into the BP Deepwater Horizon Oil Spill disaster, as well as address the usefulness of the technological disaster models presented above.

RESULTS

Based upon the common features of technological disaster models that we outlined above, we highlight key aspects of the BP Deepwater Horizon oil spill and its aftermath. The first key component of the oil spill is the precursor conditions that were in place to create the disaster.

PRECURSOR CONDITIONS

At least ten precursor conditions contributing to the disaster have been identified in media reports. One precursor condition is that the oil well enters the ocean floor at 5,000 feet below the ocean's surface. Drilling at this depth has water temperatures and pressures that prevent direct human access to the well. All access and monitoring of the well occurs indirectly through the use of robots and other technologies.

A second precursor condition is the decision to use a single-barreled (one flow channel for oil) instead of a double-barreled pipe. Use of a double-barreled pipe can decrease the amount of pressure exerted on a well's walls at any one point. It also provides a way to contain the well's contents if one barrel ruptures or malfunctions in the pipe.

A third precursor condition is that the shut-off valves were off-line, being tested at the time of the explosion. After realizing an accident had occurred, workers immediately brought the valves on-line, but they failed to function properly and stop the flow of oil through the ruptured well pipe (Times-Picayune Staff, 2010a).

Fourth, if the equipment had not been off-line, workers responsible for monitoring the pressure in the pipe delayed pressing the Emergency Disconnect

System (EDS) button. Instead, panic and arguing amongst employees ensued before any action was taken (Hammer, 2010).

A fifth precursor is surrounds the cement used to secure the well pipe in place. Workers used nitrogen-based cement, with which they had limited experience in using at great depths.

Sixth, tests on the mud in the bottom of the well were not performed (Bottom's Up test) to see if the mud was stable and if gas leaks were not present before pouring the concrete (Hammer, 2010). Other testimony at hearings indicated that seawater was used in the cement mix, instead of drilling fluid (Associated Press/NOLA. 2010a).

This suggests a lack of standard protocol being followed by workers (seventh precursor).

Eighth, warning signs that methane gas was present at the site were ignored. Several times before the April 20 explosion, workers had to discontinue any "hot work" (welding , cooking, and other uses of fire) to avoid explosions. Brief breaks from work when these "gas kicks" occurred were the only responses to indications that gas leaks were happening (Mowbray, 2010a; Kirkham, 2010a).

Ninth, BP's plan to deal with an emergency at the site was faulty. The plan had wrong phone numbers. The internet link to a supplier was not operational. One of the experts listed in the plan as a person to contact for information had died in 2005. Other aspects of the plan were based upon a potential incident happening in the conditions of the Artic region, not the Gulf of Mexico. For example, the plan describes impacts upon walruses and seals, which are not present in the waters of the Gulf of Mexico (Grace, 2010).

Tenth, there are also questions about employee misconduct and corruption in the Office of Mineral Management. The MMS is reviews and approves permits and is responsible for regulating oil well sites. Employees were found to accept gifts, vacation trips, parties and other items from oil company officials. Other employees were found to use illegal drugs and view pornography with federal email accounts (Schleifstein, 2010).

DISASTER EVENT

The Deepwater Horizon Oil Spill disaster event began when a build-up of methane gas in the oil well pipe created an explosion on April 20, 2010. The explosion ruptured the oil well's pipe, leading to a leak of oil into the water. Attempts to use the shut-off valves to stop the flow of oil into the pipe failed. The explosion also causes the oil well's rig to catch fire, leading to the deaths of 11 workers (Times-Picayune Staff, 2010a).

IMPACT OF EVENT

One immediate impact of event was that eleven workers are killed. Over 120 others had to be rescued from the burning rig or the ocean (Purpura, 2010). A second immediate impact was the leaking of crude oil and methane gas into the ocean (Times-picayune Staff, 2010b). Third, negative economic impacts on the Gulf region's tourism and seafood industries were felt. Tourists canceled bookings and oysters, fish, and such were feared to be contaminated. Meanwhile, the oil clean-up related industries experienced an economic boom (White, 2010a).

A fourth impact of the event is the psychological suffering of Gulf Coast residents, especially children. Depression, anxiety, and other stress-related illnesses increased. Also, incidences of substance abuse and violence were on the rise follow the oil spill (Barrow, 2010). Impacts on the natural environment are expected to be long-term as well (Rioux, 2010a).

RESCUE AND RECOVERY OPERATIONS

Rescue efforts initially focused on removing workers from the waters surrounding the burning oil rig. A supply ship operator heard a hissing sound before the explosion and sounded an alarm to notify crew to come to ship. After the explosion, the Coast Guard arrived on the scene to assist with moving over 120 workers to safety (Purpura, 2010). Later rescue efforts focused on removing affected birds, turtles and other marine life from the area (Kirkham, 2010b; Associated Press/NOLA, 2010c).

Recovery efforts have been varied, to include actions to remove oil from the water and plug the leaking well, as well as provide funds to those whose incomes were adversely affected by the spill (Reid, 2010; Alpert, 2010a). Intermittently, oil slicks were set on fire to burn away spilled oil (Mowbray, 2010b). Oil dispersants were also sprayed on the oil to break-down the slicks into smaller particles (Associated Press/NOLA, 2010b). Berms were placed along shore-lines in Louisiana and Alabama in an attempt to stop the oil from entering into ecologically sensitive marshes and onto public beaches (Rioux, 2010b). Skimmer ships were deployed to collect oil from the water. (Kirkham, 2010c)

Meanwhile, efforts were also underway to shut-off the flow of oil into the water. One attempt was referred to as "Top Hat." A device that physically resembled a hat or a house was lowered over the well. This device had a pipe that was inserted into the well's pipe, in an attempt to gain control over the oil flow and continue to harvest oil from the well. Because of the depth of the well, ice crystals began to form and clog the pipe, so this approach was discontinued (White, 2010b).

Afterwards, an approach referred to as "Junk Shot" was considered. It would involve pushing all sorts of rubber junk and mud into the well. There were some concerns expressed that the pressure of the well might be too much for the junk to remain in place and that debris forced back out of the pipe might produce more damage to the oil well. Instead, mud was pumped into the well (White, 2010c). As some had expected, the pressure of the well was too great and the mud was spewed back out of the well.

Another approach was to drill two relief wells, on the sides of the well, which would intersect the damaged well and be used to shut the oil flow off. As the relief wells were being drilled, efforts to replace the cap on the damaged well proceeded. Eventually, a new cap was placed on the damaged well, which stopped the flow of oil into the water. In early August 2010, one of the relief wells intersected the damaged well. Mud was pumped into the relief well (referred to as "Top Kill"), eventually plugging the damaged well. After pressure checks were made to make sure the plugging would hold, cement was forced into the well to finalize the sealing off of the damaged well (White, 2010d)

INVESTIGATION AND BLAME

A few weeks after the oil rig explosion, a series of investigations began (Times-Picayune Staff, 2010c). Oil company executives were made to testify before Congress about their plans to prevent and clean-up similar oil leaks (Associated Press/NOLA, 2010d). Tony Heyward, the executive in charge of BP North America, endured fierce questioning about his company's actions to resolve the oil disaster (Associated Press/NOLA, 2010e).

President Obama created a National Oil Spill Commission to do further interviewing and investigating of the oil spill incident (Hammer, 201b). BP also did its own internal investigation into the accident, blaming Transocean, Haliburton and other companies in the construction of a faulty oil well (Hammer, 2010c).

POLICY CHANGE AND SOCIAL JUSTICE

In response to the investigations into the disaster, the Office of Mineral Management has been divided into two sections. One section will deal with permits; the other section will deal with fee collections. These efforts were taken to interrupt the too-friendly relationships between industry and regulator (Tilove, 2010).

Another response has been the filing of both civil and criminal charges against BP and others by the U.S. Justice Department. Federal prosecutors opened offices in New Orleans to serve as a central headquarters for compiling cases. Criminal charges against BP include violation of the Clean Water Act, Endangered Species Act, and Marine Animals Act. Civil charges include violation of Occupational Safety and Health Act, Clean Water Act, Oil Pollution Act, and Outer Continental Shelf Land Act (Hammer, 2010d).

BP was also required by the Obama administration to open a trust fund to make payments to those impacted by the oil spill. Ken Feinberg is the administrator of this fund, in charge of receiving applications and distributing payments to affected individuals and businesses (Hammer, 2010e).

Congress has also raised the possibility of increasing the liability cap, amount that businesses are required to pay for damages for an incident, to $20 billion. The original cap is $75 million (Alpert, 2010b).

CONCLUSION

This chapter examined four major process models for technological disasters. We found six features common (precursor conditions, disaster event, impacts, rescue and recovery operations, investigation and blame, and social justice efforts) to these models. Using these common features, we highlighted important characteristics of the BP Gulf Oil Spill disaster.

Future scholarship on this disaster should examine in-depth the long-term impacts on populations and social institutions in society. This chapter was written only 5 months after the disaster; the impacts and adaptations documented are relatively short-term. Long-term impacts and changes in the way that the energy industry operates and society responds to similar disasters should be followed.

REFERENCES

Aini, M. S. and Fakhrul-Razi, A. 2010. "Development of Socio-Technical Disaster Model." *Safety Science*, 48: 1286-1295.

Alpert, Bruce. 2010a. "Obama administration releases fund agreement with BP in Gulf oil spill." *The Times-Picayune*, August 11.

Alpert, Bruce. 2010b. "Raising industry liabilityin wake of Gulf oil spill becomes partisan issue." *The Times-Picayune*, June 7.

Associated Press/NOLA. 2010a. "Bad cement jobs plague offshore oil rigs." *The Times-Picayune*, May 24.

Associated Press/NOLA. 2010b. "Louisiana leaders asking BP about dispersant risks in Gulf oil spill." *The Times-Picayune*, May 10.

Associated Press/NOLA. 2010c. "NASA's turtle egg rescue from Gulf oil spill is deemed a success." *The Times-Picayune*, September 8.

Associated Press/NOLA. 2010d. "Gulf oil spill testimony to Congress: Not our fault." *The Times-Picayune*, May 11.

Associated Press/NOLA. 2010e. "BP executive accused of 'stonewalling.'" *The Times-Picayune*, June 17.

Barrow, Bill. 2010. "Social fallout from Gulf of Mexico oil spill may last for years." *The Times-Picayune*, July 15.

BBC News. 2010. "Timeline: BP oil spill." September 19. http://www.bbc.co.uk/news/world-us-canada-10656239.

Cutter, S.L., Barnes, L., Berry, M., Burton, C., Evans, E., Tate, E., and Webb, J. 2008. "A place-based model for understanding community resilience to natural disasters." *Global Environmental Change*, 18: 598-606.

Grace, Stephanie. 2010. "As BP promised, the walruses are fine: Stephanie Grace." *The Times-Picayune*, June 10.

Hammer, David. 2010a. "Deepwater Horizon's ill-fated oil well could have been handled more carefully, hearings reveal." *The Times-Picayune*, May 29.

Hammer, David. 2010b. "Oil spill commission let BP exec off easy in questioning Monday." *The Times-Picayune*, July 12.

Hammer, David. 2010c. "BP internal investigation into Gulf oil spill mostly points fingers at others." *The Times-Picayune*, September 8.

Hammer, David. 2010d. "Feds establish downtown bunker to build criminal, civil cases against BP, others in Gulf oil spill." *The Times-Picayune*, September 26.

Hammer, David. 2010e. "BP cuts oil spill payments to businesses in half in August." *The Times-Picayune*, August 12.

Kirkham, Chris. 2010a. "Deepwater Horizon oil well was increasingly out of control before fatal explosion, widows testify." *The Times-Picayune*, June 7.

Kirkham, Chris. 2010b. "Bird rescue efforts filled with frustration." *The Times-Picayune*, June 14.

Kirkham, Chris. 2010c. "Red tape keeps prized oil-fighting skimmers from Gulf, coastline." *The Times-Picayune*, June 29.

Mowbray, Rebecca. 2010a. "Drilling relief wells to stop Gulf oil leak poses challenges." *The Times-Picayune*, May 30.

Mowbray, Rebecca. 2010b. 'BP did five `controlled burns' Wednesday of oil released in Gulf of Mexico oil spill." *The Times-Picayune*, May 6.

National Wildlife Federation. 2010. *Assault on America: A Decade of Petroleum Company Disaster, Pollution, and Profit.* National Wildlife Federation: Reston, VA.

Picou, J.S., Marshall, B.K., and Gill, D.A. 2004. "Disaster, litigation, and the corrosive community." *Social Forces*, 82 (4): 1493-1522.

Picou, J. S. and Marshall, B.K. 2007. 'Katrina as a paradigm shift." pp. 1-22 in *The Sociology of Katrina: First Edition.* Edited by David L. Brunsma, David Overfelt, and J. Steven Picou. Lanham, MD: Rowman & Littlefield Publishers, Inc.

Pupura, Paul. 2010. "Terror, daring rescue in Gulf of Mexico oil rig recounted." *The Times-Picayune*, May 7.

Reid, Molly. 2010. "Eager volunteers having trouble finding ways to contribute to Gulf oil spill cleanup." *The Times-Picayune*, June 26.

Rioux, Paul. 2010a. "Gulf of Mexico oil spill's environmental impact expected to be profound, long-lasting." *The Times-Picayune*, May 7.

Rioux, Paul. 2010b. "Federal government will make BP pay for all six berms in Gulf oil spill fight." *The Times-Picayune*, June 2.

Shaluf, I.M., Ahmadun, F-R., and Shariff, A.R. 2003. "Technological disaster factors." *Journal of Loss Prevention in the Process Industries*, 16: 513-521.

Schleifstein, Mark. 2010. "Gulf region MMS employees accepted gifts, food, tickets at oil and gas company expense." *The Times-Picayune*, May 25.
Tilove, Jonathan. 2010. "Comments on Mineral Management Service duties to be divided in wake of Gulf oil leak." *The Times-Picayune*, May 11.
Times-Picayune Staff. 2010a. "How the Gulf of Mexico oil spill happened: a graphic presentation." *The Times-Picayune*, May 7.
Times-Picayune Staff. 2010b. "Gulf of Mexico oil spill may be twice the size estimated two weeks ago." *The Times-Picayune*, June 10.
Times-Picayune Staff. 2010c. "8 investigations underway into Gulf of Mexico rig explosion and oil spill." *The Times-Picayune*, May 17.
White, Jacquetta. 2010a. 'Economic impact of Gulf oil spill varies by industry." *The Times-Picayune*, May 9.
White, Jacquetta. 2010b. "BP still deciding how best to deploy `top hat' to contain Gulf of Mexico oil spill. "*The Times-Picayune*, May 12.
White, Jacquetta. 2010c. "Top kill strategy to be adjusted to incorporate different clogging materials." *The Times-Picayune*, May 27.
White, Jacquetta. 2010d. "No oil found at BP relief well junction in Gulf of Mexico." *The Times-Picayune*, September 17.

CHAPTER THREE
BEYOND PETROLEUM IN THE GULF OF MEXICO: THE CHARACTERISTICS AND CONSEQUENCES OF CATASTROPHE

JOHN BARNSHAW AND LYNN LETUKAS

The failure of a blowout preventer below the offshore drilling vessel Deepwater Horizon in the Gulf of Mexico is a recent example of an occasion generating exceptional social disruption for ecologies, individuals, communities, and regions. As humans and societies encroach further upon ecological fields and develop increasingly complex technological innovations, the risk that a socio-technological exception will occur increases. This chapter explores social change associated with exceptions that generate substantial disruption to ecological and social systems. Utilizing a case study methodology based upon archival records and documents, we explore and attempt to falsify the conceptualization of catastrophe.

Catastrophe is different from a disaster in that most or all of a community or area is impacted, many local individuals are unable to undertake their usual roles, the community undergoes a period of prolonged inoperability, media plays an increasing role in the construction of the public perception of the occasion, and the political arena shifts from a local or regional issue to a national or international issue. The distinction between disaster and catastrophe is not simply an academic one, as national and international policymakers have recently become sensitized to differences between these distinctions, and emergency management professionals and policymakers have asserted that more academic research is needed to empirically falsify conceptualization distinctions. This chapter offers additional conceptual clarification on catastrophe, particularly as it relates to developments in new media technology and recent shifts in policymaking from compliance to surveillance.

For more than a half century, disaster and crisis researchers have struggled to define and conceptualize their central object of study. In 1950, when the first team of social science disaster researchers assembled at the National Opinion Research Center (NORC) at the University of Chicago, one of those scholars, Robert Endleman, defined *disaster* as:

> an event, concentrated in time and space, in which a society, or a relatively self-sufficient subdivision of a society, undergoes severe danger and incurs such losses to its members and physical appurtenances that the social structure is disrupted and the fulfillment of all or some of the essential functions of the society is prevented. (Endleman, 1952)

Although the Endleman definition was not published for nine years, it encapsulated many of the central concepts and concerns of the earliest disaster researchers (Quarantelli, 2006). Endleman's definition of disaster gained considerable attention among sociologists with the publication of Charles Fritz's chapter in Robert Merton and Robert Nisbet's widely read *Contemporary Social Problems* (1961). More than forty-five years later, Tierney (2007:505) acknowledges that the Endleman/Fritz definition of disaster remains "highly influential" in how many researchers understand disaster.

Although the Endleman/Fritz definition is the most cited definition of *disaster* in the social sciences, there are several *methodological*, *theoretical* and *substantive* issues that make its use problematic for disaster researchers (Tierney, Lindell and Perry, 2001). The Endleman/Fritz definition is methodologically problematic because the short-term "event concentrated in time and space" that disrupts the "essential functions of the society" was reinforced by a methodological approach that emphasized the importance of gathering short-term observations and data that were deemed ephemeral (Fritz, 1961:655; Natural Hazards Application and Information Center 2011; Stallings, 2002). Although some outstanding quantitative surveys (Bolin and Bolton, 1999; Brodie et al., 2006; Moore, 1958), and some long-term ethnographic work has resulted from these initial quick response research trips (Oliver-Smith, 1999; Picou et al. 1992), the vast majority has been qualitative interviews that are post-hoc and limited in generalizability beyond the initial impressions of those impacted. This is not to say that these data are not important, indeed we believe that they have been essential to the development of the field of disaster research. Our point in drawing attention to the post-hoc approach is to note how the methodological approach has influenced the findings and theorizing that conceptualizes disaster as an aberrant disruption of the social system.

Tangentially related to the methodological focus on "event" is that when findings generated from quick response or ephemeral data are reported, they are often contextualized in a theoretical approach that is often a theory of the middle range that is limited in scope and generalizability (Blau, 1995; Stallings, 1998; Tierney, 2007). Thus, by focusing on broader methodologies and more inclusive data collection processes that allow for broader theoretical development beyond theories that are narrowly tailored to explain phenomenon such as how individuals perceive warning messages or how organizations adapt in disaster may lead to more robust sociological inquiry and theory development.

Substantively, there are also two problems with the Endleman/Fritz definition of disaster. First, the use of "event" focuses both the researcher and the research on one singular event as the genesis of a particular disaster, which often, is not the case. Quarantelli (1998) and Perry (1998) have both pointed out that since disasters are social constructions, rather than purely physical phenomena, agents, social systems and societies influence disaster. Therefore, although certain moments may play a larger role in disaster than others, even what may appear to be the most singular event is actually the product of a longer history of preconditions for disaster. For example, Oliver-Smith (1999) demonstrated that the devastation resulting from the Peru earthquake on May 31, 1970 was the product of nearly five hundred years of South American settlement location, planning and building techniques.

The second substantive problem with the Endleman/Fritz definition is that one runs the risk of *ontological gerrymandering*, which is the illicit insistence on defining a term in a way that is favorable to one's own understanding of a phenomenon (Blackburn, 1996; Woolgar and Pawluch, 1985). For example, in a search of disaster definitions, Westgate and O'Keefe (1976) found that the majority of previous definitions were programmatic declarations that were useful for politicians and policymakers in declaring an area as a disaster rather than offering definitive criteria that would be useful for systematically defining or establishing boundaries of the phenomenon. Ontological gerrymandering does not appear to be simply a political problem as Dombrowsky (1998:20) notes that among disaster researchers "we see what we want to see." Thus, by focusing only on issues that are germane to the politician, policymaker, or disaster researcher, one runs the risk of creating favorable definitions that support one's own understanding of disaster while potentially excluding substantive findings that are deemed "problematic." For example, consider the classical Endleman/Fritz (1961:655) definition where disaster is seen as generating "severe danger" to a community resulting in destruction of "physical appurtenances" and "losses to its members." This definition easily conjures up images of what most people traditionally consider the domain of "natural disasters" such as floods, earthquakes, hurricanes, tornadoes, and tsunamis. However, consider for a moment research by Klinenberg (2002) on the 1995 Chicago heatwave where 521 people died with almost no destruction or devastation to physical appurtenances. Should the Chicago heatwave, clearly a "natural" meteorological occasion generating exceptional disruption to the community be excluded from the definition of a disaster because it did very little damage or destruction to physical appurtenances?

A potential solution to the methodological, theoretical and substantive problems of the Endleman/Fritz definition is to develop a *conceptualization* of disasters based upon *social occasions* and *exceptions*. In contrast to definitions that purportedly offer a definite, or final account of a phenomenon, a conceptualization is a process that sensitizes the researcher to certain ongoing aspects of the social world (Neuman, 2003).

A major advantage of sensitizing concepts is that they can be used to draw attention to important social features that otherwise may have been missed, thereby limiting the influence of ontological gerrymandering (Charmaz, 2003; Padgett, 2004; Bowen, 2006). Although sensitizing concepts are generally used in an inductive manner (Glaser and Strauss, 1967; Strauss and Corbin, 1990), we think that deductive approaches can be applied to sensitizing concepts in a study. For example, there are numerous observations and propositions developed by prior scholars about behavior in general and also specifically in disasters. We follow this convention in an attempt to determine whether or not behavior in catastrophe is similar or different from prior research.

The potential solution to the problem of the Endleman/Fritz focus on the "event" concentrated approach is to avoid the determinism associated "event" and focus on *projects, social occasions,* and *exceptions*. Dynes (1998:113) notes that "event" can imply a determinism that if an earthquake, hurricane or terrorist attack is going to take place, there are few measures agents or social systems can take to prevent it. To avoid this pitfall, Dynes (1998:113) recommends the concept of "occasion," which offers a more effective organizing concept. Occasion provides a socially defined process that is contextually bound beyond the fixed temporality of "event" into a "before," "during" and "after." Thus, the occasion serves as the catalyst that "generates" exceptions, or social disruption, that is part of a larger historical project of "before," "during" and "after" disaster (Kreps, 1998). In our view, this advancement over the traditional understanding of disaster is not without its own issues and greater clarification is necessary in what is meant by *project, exception,* and *scale*, which is where we turn next.

PROJECTS, EXCEPTIONS AND SCALE

Retrospectively constructing the past is hardly novel in sociology or disaster research. The French sociologist Emile Durkheim (1897) retroactively gathered data about suicides in France and drew inferences about his subject to better understand how social solidarity, or a lack thereof, influences the likelihood of suicide. Moreover, for more than a century in American sociology, race scholars have developed rather sophisticated conceptual schemes for retrospectively constructing the racial history of individuals and societies (DuBois, 1920; Feagin, 2010; McKee, 1993). In disaster research, the Canadian sociologist Samuel Prince (1920) retrospectively studied the individual and community response to 1917 Halifax explosion to better understand catastrophe and social change.

This current research offers an expanded conceptualization of *projects* to emphasize their social construction as well as their composition of fields, structures, agents, and arrangements to construct a conceptualization that attempts to take each of these intersecting processes into account through operational means. Central to the use of projects is to understand how they exist at a current period of time and how they change. Historically situated projects will continue into perpetuity until a social exception occurs. An *exception* is a substantial deviation from the constructed project. Exceptions differ from fluctuations in that they substantially alter the field, structures, agents and/or arrangements of the constructed project.

There are two general propositions consistent with all types of exceptions. First, *exceptions generate uncertainty*. No matter how small, large or complex an exception, there is a period where the substantial deviation is undefined or unknown. Second, irrespective of the type of exception, *some action will occur*. At any given time, some individuals, organizations or subsets of society will have an interest in developing new possibilities brought about by the exception or attempting to restore some of what was previously constituted by the exception. Thus, once an exception has occurred, it is not possible for inaction to result in a constructed project but rather, a new project is in the process of formation or reformation. This is not to say that all fields, structures, agents or arrangements must participate in some type of action, it simply indicates that some fields or structures, agents, or arrangements will engage in action in an attempt to resolve the exception.

Disaster is an ideal context for studying exceptions because they are often behaviorally observable rather than asking about lines of action constructed solely in the mind (Mead, 1934). Thus, by focusing on disasters as exceptions, researchers are able to assess social change in the historically constructed before, during and after disaster. Immediately before, during, and following a disaster, routine patterns of action are significantly disrupted as well as the social structures and built environment. In an effort to better understand what has been lost and what has been retained from a project, our view is that it is essential to understand the *scale* or size of an exception. To measure scale, we attempt to understand exceptions in their impact on the routines or rituals of individuals, organizations, communities in the social and physical fields they inhabit.

Therefore, within the context of an exception exists a continuum of based upon the scale of the exception. Figure 3.1 describes a continuum of social exceptions ranging in scale from emergency (smallest level of exception) to extinction level occasion (largest level of exception). First, an *emergency* is an exception involving first responders that is often resolved with little disruption to larger communities and social structures in the project. Second, a *mass/complex emergency* is a complex exception such as a large traffic accident or plane crash that is often resolved through complex responses that do not constitute significant disruption to communities and social structures. Third, a *disaster* is an occasion generating exceptional social and structural disruption. Fourth, a *catastrophe* is an occasion generating exceptions where most or all of the community is impacted, local officials are unable to undertake their usual work roles and the community undergoes a period of prolonged inoperability. Finally, an *extinction level occasion* is an exception generating substantial challenges to all communities and social structures in a region.

Figure 3.1—Continuum of Exception Occasions

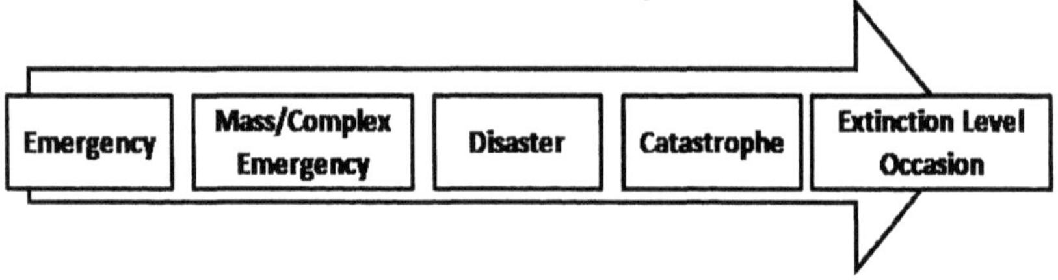

A complex emergency, such as a plane crash, may result in exceptions to routines, but not to the extent that the exceptions constitute a significant disruption to communities and social structures. Thus, authorities and emergency professionals are able to respond in such a way that most members of a community and social institutions are not adversely impacted. Conversely, a disaster is not sufficiently as large as a catastrophe. Given our understanding of disaster and the level of exceptions generated, we hypothesize that our current project of analysis, the Gulf of Mexico oil spill emanating from the *Deepwater Horizon* is a catastrophe rather than the smaller exception of a disaster. However, a better understanding of the difference in scale of exceptions between disaster and catastrophe is necessary, which is where we turn next.

CATASTROPHE AS A SOCIAL EXCEPTION

Although there are a number of recent summaries and reviews of the literature on disaster (Tierney, Lindell and Perry, 2001; National Research Council, 2006: Rodriguez, Quarantelli and Dynes, 2006) prior research by Quarantelli (1982; 1993; 2006) has suggested that catastrophes may be fundamentally different from disasters in several important ways. A *catastrophe* is a social occasion that generates exceptions where most or all of (1) the community structure is impacted, (2) the community rituals or routines are disrupted and local officials are unable to undertake their routine work roles, (3) help from nearby communities cannot be provided due to the scale of the exception (4) the community undergoes a period of prolonged inoperability, (5) the mass media socially construct the occasion, and (6) the political arena emerges as increasingly significant in dealing with the response. Figure 3.2 provides a chart of the conceptual characteristics of catastrophe which is described below.

First, in a catastrophe most or all of the community built structure is heavily impacted. For example, in the 2010 Haiti earthquake, at least 230,000 persons were killed and more than one million—nearly ten percent of the entire country—were left homeless (British Broadcasting Corporation, 2010). In contrast, in the 1985 Mexico City earthquake, considered a disaster, less than two percent of the residential housing structure stock was lost, with only 4.9 percent of the population reporting great damage to the building in which they lived in and around Mexico City (Quarantelli, 2006).

Figure 3.2—Conceptual Characteristics of Catastrophe

1. Most or all of the community structure is impacted.
2. Most or all of the community rituals or routines are disrupted and local officials are unable to undertake their routine work roles.
3. Most of all of the help from nearby communities cannot be provided due to the scale of the exception.
4. Most or all of the community undergoes a period of prolonged inoperability.
5. The mass media socially construct the occasion.
6. The political arena emerges as increasingly significant in dealing with the response.

Second, in a catastrophe most or all of the community rituals or routines are disrupted and local officials are unable to undertake their routine work roles. Sociologist Randall Collins (2004) has argued that individuals use symbols to form the basis of interaction and routines, which, in turn, form ritual chains of interaction. These micro-level interactions form the basis of social structures and much of what is found in the meso- and macro-levels of society. Thus, when a socio-historical project experiences an exception, social change is often experienced at the interactional micro-level. Related to this observation, when a catastrophe occurs, local personnel specializing in catastrophic occasions are often unable to carry out their formal and organizational work roles (Quarantelli, 2006). This is because some local workers are either dead or injured, and/or are unable to communicate the knowledge or skills they routinely provide (Barnshaw, Letukas and Quarantelli, 2008). For example, in the 2004 Indian Ocean tsunami practically all government officials were killed by the impact, fisherman were unable to fish, health professionals were unable to provide routine assistance in hospitals and thousands of foreign tourists were stranded (Letukas, 2008; Letukas, Olofsson and Barnshaw, 2009). In sharp contrast, in disasters routine patterns for large sections of society are maintained as following the 1994 Northridge earthquake in Los Angeles 12,000 people went to the horseracing track the afternoon of the earthquake (Barnshaw, Letukas and Quarantelli 2006).

Third, in a catastrophe most or all of the help from nearby communities cannot be provided due to the scale of the exception. In many catastrophes most of the residents in a particular community affected, but often, those in nearby localities are also impacted. In 1986, many areas around Chernobyl after the accident at the nuclear plant were unable to respond for risk of contamination (Porforiev, 1998).

In short, catastrophes affect multiple communities, and often have a regional character.

Fourth, in a catastrophe most or all of the community undergoes a period of prolonged inoperability. In addition to most assistance coming from distant individuals, organizations and communities, the impacted communities are often inoperable or unusable for a period of time (Quarantelli, 2006). For example, following the 1986 nuclear accident at Chernobyl almost the entire community has been displaced and most have not returned nearly twenty-five years after the occasion (Swiss Agency for Development and Cooperation, 2010).

Fifth, in a catastrophe the mass media system, especially in recent times, socially constructs the occasion. All disasters evoke at least local mass media coverage. Some disasters can attract attention from outside the local community media, to draw regional, national and at times, international attention. However, this disaster coverage rarely lasts beyond the first few days, or at most, the first month depending upon competing news cycles (Hilgartner and Bosk, 1988). Even media coverage following the September 11, 2001 attacks of the occasion dropped off considerably after the first few weeks except in the impacted metropolitan areas of New York and Washington. In contrast to disasters, mass media coverage of catastrophe differs in that there is generally much more and longer coverage by regional and national mass media than local media. This is partly because local coverage is reduced, if not totally displaced, during and following the exception.

Finally, because of the previous five processes, *in a catastrophe the political arena becomes even more important in dealing with the impact and response.* All disasters involve, at a minimum, local political considerations. However, it is a radically different occasion when the national government and elite officials become directly involved (Quarantelli, 2006). Even with the most socially prominent disasters, such as the September 11, 2001 attacks, a symbolic presence is often all that is necessary. However, in catastrophes, that symbolism is not enough, particularly for larger society. Part of this stems from the fact that catastrophes often force larger society to deal with the most fundamental constructions of the previously constituted project. For example, as the levees were breached and the floodwaters rose in New Orleans following Hurricane Katrina, television images of thousands of citizens trapped outside the Louisiana Superdome and New Orleans Convention Center, without adequate food and water, caused many Americans to "rediscover" the urban underclass as well as racial and ethnic differences that are papered over during routine occasions (Barnshaw, 2006).

Another reason the political arena is increasingly important during and following a catastrophe is that it is easy to take partisan political advantage of such times as organizational weaknesses of responding organizations becomes obvious. For example, following the 1986 Chernobyl nuclear reactor accident, the Soviet government initially tried to keep the catastrophic failure secret, but when this was not possible due to radiation monitors in Sweden, the political cost had to be paid by Soviet actors and institutions (Porforiev, 2006).

Given our understanding of catastrophe, we hypothesize that our current project of analysis, the Gulf of Mexico oil spill emanating from the *Deepwater Horizon*, is a catastrophe rather than the smaller exception of a disaster. First, however, a better understanding of our methodological approach and data utilized is necessary, which is where we turn next.

DATA AND METHODS

This chapter utilizes a case study methodology (Yin, 2009). *Case study methodology* is a research design that focuses on one or more specific instances, institutions, or occasions (cases) for the purpose of exploring a process that may not necessarily be generalizable in a statistical sense. The case study methodology is ideal for examining contemporary events when behavior cannot be manipulated because it allows for a blending of various data sources to provide greater explanatory power than the historical method (Yin, 2009). Although a case study methodology may include up to six sources of evidence (documents, archival records, interviews and surveys, direct observation, participant observation and physical artifacts), this chapter utilized documentary and archival records (Yin, 2009). Documentary sources of evidence included memoranda, administrative information, Congressional testimony transcripts, after action reports by private and public sector entities as well as annual reports of several of the parties involved such as British Petroleum, Halliburton and Transocean for a total of nineteen documents. Archival records included three newspapers including two national newspapers (*The New York Times*; *The Washington Post*) and one newspaper located in the impacted region (*The Times-Picayune*, New Orleans, LA).

These two national newspapers have the highest circulation in the United States, and in particular, the *Washington Post* provides detailed coverage of political and governmental involvement in national issues and events. Since two of our propositions of catastrophe include *the national media increasing role in the construction of the occasion*, and *the extent to which the event receives national and international political coverage*, we concluded that the *New York Times* and the *Washington Post* would be ideal sources to examine the extent to which the oil spill was the subject of national media attention. We also examined one local newspaper *The Times-Picayune* because it provided detailed coverage of how local communities were impacted by the oil spill. Our time period of analysis included April 20, 2010, the day of the Deepwater Horizon oil rig explosion in the Gulf of Mexico, to September 10, 2010, two days following the release of the internal report from British Petroleum (BP) detailing the cause of the explosion, which allowed for a detailed analysis of the media construction of the short-term response and recovery efforts as well as coverage of the BP investigation (Tierney, Lindell, and Perry, 2001).

We used *Lexis-Nexis Academic* to conduct a full-text database search using two search terms that were determined to be the most frequently used to describe the oil spill by the mass media, after preliminary examination of articles in all three newspapers: "BP oil spill," "BP," and "Deep Water Horizon." The Lexis-Nexis search identified a total of 2,342 newspaper articles, all of which were subjected to three selection criteria (duplication, relevance, content), following which a combined total of 2,252 articles remained. Finally, we selected a sub-sample of every tenth article resulting in 62 *New York Times*, 45 *Washington Post* and 119 *Times-Picayune* articles (N=229).

In order to explore and attempt to falsify the conceptualization of catastrophe as developed by Quarantelli (2006), this research utilized a deductive methodological approach in its analysis of news content. The software program ATLAS.ti 6.0 was used for the coding and analysis of these articles. After the initial coding, all codes, notes and memos were read through and analyzed. Representative quotations were then selected, some of which have been used in the presentation of our findings, which is where we turn next.

FINDINGS

Previously, Quarantelli (1993; 2006) has asserted that *catastrophe* is a social occasion that generates exceptions where most or all of (1) the community structure is impacted, (2) the community rituals or routines are disrupted and local officials are unable to undertake their routine work roles, (3) help from nearby communities cannot be provided due to the scale of the exception (4) the community undergoes a period of prolonged inoperability, (5) the mass media socially construct the occasion, and (6) the political arena emerges as increasingly significant in dealing with the response. Based upon our analysis of archival sources and documentation we generally find support for these claims.

First, in a catastrophe most or all of the community built structure is heavily impacted. The *Deepwater Horizon* explosion is relatively unique to the study of catastrophe in that in most catastrophes the heavy physical loss of community structure is readily observable. However, in some catastrophes such as the 1986 nuclear reactor accident at Chernobyl the community built structure was heavily impacted in that it was deemed inoperable or unusuable although still largely intact. Similarly, following the *Deepwater Horizon* explosion, we find substantial evidence of community built structures being heavily impacted, most notably along the Gulf of Mexico, with the largest economic costs in the urban centers such as New Orleans.

In our analysis of the *BP Deepwater Horizon Accident Investigation Report* (2010) there is no mention of individuals, organizations or communities being impacted or heavily impacted by the accident (British Petroleum, 2010). However, in the *Report to the President by the National Commission on the BP Deepwater Horizon Oil Spill and Offshore Drilling* (2011), we found numerous instances of community destruction and impact. First, and most directly impacted was the community on the oil rig *Deepwater Horizon*, where eleven of the 126 crew were killed and sixteen others were seriously injured (National Commission on the BP/Deepwater Horizon Oil Spill and Offshore Drilling, 2011). Approximately 36 hours after the initial exception that led to the explosion on the *Deepwater Horizon* intensive fire-fighting efforts were unable to save the $560 million oil platform, which sank to the bottom of the Gulf of Mexico (Transocean, 2010). The explosion on the *Deepwater Horizon* set off a cascading exception resulting in an estimated 52,700 to 62,200 barrels of oil spilling per day into the Gulf for a total of approximately 4.9 million barrels, by far the largest maritime oil spill in United States history (National Commission on the BP/Deepwater Horizon Oil Spill and Offshore Drilling, 2011). The spill is estimated to cover a concentrated area of 2,500 miles and a diffuse area of 68,000 square miles with more than 600 miles of shoreline being fouled in five states (Norse and Amos, 2010). This tremendous oil spill has long-term implications for the ecological sustainability of the Gulf of Mexico and similar to Chernobyl, the full impact will not be realized for years.

Second, in a catastrophe most or all of the community rituals or routines are disrupted and local officials are unable to undertake their routine work roles. In the afore mentioned commission report (2011), there was evidence of substantial disruption to environmentally sensitive ecologies that have dramatically impacted the seafood and tourism industries, two of the largest components of the regional economy (National Commission on the BP/Deepwater Horizon Oil Spill and Offshore Drilling, 2011). In 2008, Gulf Coast fisheries harvested 1.27 billion pounds of fish and shellfish generating $659 million in revenue (National Oceanic and Atmospheric Administration, 2010). In addition to the commercial fishing industry, there were an estimated 3.2 million fishers in the Gulf Coast region who took 24 million trips (National Oceanic and Atmospheric Administration, 2010). Following the *Deepwater Horizon* spill, approximately 33 percent of all federal waters were closed extending as far away from Louisiana as Panama City Beach Florida.

Tourism generates an estimated $19.7 billion annually in the Gulf Coast region, with approximately 50 percent of the total coming from Florida (United States Census Bureau, 2007). In addition to the disruption of employees of the fishing and tourism industries the impact of the oil spill has also adversely impacted the health of many Gulf residents. While the long-term effects of contamination may not be known for years, the disruption in loss of jobs and income due to the spill is salient in the 25 percent increase in clinical depression among Gulf facing counties since the oil spill (National Commission on the BP/Deepwater Horizon Oil Spill and Offshore Drilling, 2011).

Third, in a catastrophe most or all of the help from nearby communities cannot be provided due to the scale of the exception. Within moments of the Deepwater Horizon explosion it was apparent that the necessary resources to resolve the exception were not readily available. Within hours, BP began hiring vessels to skim oil off the surface of the water and began ordering boom in an attempt to surround and contain oil that was rapidly gushing from the blown out well (Resnick-Ault and Klimasinska, 2010). Although BP was initially reluctant to use outside international assistance, four weeks after the explosion, they began accepting skimmers, boom and technical assistance from international partners with experience in cleaning up oil disasters (Eilperin and Kessler, 2010).

Fourth, in a catastrophe most or all of the community undergoes a period of prolonged inoperability. Following the *Deepwater Horizon* explosion, large areas of the Gulf Coast were deemed inoperable for human activity. On May 2, 2010, the National Oceanic and Atmospheric Administration (NOAA) National Marine Fisheries Service began closing an area of approximately 6,817 square miles, roughly three percent of the entire Gulf of Mexico federal fishing zone and continued closing areas and by one month later, on June 2, 2010, it prohibited 37 percent of the entire federal Gulf fishing zone (National Commission on the BP/Deepwater Horizon Oil Spill and Offshore Drilling, 2011). Later, in May and July 2010, President Barack Obama issued a moratorium on offshore drilling for an indefinite term that was later clarified as six months (National Public Radio, 2010). At the local and state level, much of the fishing and tourist communities have remained closed due to the impact of the Deepwater Horizon spill and some jobs, businesses will never recover (Alpert, 2010b).

Fifth, in a catastrophe the mass media, especially in recent times, socially constructs the occasion. With the proliferation of twenty-four hour cable news, the Internet, blogs and citizen reporters, perhaps the greatest changes have come in how catastrophe are understood by those who experience the impact and the broader public. Since the *Deepwater Horizon* oil spill was constructed as an exception that was facilitated largely through rational-technical means, this sensitized the media, the public, and policymakers to hold BP accountable for all damages related to the impact.

A frequent source of discussion in the media was the flow rate of oil into the Gulf of Mexico emanating from the Macondo oil well. Although initial estimates from BP were that 1,000 barrels per day might be flowing into the Gulf of Mexico, at a press conference on April 28, 2010, Coast Guard Rear Admiral Mary Landry stated that the flow of oil entering the Gulf could be *as much as* 5,000 barrels per day, a number that was later deemed highly speculative (National Commission on the BP/Deepwater Horizon Oil Spill and Offshore Drilling, 2011). As was later acknowledged in the national commission report to the President, over the next four weeks, the media frequently repeated the number and it remained the official estimate of the spill size. However, once video from the feed of the blowout became publically available, it soon became apparent that the size of the exception was far larger than reported. More recent official estimates place the low number at more than ten times those initial estimates with 52,700 to 62,200 barrels of oil spilling per day into the Gulf for a total of approximately 4.9 million barrels, by far the largest maritime oil spill in United States history (National Commission on the BP/Deepwater Horizon Oil Spill and Offshore Drilling, 2011).

Another aspect of the *Deepwater Horizon* oil spill that was socially constructed by the media, albeit to a lesser extent, was the lack of specificity of who, and where was being impacted. Many Floridians, particularly those living and working south of the Panhandle, expressed frustration that national media was not informing the public that their beaches and tourist attractions were open and that little tar or oil had washed ashore, especially during the peak of tourist season (National Commission on the BP/Deepwater Horizon Oil Spill and Offshore Drilling, 2011). These sentiments were perhaps best expressed by Keith Overton, Chairman of the Florida Restaurant and Lodging Association who stated:

> These losses have occurred in our area in the Tampa Bay area, without a single drop of oil ever reaching our beach and that is true for most of Florida. Pensacola has had some oil but the rest of the panhandle is in pretty good shape right now. But you wouldn't know that if you looked at the national news media or you read the newspaper each day.

Finally, because of the previous five processes, *in a catastrophe the political arena becomes even more important in dealing with the impact and response.* As noted previously, all disasters of course involve, at a minimum, local political considerations. However, it is a radically different occasion when the national government and the elite officials become directly involved in the process (Quarantelli, 2006). Following the *Deepwater Horizon* oil spill, local, state, national, and international media played an important role in attempting to deal with response. At the local level, particularly in Louisiana, government officials frequently reported being stretched beyond capacity, which is common in catastrophe, given the scale of the exception as evidenced by Jefferson Parish (Louisiana) President Steve Theriot who stated, "We don't have enough assets in the world to take care of the Gulf Coast (Rainey, 2010:1)."

At the state level, politics played a very important role in shaping responsibility for the recovery. Louisiana Governor Bobby Jindal repeatedly blamed BP for the oil spill and repeatedly requested funds to protect the Louisiana shorelines (Alpert 2010a). Jindal told national media that "We're in a war here to fight to protect our way of life" and repeatedly requested for more help from the federal government, specifically Jindal asking that Coast Guard officers be assigned to each affected parish who could immediately OK requests for additional boom or other assistance (Alpert, 2010a). At the federal level, President Barack Obama made numerous trips to the impacted Gulf Coast region and held a high profile prime-time press conference, the first in months, to address problems with the federal response to the BP oil spill (Obama, 2010). On cable news networks, television pundits such as James Carville critiqued President Obama calling him "naïve" for entrusting BP with command over a spill they created (Basset, 2010:1). Many in the national media began to wonder if the lackluster response amounted to "Obama's Katrina" (Macmanus, 2010). Such framing of responsibility and failure is indicative of the problem partisan politicians in facing a catastrophe, a scale of exception that adequate planning and preparation is often too complex to execute prior to the exception.

for crisis policies and emergency management procedures. (Working Paper No. 90). Newark, DE: Disaster Research Center

Basset, L. (May 21, 2010). James Carville takes on Obama on Oil spill: He's risking everything with go along with BP strategy. *The Huffington Post*, Retrieved January 10, 2011, from http://www.huffingtonpost.com.

Blackburn, S. (1996). *Oxford dictionary of philosophy*. New York: Oxford.

Blanchard, W. (2008). *FEMA emergency management higher education project report.* Emmitsburg, Maryland: Emergency Management Institute.

Blau, P. (1995). A circuitous path to macrostructural theory. *Annual Review of Sociology* 21, 1-19.

Bolin, R. and Bolton, P. (1999). *The Northridge earthquake: Vulnerability and disaster.* New York: Routledge.

Bowen, G. (2006). Grounded theory and sensitizing concepts. *International Journal of Qualitative Research* 5, 1-10.

British Broadcasting Corporation. (February 12, 2010). Haiti will not die, President Rene prevail insists. *British Broadcasting Corporation.* Retrieved January 14, 2011, from http://www.bbc.com.

British Petroleum. (September 2010). Deepwater horizon accident investigation report. Retrieved September 8, 2010 from the British Petroleum Web site: http://www.bp.com/liveassessments

Brodie, M., Weltzien, E., Altman, D., Blendon, R., and Benson, J. (2006). Experiences of hurricane katrina evacuees in houston shelters: Implications for future planning. *American Journal of Public Health* 96, 1402-08.

Charmaz, K. (2003). Grounded theory: Objectivist and constructivist methods. In N. Denzin and Y. Lincolm (Eds.), *Strategies for qualitative inquiry.* Thousand Oaks, CA: Sage, pp. 249-91..

Collins, R. (2000). *Interaction ritual chains.* Princeton, NJ: Princeton University Press.

Dombrowsky, W. (1998). Debate-test-dummy: Reaction to Hewitt. In E.L. Quarantelli (Ed.), *What is a disaster*. New York: Routledge, pp. 99-100).

DuBois, W.E.B. (1893). *Suppression of the African slave trade to the United States of America 1638-1870.* Baton Rouge, LA: Louisiana State University Press.

Durkheim, E. (1897). *Suicide.* New York: The Free Press.

Dynes, R. (1998). Coming to terms with community disaster. In E.L. Quarantelli (Ed.), *What is a disaster*. New York: Routledge, pp. 109-26.

Eilperin, J. and Kessler, G. (June 14, 2010). After delays, U.S. begins to tap foreign aid in gulf spill; Domestic interests in cleanup keep some offers on hold. *Washington Post*, Retried September 10, 2010, from http://www.washington post.com

Eisenhardt, K. (1989). Agency theory: An assessment and review. *Academy of Management and Review,* 14, 57-74.

Endleman, R. (1952). *An approach to the study of disasters.* Unpublished manuscript, National Opinion Research Center, University of Chicago.

Feagin, J. (2010). *Racist America: roots, current realities and future reparations*. New York: Routledge.
Fritz, C. (1961). Disaster. In R. Merton and R. Nisbet (Eds.), *Contemporary Social Problems*. New York: Harcourt, Brace & World, pp. 651-94.
Glaser, B. and Strauss, A. (1967). *The discovery of grounded theory: strategies for qualitative research*. New York: Aldine de Gruyter.
Granovetter, M. (1973). The strength of weak ties. *American Journal of Sociology*, 78, 1360-80.
Hilgartner, S. and Bosk, C. (1988). The rise and fall of social problems: A public arenas model. *American Journal of Sociology*, 94, 53-78.
Klinenberg, E. (2002). *Heatwave: The social autopsy of disaster in Chicago*. Chicago, IL: University of Chicago Press.
Kreps, G. (1998). A reply to Hewitt's critique. In E.L. Quarantelli (Ed.), What is a disaster. New York: Routledge, pp. 101-03.
Krippendorf, K. (1980). *Content analysis*. Beverly Hills, CA: Sage.
Letukas, L. (2008). *Is aid a social problem? Cross-national media constructions of relief efforts following the Indian Ocean tsunami*. Unpublished master's thesis, University of Delaware, Newark, DE.
Letukas, L., Oloffson, A., and Barnshaw, J. (2009). Solidarity trumps catastrophe? An empirical and theoretical analysis of post-tsunami media in Sweden and the United States. In L. Nord (Ed.) *Scandinavian media coverage of the Indian Ocean tsunami* (pp 105-124). Stockholm, Sweden: Nordicom.
McKee, J. (1993). *Sociology and the race problem: The failure of a perspective*. Champaign, IL: University of Illinois Press.
MacManus, D. (2010, May 30). The BP oil spill: Obama's Katrina? *The Los Angeles Times*, Retrieved January 5, 2011, from http://articles.latimes.com/.
Mead, G. (1934). *Mind, self, and society*. Chicago, IL: University of Chicago Press.
Moore, H. (1958). *Tornadoes over Texas*. Austin, TX: University of Texas Press.
Morgenstern, R. and Barrett, N. (1974). The retrospective bias in unemployment reporting by sex, race and age. *Journal of the American Statistical Association*, 69, 355-57.
National Commission on the BP/Deepwater Horizon Oil Spill and Offshore Drilling. (2011). *Deepwater: The gulf oil disaster and the future of offshore drilling*. Retrieved January 14, 2011, from http://www.oilspillcommission.gov/final-report.
National Oceanic and Atmospheric Administration. (2010, May 12). *Fish stocks in the Gulf of Mexico*. (pp 1-3). Washington, D.C. U.S Department of Congress
National Public Radio. (2010, July 12). U.S. issues new moratorium on offshore drilling. *National Public Radio*. Retrieved January 5, 2011, from www.npr.org.
National Research Council. (2006). *Facing hazards and disasters: Understanding human dimensions*. Washington, D.C.: The National Academies Press.
Natural Hazards Applications and Information Center. (2011). *Quick response research program*. Retrieved January 14, 2011, from University of Colorado

at Boulder, Natural Hazards Center Website: http://www.colorado.edu/hazards/research/qr/.

Norse, E. and Amos, J. (2010). Impacts, perceptions and policy implications of the deepwater horizon oil and gas disaster. *Environmental Law Institute,* 40, 11058-73.

Obama, B. (2010, May 27). Deepwater BP oil spill: presidential press conference [Television press conference]. Washington, D.C.: White House.

Oliver-Smith, A. (1999).*The angry earth: Disaster in anthropological perspective.* New York: Routledge.

Padgett, D. (2004). Coming of age: Theoretical thinking, social responsibility and a global perspective in qualitative research. In D. Padgett (Ed).*The qualitative research experience* (pp 297-315). Belmont, CA: Wadsworth.

Perry, R. (1998). Definitions and the development of a theoretical superstructure for disasters. In E.L. Quarantelli (Ed). What is a disaster. New York: Routlege, pp. 197-216.

Porforiev, B. (1998). Issues in the definition and delineation of disasters and disaster areas. In E.L. Quarantelli (Ed). What is a disaster. New York: Routlege, pp 56-72.

Prince, S. (1920).Catastrophe and social change; Based upon a sociological study of the Halifax disaster. Unpublished doctoral dissertation, University of Chicago, Chicago, IL.

Picou, S., Gill, A., Dyer, L., and Curry, E. (1992). Stress and disruption in an Alaskan fishing community: Initial and continuing impacts of the Exxon Valdez oil spill. *Industrial Crisis Quarterly,* 6, 235-57.

Quarantelli, E.L. (1993). Community crises: An exploratory comparison of the characteristics and consequences of disasters and riots. *Journal of Contingences and Crisis Management,* 11, 15-39.

Quarantelli, E.L. (1998). Epilogue: Where we have been and where we might go. In E.L. Quarantelli (Ed). *What is a disaster.* New York: Routledge, pp. 236-73.

Quarantelli, E.L. (2006). *Catastrophes are different from disasters: Some implications for crisis planning and managing drawn from Katrina.* Retrieved January 14, 2011, from Social Science Research Council Web site: http://www.understandingkatrina.ssrc.org/Quarantelli/.

Rainey, R. (June 10, 2010). Jefferson fretting over spill costs. *Times-Picayune,* Retrieved January 10, 2011, from http://www.timespicayune.com.

Resnick, A. and Klimasinska, K. (April 22, 2010). Transocean rig sinks in Gulf of Mexico as Coast Guard looks for survivors. *Bloomberg,* Retrieved January 10, 2011, from http://www.bloomberg.com.

Rodriguez, H., Quarantelli, E.L., and Dynes, R. (Eds), (2006). *Handbook of disaster research.* New York: Springer.

Shachar, R. and Zvi, E. (2007). Correcting for bias in retrospective data. *Journal of Applied Econometrics, 22,* 657-75.

Stallings, R. (1998). Disaster and the theory of social order. In E.L. Quarantelli (Ed.). In E.L. Quarantelli (Ed). What is a disaster. New York: Routlege, pp. 127-45.

Stallings, R. (Ed.), (2002). *Methods of disaster research*. Philadelphia, PA: Xlibris.

Strauss, A. and Corbin, J. (1990). *Basics of qualitative research: Grounded theory procedures and techniques*. Newbury Park, CA: Sage Publications.

Swiss Agency for Development and Cooperation. (2010). *Fuel Unloaded from Chernobyl reactor*. Retrieved from: http://www.chernobyl.info/index.php?userhash=&navID=534&lID=2

Tierney, K. (2007). From the margins to the mainstream? Disaster research at the crossroads. *Annual Review of Sociology*, 33, 503-25.

Tierney, K., Lindell, M., and Perry, R. (Eds.), (2001). *Facing the unexpected: Disaster preparedness and response in the United States*. Washington, D.C. John Henry Press.

Times-Picayune. (June 17, 2010). The oil spill in brief. *Times-Picayune*. Retrieved September 10, 2010, from http://www.timespicayune.com/.

Transocean. (2010, April 26). *Transocean limited provides deepwater horizon update*. Retrieved from http://finance.yahoo.com/news/Transocean-Ltd-Provides-iw-533460218.html?x=0&.v=1.

Washington Post. (2010, September 6). Louisiana beach town laments the summer that never was. *Washington Post*, Retrieved September 30, 2010, from http://www.washingtonpost.com

Westgate, K. and O'Keefe, P. (1976). *Some definitions of disaster*. Bradford, England: Disaster Research Institute, University of Bradford.

Wisner, B., Blakie, P., Cannon, T., and Davis, I. (2004). *At risk: Natural hazards, people's vulnerability and disasters*. second edition. New York: Routledge.

Woolgar, S. and Pawluch, D. (1985). Ontological gerrymandering, *Social Problems*, 32, 214-27.

Yin, R. (2009). *Case study research: Design and methods*. 4th ed. Thousand Oaks, CA: Sage.

CHAPTER FOUR
CORPORATE CATASTROPHES FROM UC BHOPAL TO BP DEEPWATER HORIZON: CONTINUITIES IN CAUSATION, CORPORATE NEGLIGENCE, AND CRISIS MANAGEMENT

TOMÁS MAC SHEOIN AND STEPHEN ZAVESTOSKI

INTRODUCTION

In 1984 a leak of toxic gases from a pesticides factory owned by a subsidiary of the US transnational corporation Union Carbide (UC) devastated the Indian city of Bhopal, killing thousands and unleashing an environmental disaster whose effects are still being felt today. In April 2010, the blowout on the Deepwater Horizon, an offshore oil rig in the Gulf of Mexico under contract to British transnational corporation BP, killed eleven workers and led to an environmental catastrophe whose effects are still unknown. This chapter will look at similarities between these two process industry disasters in terms of causation (failures of operating and safety equipment and inadequate state regulation), corporate negligence (prioritizing production over occupational and environmental safety, failure to plan for worst-case scenarios) and crisis management (control over information, transfer of blame, legal tactics). The chapter will argue that both catastrophes were the product of negligence on the part of both capital and the state which prioritized production, profit and capital accumulation over questions of community, environmental and worker safety. The chapter concludes with a brief discussion of the similarities and differences, including social movement responses, to these two corporate catastrophes.

In this chapter we aim to compare two dissimilar corporate catastrophes–the 1984 Bhopal gas leak disaster and the 2010 Deepwater Horizon oil release disaster.[1] The dissimilarities, we contend, assist our analysis rather than limit it. Although both were disasters of the process industry, one resulted in unprecedented loss of human life with relatively transient environmental impacts, while the other resulted in relatively little loss of human life but unprecedented and as yet uncertain public health and ecological impacts. The similarities in the causes of, and corporate responses to these two disasters, therefore, are all the more striking. We would be unsurprised if comparison of the Exxon Valdez or another oil spill disaster to the Deepwater Horizon disaster found similarities in the causes and corporate responses. By comparing the Bhopal disaster to the Deepwater Horizon disaster, however, we can begin to uncover specific patterns in the causes of industrial disasters and the responses of the implicated corporations. These patterns, we argue, point to the far-reaching powers of corporate actors in a capitalist system–powers that frame cost-cutting and other profit-seeking measures, even when made in the context of inherently dangerous technological systems, as logical and justified. The very same powers are then invoked to downplay the severity of disasters and minimize the liabilities of culpable corporate actors.

BACKGROUND TO DISASTER

In December 1984, a petrochemical factory in Bhopal, India suffered a runaway reaction which spewed toxic gases over the sleeping population of the Indian city, killing thousands and unleashing an environmental disaster whose effects are still being felt today. The factory was owned and operated by Union Carbide India, Ltd., (UCIL), a majority-owned subsidiary of US transnational corporation (TNC) Union Carbide Corporation (UCC). In response to Bhopal, op-ed writers and corporate apologists writing in the press hastened to assure their readers "It can't happen here," citing the highly trained workers, advanced technology and safety culture of the American petrochemical industry, stringent state and federal regulations, and the North American concern for human life, implicitly blaming the incident on the Indian managers and Indian workforce. This position was fatally undermined the following year when an accident at Bhopal's sister plant at Institute, West Virginia, operated by American workers under the full control of American managers, reproduced the Bhopal accident in every respect but the casualty rate (Jones, 1988). Indeed, what has become increasingly obvious over the quarter century since the Bhopal catastrophe–a quarter century that included, among others, "accidents" in Texas City, Texas (March 2005, fifteen dead, eighteen injured); Sterling, Louisiana (May 1991, eight dead, 140 injured); Channelview, Texas, (July 1990, seventeen dead); Pasadena, Texas (October 1989, twenty-three dead)—is not only that it can happen here, but that it will continue to happen here regularly as long as the production, use and transport of highly toxic substances is controlled by TNCs whose first priority is profit and capital accumulation. (Mac Sheoin, 2009, 2010).

The April 20, 2010 blowout on the Deepwater Horizon oil rig, owned and operated by Transocean and on lease to BP, represents the latest fatal incident to strike the North American process industry. The explosion on the rig killed eleven workers and seriously injured seventeen others.[2] It also resulted in the sinking of the rig two days later and the ensuing four-months of oil gushing from the well into the Gulf of Mexico. The roughly 4.9 million barrels released stands as the greatest spill in North American history. Just as Bhopal created a crisis for UCC and the global chemical industry, the Gulf oil disaster created a crisis for BP and the oil industry as a whole.

In what follows, we discuss the Bhopal and Deepwater Horizon disasters in terms of their proximate and ultimate causes. Proximate causes include the failures of operating and safety equipment, while ultimate causes can be traced to inadequate regulatory frameworks and negligence in enforcing existing laws, which in turn facilitate the prioritization of production over occupational and environmental safety and the failure to plan for worst-case scenarios. We focus on the parallels between the ways in which Union Carbide and BP responded to their respective disasters, highlighting crisis management strategies that emphasize information control and the transfer of blame. As the causes, consequences, and crisis management of Bhopal are already well-documented in the literature (see, for example, Eckerman, 2005; ICFTU-ICEF, 1985; Hannah, Morehouse, and Sarangi, 2005; Jones, 1988, Shrivastava, 1987), we will deal with Bhopal only in summary and deal with Deepwater Horizon in more detail.

BHOPAL AND DEEPWATER HORIZON AS NORMAL ACCIDENTS

The Bhopal and Deepwater Horizon disasters were both examples of what sociologist Charles Perrow calls "normal accidents" (Perrow, 1984): accidents that become normal, or inevitable, when systems are defined by tight coupling of components and high interactive complexity of processes. Post-disaster reports in Bhopal and the Gulf pointed to combinations of failures in different systems as causes. In both cases safety devices and alarms failed. The fact that previous problems with these systems were known to management but not dealt with is typical of highly complex systems (Perrow, 1984). Behind these are the broader issues of system design problems and lack of adequate maintenance. Finally, by virtue of a tendency towards unwarranted confidence in highly complex systems designed to manage risk, organizations responsible for high-risk technological systems systematically fail to have adequate plans in place to deal with worst case scenarios. This was clearly the case in both the Bhopal and Deepwater Horizon disasters, where neither Union Carbide nor BP had adequate disaster response plans, much less the necessary equipment in place to respond to disasters of such scale.

Normal accidents are also characterized by a response from potentially liable parties that frames the disaster as an unforeseeable and unpredictable act of God, with odds so remote no amount of planning could have prevented the disaster or prepared an effective response. Union Carbide and BP both pursued this line of reasoning while also pointing to their past safety records as evidence that the present disasters were clearly not a result of lack of concern for safety. To prevent evidence from emerging that could erode the "innocent victims" images of themselves that these corporations attempted to create, both corporations attempted to control information available to the media, government officials, and the victims of the disasters. Both corporations also aimed to create images of themselves as responsible corporate actors by accepting moral responsibility, meanwhile working to avoid legal liability by transferring blame. As Perrow points out, the causes of the vast majority of industrial disasters are attributed to operator error. In Bhopal, Union Carbide blamed both workers for failing to follow safety protocols and local management for failing to detect that protocols were not being followed. In the Deepwater Horizon case, BP placed blame on work done not by its own employees, but the employees of its contractors. As we will argue later, each of these corporate strategies was facilitated by a political economy in which TNCs, headquartered outside the countries in which the incidents occurred, exploit aspects of the global economic order to prioritize production and profit over safety while escaping the real costs of the industrial disasters for which they are responsible.

CAUSES OF THE DISASTER AT BHOPAL

PREVIOUS ACCIDENTS

Union Carbide has an extensive history of industrial accidents and employee deaths. For the sake of brevity, we will focus on incidents at the Bhopal plant where at least six major accidents occurred between 1978 and 1984 (Eckerman, 2005). In 1974, residents of a neighborhood adjacent to the plant, discovered that a well had been contaminated when some cattle died. According to former UCIL worker T.R. Chouhan (2004), in 1978 when a fire broke out at the coke yard the plant's only fire truck was up on jacks with all four wheels removed. In December 1981, phosgene leaked, injuring three workers, one of whom died the following day. The phosgene release had resulted from a malfunctioning valve that should have been repaired. In January 1982, another phosgene leak overcame 24 workers; a February 1982 methyl isocyanate (MIC) leak affected 18 people; an August 1982 incident involved an engineer receiving burns over 30 percent of his body when he was exposed to MIC; an October 1982 incident resulted in the release of MIC, methylcarbaryl chloride, chloroform and hydrochloric acid, the evacuation of the plant, and injuries to three workers. Finally, a 1982 safety audit identified 61 hazards, 31 of them "major" (Bhopal People's Health and Documentation Clinic, 1998). Many of the items in the 1982 report were temporarily fixed, but had deteriorated once again by the time of the disaster in 1984. Since 1976, unions representing Bhopal workers had regularly drawn the attention of management and state authorities to safety problems at the factory. Despite this, little or nothing was done by factory management or state regulators to address the problems (ICFTU-ICEF, 1985, p. 11).

EQUIPMENT FAILURES

The refrigeration unit was shut down in June 1984 to save money. The vent gas scrubber (VGS) was turned off in October 1984 by a supervisor who believed it was necessary only when MIC was being produced. The flare tower was also out of service as a section of corroded pipe had been removed and not replaced in mid-October 1984. Such failure to undertake maintenance was common in the factory, with leaking valves and broken gauges a serious problem. Malfunctioning gauges made it difficult for workers to know what was happening in the production and storage systems. The pressure indicator/control, temperature indicator and level indicator for the MIC tanks were reported to have been malfunctioning for over a year. In the case of the VGS, the caustic flow indicator was malfunctioning, leading to difficulties in knowing whether the system was operating (ICFTU-ICEF, 1985).

DESIGN FAULTS

As well as these equipment failures, the Bhopal factory showed design failures in safety and production systems. The emergency relief system was too small to deal with the amount of escaping gas, a well-known problem in the chemical industry. A further design problem related to the water spray system, which did not reach the height of the gas leak. A new larger spray system had been recommended after the 1982 inspection, but was not installed due to constraints on investment in a loss-making plant. The major design fault involved the storage of large amounts of MIC, which resulted from a design decision UCC had insisted on, overruling objections by UCIL engineers (ICFTU-ICEF, 1985).

STAFF CUTS AND TREATMENT OF STAFF

In pursuit of cost savings, staff at the factory had been cut and training reduced. The MIC production crew was cut from 12 to 6 and the MIC maintenance crew from six to two. Staff training was also deficient and workers were assigned work for which they were not qualified or trained. Health and safety training in particular was lacking, with workers saying they were given little or no information on the toxicity of chemicals used in the factory (ICFTU-ICEF, 1985). Overworking employees in unsafe conditions and then threatening whistleblowers and union

interest of 50.9 percent of shares in UCIL, UCC's lawyers claimed control of the Bhopal factory rested with the Indian company, with its Indian counsel telling *The American Lawyer* in 1985 "It is again wrong to say that the American company is the parent company. It is not the parent company. It holds some shares . . . The Indian company has nothing to do with the US company" (as cited in Jones, 1988, p. 45).

As well as denying it controlled its Indian subsidiary, UCC now claimed that the Bhopal plant was designed by Indians, denying previous statements to the *Guardian* (1984, p. 11) that "[t]he Indian plant was designed and built by American nationals." UCC later admitted a large number of Americans were on site when the plant was being constructed and started production. As for safety standards, UCC insisted that it had provided general guidelines but implementation was left up to the Indians. At the launch of UCC's report on the disaster in March 1985, Anderson claimed that "Non-compliance with safety procedures is a local issue," and foisted blame on Indian managers: "It's their company, their plant, their people" (Jones, 1988, p. 36-37).

LEGAL TACTICS

UCC's denial of responsibility for the Bhopal plant's design and operation and denial of control of the Indian company were central to its legal tactics in refusing to accept liability for the disaster. UCC's legal tactics belied the image of a caring corporation that its public relations were attempting to create. While in December 1984 it was saying "[e]very effort will be made to mitigate against the deep sorrow of the people of Bhopal," by February 1985 (when the media gaze had moved on to other issues) it was declaring "We will fight [the legal battle] right to the end." By November 1985, few signs remained of UCC's previous "caring" approach when, according to CEO Anderson, "Maybe they, early on, thought we'd give the store away. (Now) we're in a litigation mode. I'm not going to roll over and play dead" (as cited in Jones, 1988, p. 126).

One of UCC's central legal tactics was to oppose hearing of the cases against it in the USA where punitive damages and American disclosure rules would likely raise the stakes for the TNC. Anderson strategically announced in November 1985 that UCC would take the case all the way to the US Supreme Court, noting "This could drag on for five or more years." In May 1986, on the legal premise of *forum non conveniens* (an inconvenient forum), Judge Keenan of the Federal District Court of New York returned the litigation to India.

When it appeared after several more years that the Indian Courts were nowhere near a settlement or trial, survivors applied to the courts for interim relief. UCC fought the provision of interim relief and dragged on the proceedings so long that the Indian Supreme Court gave as the main reason for its imposition of a settlement in 1989 the need to begin disbursing compensation to survivors. UCC's Indian lawyers were condemned for using a variety of tactics to impede and delay resolution of the case. Over the years, the victims of Bhopal tried a number of strategies for bringing new cases before U.S. judges, each time being denied. Even 26 years later, with appeals and other delays, a number of cases are still open in both U.S. and Indian courts.

THE INDIAN GOVERNMENT'S RESPONSE

The Indian government also needed to manage the Bhopal crisis. There were two major motivations behind the state's behaviour. First, regulatory failures by the Indian state had contributed to the disaster. Environmental and industrial health and safety regulation in India were notably lacking. Extremely understaffed and under-resourced state Pollution Control Boards were tasked with enforcing pollution-related laws. Inspections at the plant in Bhopal were infrequent, and violations almost never responded to. Matters were no better in terms of labor standards. With a small number of inspectors and hundreds of factories to inspect, companies like UCIL knew they could act with relative impunity. Responsibility for inspection of health and safety lay with the Madhya Pradesh state authorities who employed two factory inspectors for the Bhopal area. These inspectors were mechanical engineers with neither equipment nor training to assess chemical hazards. The state authorities undertook no action in response to warnings from trade unions as to the hazards the factory represented. The second problem for the Indian state was the close connections between the Indian state and UCIL. Since the Indian state held 23 percent of the equity in UCIL, the failures that led to the disaster were in part the responsibility of the Indian state as an investor in the responsible company. Furthermore there were close connections between members of the Indian political elite and the company. UCIL offered the use of its luxurious corporate guesthouse in Bhopal to the elites of local government and national political parties (Hannah et al., 2004, p. 115).

For these reasons, the state shared with UC a desire to control the crisis Bhopal represented. Government response began with a reassurance operation, characterized by declarations that local food supplies were safe and water supplies unpolluted. Rather than institute a public inquiry the state handed investigation over to the Central Bureau of Investigation, the Indian equivalent to the FBI. Meanwhile, the Government treated research by its medical and scientific arms as state secrets, paralleling UCC's trade secrets. It also marginalized the victims from the litigation through the Bhopal Gas Leak Disaster (Processing of Claims) Act of 1985, which consolidated all claims under the auspices of the Indian government.

MANAGING THE SAFETY CRISIS AFTER BHOPAL

Bhopal also represented a crisis for the global chemical industry whose response was spearheaded by the chemical industry associations. On one hand, the chemical industry needed to distance itself from UCC so that the reputations of other companies in the industry were not tarnished by UCC's disaster. On the other hand, the industry needed to step in to reassure people that it took safety seriously and would work with members of the industry to help them operate more safely. The solution to this dilemma came in the development of the Responsible Care program. Introduced by the US Chemical Manufacturers Association, the voluntary program asked industry members to implement certain practices and report information on its releases. It also served as a means of convincing regulators that the industry could regulate itself (Mac Sheoin, 2009; Givel, 2007).

CAUSES OF THE DEEPWATER HORIZON DISASTER

Academic analysis too often exists at the level of pure abstraction. Even when our analysis engages with real events, they may fail to recognize the life and death consequences of the events we discuss. In the case of the Deepwater Horizon the media, which usually does a better job at highlighting the human elements of a news event, has all but ignored the deaths that resulted from the catastrophe, emphasising the ensuing environmental catastrophe, while downplaying and ignoring the workplace disaster. There were 126 workers on Deepwater Horizon, 79 hired by Transocean, six by BP, while the rest were casual workers. Those who died were skilled, semi-skilled and unskilled workers: two mud engineers (Gordon Jones, 28; Keith Blair Manuel, 56) one driller (Dewey Revette. 48), one toolpusher (Jason Anderson, 35), one crane operator (Aaron Dale Burkeen, 37), two assistant drillers (Donald Clark, Stephen Ray Curtis), three floorhands (Kark Kleppinger Jr., 38; Shane Roshto, 22; Adam Weise, 24) and one derrick hand (Roy Wyatt Kemp, 27) (Blackmon, O'Connell, Berzon, and Campoy, 2010). We hope that our analysis of this event not only helps us remember these lives lost, but also provides an analytical basis for action that can protect the health of workers and communities by preventing industrial disasters in the future. Observing patterns across a range of corporate catastrophes is one way of developing such an analytical framework for understanding the causes of, and corporate responses to, industrial disasters. It is towards just such a set of observations that we now turn.

OFFSHORE OIL'S CATASTROPHIC HISTORY

Normal accidents are, by definition, not rare. Corporate accounts of disasters as unforeseeable acts of God, therefore, can be easily tested against the historical record. The offshore oil industry has a long history of blow-outs, explosions, fires and other incidents. The extraction of oil is a deadly business. Hirschberg, Spiekerman and Dones (1998) estimate that severe accidents in oil extraction from 1969 to 1996 killed 1,598 people and injured 4,366 people. Clemmer and Diem (1985) attribute 344 deaths to "major mishaps" in offshore drilling globally from 1955 to 1981, noting this figure does not include deaths in the course of ordinary operations. LeBlanc (1981) gives total deaths in 140 offshore incidents from 1955 to 1980 as 316. In the Gulf of Mexico since 2001, sixty-nine people died, 1,349 were injured and there were 858 fires and explosions, according to reports from the Minerals Management Service (MMS) (Crace, 2010), while 116 workers died and 294 were injured in thirty-three incidents on Pemex offshore platforms between January 2005 and May 2009 (News wires, 2009). For the earlier period, Priest's excellent account of the offshore oil industry's safety in the Gulf of Mexico stresses both the unavailability of statistics and the unreliability of such statistics as are available due to non-reporting and under-reporting by the industry (Priest, 2008, p. 140, 142, 149, 151). Priest estimates that from 1954 to 1966 at least ninety people died in drilling on the Outer Continental Shelf (2008, p. 14) and concludes that, despite improvements, "accidents, injuries and fatalities, although not as frequent as in the early years of the industry, unfortunately are still a routine aspect of working offshore" (2008, p. 152). After the world's largest oil rig sank off Brazil in 2001, Reuters reported that ninety-one Petrobras workers, including ten killed in that disaster, had died over the previous three years, while in the ten years after the Piper Alpha disaster in 1988, at least another fifty workers were killed in the British sector of the North Sea (Weston 1989, p. 21). Table 4.1 reports a total of 1,706 deaths for incidents in the industry from 1956 to 2010, but does not include deaths due to crashes by helicopters carrying workers to, from and between rigs, for which IAOGP (2010) reports 362 deaths.

Table 4.1: Fatal Offshore Incidents

Year	Rig	Location	Deaths	Company	Cause
12/56	Quatar 1	Arabian Gulf	20	Shell	Sinking
4/57	Mister K	Gulf of Mexico	9	Golden Meadow Drilling Co.	Capsize
10/58		Gulf of Mexico	7	CATC Group	Explosion
1962	CP Baker	Gulf of Mexico	22	Reading and Bates	Blowout/ fire
1965	Bruyard	South China Sea	13	Shell	Sinking
12/27/65	Sea Gem	Britain	13	BP	Collapse
8/68	Little Bob	Gulf of Mexico	7	Fluor Drilling	Blowout
1969	Bohai II	China	72	PRC government	Capsize
1971	Big John	Off Brunei	9	Atwood Oceanics	Blowout
1971	Wodeco II	Off Peru	7	Fluor Drilling	Blowout
10/13/71	Western Offshore 2		16		
6/15/74	Main Pass 69A	USA	7		
10/9/74	Gemini	Gulf of Suez	18	Offshore Co.	Capsize
8/15/75		Gulf of Mexico	6		
3/76	Deep Sea Driller	Norway	6	Deep Sea Drilling	
4/16/76	Ocean Express	Gulf of Mexico	13	Odeco	Capsize
12/8/77		Gulf of Mexico	17		
2/25/78		Norway	5		
3/5/79		USA	8		
5/11/79	Ranger I	Gulf of Mexico	8	Atlantic-Pacific Marine	Capsize
11/25/79	Bohai II	China	72	PRC government	Sinking
1980	Bohai III	South China Sea	70		Fire/ explosion
1980		Alaska	51		
1/17/80	Funiwa-5	Nigeria	180	Chevron, NNPC. Texaco	Blowout
2/21/80		USA South Timbal	6		
3/27/80	Alexander L Kielland	Off Norway	123	Phillips Petroleum	Capsize
6/15/80		China	70		

Year	Rig	Location	Deaths	Company	Cause
8/30/80	Ocean King	Gulf of Mexico	5	Ocean Drilling and Exploration Company	Explosion/ fire
10/2/80	Ron Tapp-meyer	Arabian Gulf	19	Reading and Bates	Blowout
11/3/81		Off Brazil	5		
2/15/82	Ocean Ranger	Off Canada	84	Odeco	Capsize
3/83	Nowruz Platforms	Persian Gulf	20		Fire
3/20/83	BOS 355	West Africa	13		Explosion/ fire
9/9/83	60 Years of Azerbaijan	Caspian Sea	5		Sinking
10/25/83	Glomar Java Seas	South China Sea	81	Arco, CNOOC	Capsize
11/5/83	Byford Dolphin	Norway	5		Decompres-sion
8/16/84	Enchova I	Bay of Campos, Brazil	61	Petrobras	Explosion
5/20/85	Tonkawa	Gulf of Mexico	11	Temple Drilling Co.	Capsize
10/17/85	Trintoc Atlas	Central America	14		Explosion
11/4/85	Concern	North Sea	10		Capsize
12/14/85	Ciudad del Carmen (?)	Mexico	33		
7/7/88	Piper Alpha	North Sea	167	Occidental	Explosion
4/28/89	Al Baz	Nigeria	5		
7/31/89	Avco 5	Gulf of Mexico	10		Capsize
10/3/89	High Island Pipeline	Gulf of Mexico	11		Collision/ explosion
11/3/89	Seacrest	Gulf of Thailand	91	Unocal	Sinking
7/25/90	Brent Spar	North Sea	6		
8/15/91	McDermont lay barge 29	Hong Kong	22		Sinking
3/14/92		North Sea, off Shetlands	17		
3/25/93		Venezuela	11		Explosion
1/18/95	Ubit	West Africa	10		Fire/ explosion

Year	Rig	Location	Deaths	Company	Cause
10/15/95	DLB 269	Mexico	26		
1996	Ubit	Nigeria	18		Fire/ explosion
11/2/99	Mighty Servant 2	Indonesia	5		Sinking
3/15/01	P-36	Brazil	11	Petrobras	Explosion
7/27/05	Bombay High North	Indian Ocean	22	ONGC	Fire
4/12/07	Bourbon Dolphin,	Off Shetland	8		Capsize
10/23/07	Usumacinta, Kab 101	Gulf of Mexico	22	Pemex, Perforadora Central	Blowout
4/10	Deepwater Horizon	Gulf of Mexico	11	BP	Blowout

Note: This table has been compiled from Hirschberg, Spiekerman and Dones (1998); Offshore Data Services (1983), LeBlanc (1981), IAOGP (2010), Priest (2008) http://home. versatel. nl/the_sims/rig/losses.html and LesixNexis. Five deaths were required for inclusion.

PREVIOUS INCIDENTS AT BP AND TRANSOCEAN

Corporate claims of pristine industry safety records are further eroded by evidence of safety transgressions within the companies making the claims. As was the case with UCC prior to Bhopal, BP and Transocean had histories of incidents. BP's record includes a number of incidents at its onshore facilities, the most serious of which was an explosion in 2005 at its Texas City refinery that killed fifteen workers and injured eighteen. In response to this the corporation claimed to have made great efforts to improve safety, including investing $1 billion, yet in October 2009 BP was fined $87 million for failing to correct hazards at the refinery, whose condition "could lead to another catastrophe" according to Labor Secretary Hilda Solis (Mac Sheoin, 2010, p. 31). However, despite this new emphasis on safety, BP continued to emphasize cost-cutting, a major contribution to the lack of safety of their operations. In February 2009, for example, BP's CEO Tony Hayward told reporters "The mantra in BP today is 'Every dollar counts'" (quoted in Crooks, 2010a). Pressure on BP to deliver on time was the cause of BP's worst incident in the Gulf when its *Thunder Horse* platform nearly sank in 2005 because a valve had been installed backwards. This mistake, and others related to welding, were caused because BP rushed construction of the platform and moved it to the Gulf before it was ready "to demonstrate to their shareholders that the project was on time and on schedule." Construction was then frantically completed at sea, according to an engineering consultant who worked on the project (Lyall, 2010). Table 4.2 reports on the history of BP's safety issues in the country in which it has its headquarters and in the USA.

Table 4. 2: Safety Incidents Involving BP in England, Scotland and the USA—1999-2010

Date	Incident
4/99	BP prosecuted after two contract workers overcome by nitrogen fumes at refinery in Coryton, Essex, England.
Summer/99	BP's main oil platform in the North Sea shut down by regulators after two near-catastrophic gas leaks in two days
2000	BP pays $22 million fine for contractor dumping toxic waste rather than shipping it from North Slope. BP denied knowledge of dumping.
1/02	BP fined £1 million after pleading guilty to breaches of health and safety legislation at its petrochemical plant in Grangemouth, Scotland in June 2000.
8/16/02	Explosion in BP well in Alaska injured and burned worker
12/21/02	One contract worker killed and two injured while working at BP Alaska's North Slope.
9/04	Two killed and one injured in explosion at Texas City refinery
2004	Worker dies at Whiting, Indiana refinery after falling through a rotted railing
2005	Worker dies at BP refinery in Cherry Point, Washington
3/23/05	15 dead and over 170 injured in explosion at BP refinery in Texas City
12/05	BP announces plan to invest $1 billion in TC refinery
3/06	267,000 gallons of oil leak from corroded BP pipeline in Prudhoe Bay, Alaska
4/06	BP fined $2.4 million by OSHA for violations at Ohio refinery
5/06	Contract worker severely burned at BP refinery in Ohio
7/21/06	Contract worker killed at Texas City refinery
11/05	OSHA proposes $384,000 fine for violations at BP's refinery at Whitney, Indiana
11/13/06	Worker dies at BP drilling installation in Alaska
11/17/06	Contract worker dies in eastern Oklahoma after drilling of oilwell for BP.
5/07	Health and Safety Executive orders BP to improve safety on North Sea oil rigs after issuing 14 improvement notices in 12 months.
8/7/07	Contract worker dies at BP refinery in Cherry Point, Washington

Date	Incident
6/5/07	**Worker dies from electrocution at Texas City refinery**
10/07	**BP agrees to pay $373 million in fines for Texas City refinery explosion, Alaska oil leak and propane price manipulation**
1/14/08	**Worker dies in TC refinery from head injuries**
10/30/09	**OSHA proposes $87.4 million fine for BP's failure to correct problems at TC refinery**
11/09	**Spill of 46,000 gallons of oil at North Slope, Alaska.**
8/10	**BP accepts fine of $50.6m for safety violations at Texas City refinery**

Deepwater Horizon itself had previously suffered numerous spills and fires and the US Coast Guard had issued citations eighteen times between 2000 and 2010. Deepwater Horizon also suffered other serious incidents, including one in 2008 where the platform listed after the accidental removal of a pipe section from the rig ballast system, leading to the evacuation of seventy-seven people (Grayson 2010). The month before the disaster Transocean commissioned Lloyd's Register to investigate and assess safety culture on four of its rigs in the Gulf, including the Deepwater Horizon, in response to "a series of serious accidents and near-hits within the global organization" (Urbina, 2010a). The summary report said "Without a doubt, previous incidents and near-hits within the organization were as a result of multiple causes and many contributing factors." In complaints that echoed those made by BP workers at Texas City (Mac Sheoin, 2010, p. 28), some 43 percent of workers said they feared reprisals if they reported problems, while some workers charged maintenance was deferred by the company to save money and nearly 40 percent of workers said management investigated safety problems purely to assign blame.

Transocean had been involved in a rising number of incidents in the Gulf since it took over GlobalSantaFe in November 2007. From 2005 to 2007, thirteen of thirty-nine deepwater drilling incidents in the Gulf investigated by the MMS involved Transocean rigs. This 33 percent was roughly equivalent to Transocean's 30 percent share of rigs operating in the deepwater Gulf. Since the merger Transocean has been responsible for 73 percent of incidents investigated by the MMS (24 of 33 incidents) despite owning fewer than 50 percent of operating rigs (Casselman, 2010).While Transocean's safety record for 2009 is better than the industry average, four workers died on board Transocean rigs in separate incidents that year. Transocean has also suffered previous problems with the two issues at the centre of Deepwater Horizon investigations, the BOP and the cement seal on the well. For the BOP, a failure on the *Discoverer Enterprise*, which in 2006 was drilling for BP, was caused in part by "extended use. . . without inspection/ maintenance" (Casselman, 2010a). Nor are Transocean's safety problems confined to the Gulf. In 2007 eight people died off the coast of Scotland when a support vessel towing a chain for a Transocean rig capsized. A Norwegian government investigation said mistakes made by Transocean and the vessel's owners contributed to the incident. A specialist inspection report by the British HSE of four rigs operated by Transocean in 2009 accused the company of "bullying, harassment and intimidation" of staff, saying its organizational culture was based on blame and intolerance (Macalister, 2010a).

EQUIPMENT FAILURES AND DESIGN FAULTS

As in Bhopal, alarms and other equipment, including production and safety equipment, failed. BP said as many as seven different problems could have contributed to the catastrophe, relating to the cementing of the well, tests run on the well and the BOP. What was involved was the failure of a number of different subsystems in a tightly coupled system—a normal accident in Perrow's terms. In what seems an unbelievable decision, Transocean senior officials had "inhibited" alarms on the Deepwater Horizon as "they did not want people woken up at three o'clock in the morning due to false alarms," Mike Williams, who was in charge of the rig's electronic systems, told federal hearings in July. While sensors monitoring conditions on the rig and in the well were still working, the computer had been instructed not to trigger alarms in response to adverse readings. Thus on the day of the explosion, visual and sound alarms which should have gone off when sensors detected fire or combustible or toxic gases did not warn workers of impending danger. (Pilkington, 2010). About a month before the explosion, workers discovered a leak in the hydraulic system that controls the BOP, a five storey safety device that squats on top of the wellhead on the sea bottom, and contains the last line of defense, the blind shear ram, whose job is to cut the drilling pipe and seal the well. The problem was reported to BP officials in Houston, who appear not to have mentioned it in drilling reports to the MMS. An audit in April found the BOP was "well past" the date for its major inspection, due every three to five years. (Casselman, 2010b).

These were not new problems unknown to the industry. Eighteen of thirty-nine blowouts between 1992 and 2006 in a MMS study of 2007 were associated with cementing of wells. There had also been a history of failures of the BOP's blind shear ram. The BOP had failed during a well-sealing operation, causing what had been, before Deepwater Horizon, the world's largest rig spill on the Ixtoc 1 rig in 1979 in the southern Gulf. This disaster released 138 million gallons of oil over nine months (Johnson and Torrice, 2010). Furthermore, a confidential analysis of BOP reliability performed for Transocean found they had a failure rate of 45 percent. Yet despite this BP and other oil companies wanted to reduce government-mandated testing of BOPs, helping fund a study arguing these tests should be run every thirty-five days instead of every fourteen days. Finally there were doubts as to whether BOPs, developed for shallow water, would actually work in deepwater conditions (Barstow, Dodd, Glanz, Saul, and Urbina, 2010).

PROFIT PRESSURES

Priest's (2008, p. 144) report on safety in offshore operations in the Gulf of Mexico notes that:

> [S]afety was compromised not only by the sheer technological challenges, but by the necessity to complete work as quickly as possible. Offshore installations were expensive to build and operate. The profitability of projects often depended on how soon production could be brought online. Drilling vessels were contracted on day-rates, increasing time-cost pressures. Production processes were highly interdependent, and delay in one section could cause delays elsewhere. And delays cost money. So there was incredible time pressure to drill the wells, install the platforms and get the oil and gas flowing.

Similar pressures have been reported in the onshore oil and chemical industry, where many accidents occur during start-up of production processes that have been offline for repair or maintenance. Pressure to complete work speedily may have been a factor in the Deepwater Horizon catastrophe. Transocean was charging BP $533,000 a day for the Deepwater Horizon, and with payments to other contractors and other costs, the daily cost of the Deepwater Horizon was over $1 million. Deepwater Horizon was already behind schedule and was due to start drilling at another location. According to a MMS official quoted in the *International Herald Tribune* (Urbina, 2010b), BP had applied to drill with the Deepwater Horizon in another field on March 8. The official estimated (at $500,000 a day) that the 43-day delay had, by the day of the explosion, cost BP over $21 million (and, if the quoted costs of $1 million a day are used, could have cost $43 million). This is serious money for a TNC whose mantra is "every dollar counts." Thus senior and middle management were under severe pressure to finish the overdue operation.

USE OF CONTRACTORS

In terms of using contractors, Union Carbide treated its own subsidiary as if it were a contractor while BP was contracting out work. A MMS survey in 1995 found 66 percent of Gulf offshore operators reported an increase in the use of contractors (Priest, 2008, p. 150). The increased use of contractors has been noted in onshore production as one factor relating to incidents (Mac Sheoin, 2010, pp. 28-29), and may be a corporate strategy to avoid the legal complications of having partially or fully owned subsidiaries handle this kind of work. The Oilworkers' Federation (FUP) reports that between 1999 and 2001, 67 of 93 workers who died in Petrobras facilities, which are mainly oil rigs, were contract workers (Platt's Oilgram News, 2001). One result of the increasing use of contractors has been increasing complexity in the workplace which "has created uncertainty over the location of legitimate authority and decision-making, and bred ignorance about work conditions and the responsibilities of personnel and confusion over safety procedures" (Priest, 2008, p. 151). One of the main arguments now raging over the Deepwater Horizon is exactly who was in charge and who made specific decisions. While this confusion has partly resulted from attempts by companies to deny responsibility, there does seem to be some factual basis to the difficulty in identifying those in charge. Employees of twelve different companies were working on the Deepwater Horizon at the time of the disaster. The squabbling among companies over responsibility reflects the problem of complexity in the division of labor on the rig and the question of overall control.

TREATMENT OF STAFF AND HOSTILITY TO TRADE UNIONS

Both contractors and direct company employees were also suffering from long shifts and the pressure of work. Workers exhausted from periods of intense work involving extensive overtime have been associated with recent incidents in the process industry (Mac Sheoin, 2010, p. 28, 32). The lack of unionisation in the off-shore workforce, combined with understaffing meant workers worked long hours under pressure. Some indication of these pressures was given by an anonymous captain (who could not be named for fear of reprisals) to an International Transport Workers' Federation (ITF) group visiting Louisiana in 2001:

> I've been a captain for ten years and in my time in the oil industry, it rarely (never) happens that you work 12 hours and you rest 12 hours. It's more like 16 hours work and 8 hours rest. This is due to not having qualified reliefs 80 per cent of the time. So in order for you to save your job and be able to pay your bills and feed your family, you do what you have to–and that is "break the law" and make the company happy. (ITF, 2002)

This exploitation is enabled by a lack of unionization in an industry which is hostile to unions. Donn (2002, p. 11) reports

> A group of US maritime unions jointly established the OMU in 2000 which has tried to organize workers in the Gulf of Mexico. However the employers have responded by firing vessel masters and other employees who support the Offshore Mariners or even those who refuse to reveal who its members are.

The delegation from the ITF examining the situation in the Gulf of Mexico in 2001 was led by an ITF Assistant Secretary with a background in the North Sea who expressed shock at the absence of regulation and the lack of a "health and safety culture" which governs working practices in the North Sea since the Piper Alpha disaster.[3] The visiting delegation found themselves subject to some of the harassment common in the industry:

> the delegates had first-hand experience of the police harassment that has become rife in the area. At the company gates of Trico Marine, they found the local

> Sheriff's Deputy already there, barring their way. He even refused to let them deliver a letter to the company. Elsewhere they were made to show their passports and detained between two sets of squad cars for about an hour. Throughout the rest of the day they were trailed by no less than nine police cars. The ITF is considering legal action for unlawful detention. (ITF 2002)

Lack of unionisation of workers in the offshore oil industry has been condemned in Canada, England, Mexico as well as in the USA, while there have been consistent reports of bullying and harassment throughout the industry. The lack of union protection in the sector undermines safety, not just of workers but also of the environment. One Transocean employee told *Blowout* (2010a) "If you ask a lot of questions out there, you lose your job."

CORPORATE NEGLIGENCE AND RESPONSIBILITY

According to Congresspeople Henry Waxman and Bart Stupak, "BP repeatedly chose procedures in order to reduce costs and save time and made minimal efforts to contain the added risk" (Lyall, 2010). In its race to complete the well BP

> took what numerous industry experts and fellow oil executives said were highly questionable shortcuts. These included using a well design that presented few barriers to high-pressure gas rising up; skipping a crucial $128,000 test of the quality of the cementing; and failing to install capping devices at the top of the well that could also have kept gas from lifting a critical seal (Barstow et al., 2010).

A BP document shown to *New York Times* reporters by a Congressional investigator said BP chose to use the riskier of two methods of casing the rig, at least partly for financial reasons. The risk of this method was that if the cement around the casing pipe did not seal correctly, gases could leak to the wellhead, where there was only a single seal as a barrier, while the method BP did not adopt would have provided two barriers (Urbina, 2010b). BP was also strongly criticized for decisions it made regarding the lining of the well and how many centralizers it used to stabilize the well before it was cemented. On April 16 BP decided to use only six centralizers despite advice from Halliburton to use twenty-one centralizers due to the challenging nature of the well (Hoyos, 2010). For the lining BP chose to use a "long string" method which, while cheaper, gives less protection against gas leaks than alternative methods using multiple layers. The chairpeople of the Committee on Oversight and Investigations charged in a letter to BP's CEO on June 14 that "BP chose the more risky casing option, apparently because the liner option would have cost $7m to $10m more and taken longer" (Hoyos, 2010). Several reports said the BOP had been modified to speed the drilling process (Johnson and Torrice, 2010).

BP was also negligent in its disaster planning. When it applied for drilling permits BP claimed it could handle a leak of 250,000 barrels per day (bpd) (Crooks, 2010, p. 20). The company's plan was so inadequate and out of date that it listed a wildlife expert who had died in 1985 and promised to protect walruses. The limitations of BP's disaster plan became immediately obvious in the aftermath of the accident, when it struggled to mobilize enough vessels and specialist equipment to respond adequately. This should not have been surprising. In 2004, two BP managers wrote in a trade journal that BP wasn't adequately prepared for dealing with a deep-sea spill (Casselman and Chazon, 2010).

FAILURES IN REGULATION

Cost-cutting to the degree practiced by BP and Union Carbide is only feasible in a political economic context where regulations are nonexistent, weak, or poorly enforced. And while no direct causal link can be established between a weak regulatory environment and these two disasters, such an environment at least represents an ultimate cause by facilitating the kinds of practices discussed previously. For the Gulf's 3,500 rigs and platforms, the MMS had the sum total of fifty-five federal inspectors, while the total number of MMS inspectors on all drilling rigs in the Gulf, Pacific and Alaska waters was only sixty-two. Nor could the MMS inspection of rigs be considered hands on. According to the *International Herald Tribune* (Urbina, 2010c) "Its safety inspections usually consist of helicopter visits to offshore rigs to sift through company reports of self-administered tests." At Congressional hearings on May 26, Nick Rahall, Chairperson of the House Committee on Natural Resources, said "To now learn that certain agency personnel allowed industry to fill out their inspection reports in pencil, with MMS inspectors then writing on top of the pencil in ink prior to turning in their reports, is truly reprehensible (Fifield, 2010).

As well as this example of the virtues of industry self-regulation, MMS also showed the traditional operation of the "revolving door" between industry and regulator. Details of exactly how close the regulators were to the industry were revealed in a report released by the Interior Department's Inspector General in late May on staff at the MMS office in Lake Charles, Louisiana, prior to 2007, some of whose duties included inspecting oil rigs. Confirming the existence of a revolving door by noting staff moved with "ease" between jobs in industry and government, the report said officials took part in hunting and fishing trips and golf tournaments organized or partly funded by offshore oil companies. (Power, 2010). Other reports claimed some regulators literally got into bed with some industry staff. More is involved here than the private sexual practices of state employees. According to *Blowout* (2010b, p. 8), the MMS's internal inspectors reported in 2008 that "key MMS personnel involved in sexual liaisons with oil company management had discouraged the MMS in 2003 from demanding better systems to prevent well blowouts."

The MMS was also criticized for issuing BP a "categorical exclusion" on April 6, 2009 from carrying out an environmental analysis of the Deepwater Horizon project. MMS, which routinely granted such exemptions, was doing BP no special favor in this case. Coincidentally, eleven days before the disaster, BP wrote to the Council on Environmental Quality supporting such exclusions as helping "avoid unnecessary paperwork and time delays" and allowing "an expedited review of projects that are considered routine and have little or no environmental impact" (Kirchgaessner, 2010a).

These regulatory failures stemmed from the policies of their political masters. In the US the neoliberal assault on regulation, begun under Reagan, mildly rebuked under Clinton, and continued under Bush, was not rolled back by Obama, who, in an attempt to get climate change regulation passed, had shortly before the Deepwater Horizon disaster announced permission for the oil industry to drill in a widely expanded offshore area. Thus part of the problem with MMS came from its contradictory mission: like many regulatory organisations, its job included both regulating and promoting the industry. It not only regulated the oil industry's activities offshore, but also collected substantial revenues from the industry for the US government, some $13 billion in 2009. MMS was also hobbled by a Congressional provision which forced the agency to process all exploration applications within a period of thirty days, which precluded the agency from undertaking an adequate environmental review of the applications (Johnson and Torrice, 2010).

The closeness of regulators to industry was mirrored by the closeness of politicians to industry which echoes the situation in Bhopal. In the USA, under the Bush regime it appeared the oil industry had taken over the state, as oil industry executives not only changed places with those regulating the industry in the state bureaucracy, but moved into the executive suites themselves. While the Bush administration was noted for its closeness to the oil industry, the industry is strongly involved in the American state whichever party is in power. From 1998 to 2004 the five largest oil companies spent $175 million on lobbying (Faber, 2009, p.104).

This closeness of politicians to industry isn't simply about individual corruption, but related to the enthusiastic embrace by politicians of industry positions. Thus, on April 2, 2010, when Obama announced a new policy extending offshore drilling, he said "It turns out, by the way, that oil rigs today generally don't cause spills. They are technologically very advanced" (Obama 2010). These statements were a pale reflection of the hubris corporate officials were displaying. BP's head of Exploration and Production, Andy Inglis, speaking after the company had drilled the deepest well ever for a commercial operation said "We don't do simple things. We are prepared to work at the frontier, and manage the risks" (quoted in Crooks, 2010b, p. 19).

TNC INVOLVEMENT AND EXTRA-TERRITORIALITY

In many cases, the corporate strategies discussed above are facilitated by the position of TNCs which are not bound to the labor, environmental or other standards of the countries in which they are headquartered. This allows them to exploit aspects of the global economic order–in which heightened interdependency of nations, access to cheap natural resources spread across the globe, and the global movement of goods, privileges TNCs that may theoretically have physical places they inhabit, but practically speaking are more accurately defined by their ability to exist in a range of forms and in virtually any place on earth. TNCs use their nimble footedness to prioritize production and profit over safety while escaping the real costs of the industrial disasters for which they are responsible. For Deepwater Horizon, two of the three major TNCs involved are headquartered outside the USA: BP in London and Transocean in Switzerland. Transocean, originally founded in Louisiana in 1926 relocated to Switzerland in 2008, partly for tax reasons (Casselman, 2010a, p. 15).

A failure of regulation can be laid at the door of an extraterritorial power, no less than the shipping registry of the remote Marshall Islands in the north Pacific, under which the Deepwater Horizon and apparently another 33 of Transocean's oil rigs were registered, as a result of which the Marshall Islands were technically responsible for inspecting safety standards on the Deepwater Horizon. The Marshall Islands registry, administered by a private firm, International Registries Inc., is a noted flag of convenience.[4] Transocean has admitted it registered its rigs in the Marshall Islands partly to gain a "financial tax benefit" but Democrat Members of Congress alleged it was also intended to avoid safety regulations and inspections, one claiming "Coast guard inspection of a US-flagged mobile offshore drilling rig takes two to three weeks, but the safety examination of a foreign flag offshore drilling unit, such as Deepwater Horizon, takes four to eight hours" (MUA, 2010).

MANAGING THE CRISIS RESPONSE

BP worked with various government agencies to manage its response. The primacy of BP in dealing with the spill, as well as its integration with the state, is shown in the personnel working in the response control centre in a BP building in Houston where some 400 BP personnel were assisted by 150 from other companies, while some thirty US officials from agencies such as the Coast Guard and the MMS worked alongside them. A total of 160 private companies were involved in the response (Crooks and Morris, 2010). While BP's attempts to deal with the spill required approval and permits from federal officials, the operation was in private hands. The state has neither the technological capability, the trained cadre, nor the knowledge to direct the response. Only TNCs like BP and its industry peers had the technological capacity to work at such deepwater levels. This meant that the state was dependent on industry not only to control the spill, but also for information on the size of the spill. This echoed the situation in Bhopal when the Indian state had to deal with the toxic MIC remaining in the factory. The resulting "Operation Faith" was a co-production of capital and state, with the Indian state allowing UC to process the toxic chemical as the state had neither cadre nor capacity to do so, nor any alternative proposal for disposing of the MIC.

Rebecca Solnit (2010, p. 29) describes the government ceding of control over the oil spill response to BP:

> Why BP has been using dispersant at all is a question whose answer seems to be about a policy of disguising, repressing and hiding the damage. One clean-up worker quit because she said they were told to remove the surface sand; the BP supervisors just wanted the beach to look clean. . . BP has created a no-fly zone; the co-operative Coast Guard keeps boats at bay, their captains afraid of huge penalties if they cross into restricted waters; workers were obliged to sign non-disclosure contracts, others had all recording technologies confiscated, data on worker exposure were suppressed; scientists were not allowed data; birders were allowed to band but not to put transmitters on rescued birds, according to Drew Whelan [of the American Birding Association]; and data on the spill were constantly spun so the volume of the gas upwelling was smaller, the impact was less, the facts were unavailable . . . BP rules the waves and a bit of Louisiana. I met one boat's captain who'd been trying to get information from the Coast Guard but was repeatedly passed on to BP, which also seemed to be calling the shots on land about who could go where, controlling media access and even air traffic to the area of the spill. Police, sheriffs, National Guardsmen and politicians seemed to be taking their orders from the corporation.

BP's concerns with the visual aspects of the clean-up operation were alleged by some to have put workers' health and safety at risk. In late May, Clint Guidry of the Louisiana Shrimp Association accused BP of threatening to sack cleanup workers who turned up at work with respirators. BP sidestepped the accusations by saying it was the Obama administration's decision whether such gear was needed. Hugh Kaufman, chief investigator with the EPA ombudsman, drew comparisons with the situation of workers cleaning up after the attacks on US military and financial targets in September 2001: "If people are wearing respirators, it scares people because they realize how toxic it is. The administration is downplaying the problem because it saves them money down the line. It was the same at Ground Zero" (Goldenberg, 2010b). Local environmentalists have accused the EPA of failing to release the results of air and water tests, while the threat to public as well as workers' health from exposure to the crude oil, chemical dispersants and mixture of both was being minimized or ignored:

> Seven workers trying to scrub the oil from Breton Sound were taken to hospital on Wednesday, complaining of burning eyes, headaches, nausea, dizziness and chest pains. Administration officials suggested on a conference call with reporters that the workers could be suffering from sunstroke" (Goldenberg, 2010b).

CONTROL OF INFORMATION

Initially, BP and the U.S. government cooperated to minimize the extent and effects of the oil spill. This was in part caused by government needing to rely on the corporation for information and by a common desire to avoid crisis in an intensely profitable industry, which was also of great importance to energy provision and state finances in the USA. Rather than attend to the spread of the polluting oil, the media was able to concentrate on the spectacle of the various technical fixes with which BP attempted to stem the oil. Thus a narrative of heroic technologists responding to crisis was provided. But the happy ending was late in coming. As BP failed again and again, the state began to separate itself from BP, at least partly in an attempt to direct at BP public anger that threatened to target the state. US officials began to describe BP as "British Petroleum," a name the company had abandoned a decade previously, emphasizing its status as a foreign company, while a rhetoric of war and invasion was increasingly used by US officials.

BP followed a strategy of attempting to minimize the extent and effects of the spill, as seen in CEO Hayward's statement on May 13 that "The Gulf of Mexico is a very big ocean. The amount of volume of oil and dispersant we are putting into it is tiny in relation to the total water volume" (Webb, 2010a). From the beginning BP seriously underestimated the flow from the well, underestimates accepted unquestioningly by the US state. Unsurprisingly public BP estimates were on the low side of 1,000 to 5,000 bpd, at a time when BP's internal worst-case estimate was 20 times that figure. In June, government scientists estimated the rate to be 12,000 to 19,000 bpd (Johnson and Torrice, 2010), a revision based on video footage of the seabed leak BP released only after pressure from the US Congress and ocean scientists. By July the *Financial Times* was printing estimates of 35,000 bpd, while noting the rate could be as high as 60,000 bpd. (Crooks, 2010b).

BP also denied claims by scientists to have detected huge underwater plumes of oil spreading from the spill site, with Haywood responding "The oil is on the surface. Oil has a specific gravity that's about half that of water. It wants to get to the surface because of the difference in specific gravity" (McGreal, 2010). The chairperson of the House of Representatives Environmental Committee accused BP of downplaying the disaster for economic motives: "The fine that can be imposed on them is based on how many barrels [pour into the sea]. It could wind up in billions of dollars of fines. They had a stake in low-balling the number right from the beginning. They were either lying or they were incompetent" (McGreal, 2010). In Bhopal a similar financial logic underlay attempts to contain the disaster's health effects to limit possible claims.

Controlling the flow of information included controlling access to the polluted area, as well as embedding reporters with official response teams to ensure the right message was reported.[5] Elements of the US state introduced new regulations to keep people at a distance from the spill and its effects. Rebecca Solnit (2010, p. 31) quotes Drew Whelan regarding colonies of royal terns on Queen Bess Island: "the Coast Guard has imposed new rules on these colonies to keep people out, which includes criminal trespass, a felony which could be punishable by 15 years in jail and up to a $45,000 fine." While such rules were justified on health and safety grounds they had the convenient side-effect of ensuring the official line wasn't queried. This was added to by getting workers to sign non-disclosure forms and using prison labour. The New York Times (Peters, 2010) reported journalists "have repeatedly found themselves turned away from public areas affected by the spill, and not only by BP and its contractors, but by local law enforcement, the Coast Guard and government officials," while a spokesperson for the Gulf Remediation network (GRN) said "Since this catastrophe began almost a month ago, BP has attempted to restrict independent monitoring of the spill site and its impacts, and has worked to restrict data from the public and independent experts. For example, on the Saturday after the spill officials refused to allow GRN to fly over the site in order to survey the extent of the slick and cleanup efforts" and scientists complained of the state's failure to publish even one test result on water from the deep ocean (Edwards and Kane, 2010).

A final part of BP's information strategy involved a massive increase in advertising spending, with BP devoting its total advertising budget to the spill. Between April 20 and the end of July, BP spent $93.4 million on advertising, over $6 million a week, more than three times the amount it spent in the corresponding period of 2009. BP justified this increased spending as an attempt to keep residents of the Gulf coast informed, though it was obviously also intended to help diminish criticism of the company.

MISSING INFORMATION AND TRADE SECRETS

Twenty-five years after Bhopal, UCC has still failed to provide detailed information on the composition of the gases that engulfed the city's inhabitants (Mac Sheoin, 2009, p. 428, n.6). Similarly, BP withheld information about the toxicity of the 2 million gallons of dispersants that had been applied by the end of August (Goldenberg, 2010c). When concerns were expressed about their toxicity and the EPA demanded on May 20 that BP reveal their chemical composition, BP refused, claiming the dispersants' composition was proprietary information of their manufacturer Nalco, in essence "trade secrets." On May 27, Nalco made some of this information available and the EPA published it on its website (Johnson and Torrice, 2010). By June, complaints arose that BP was still refusing to disclose publicly the full formula for Corexit. Gregor Eberli, professor of marine geology at University of Miami told the *Financial Times* (Cookson, 2010) "It is astounding that the exact chemical composition of Corexit is not publicly available, given the magnitude of this disaster. I don't understand why there hasn't been more pressure on the companies to disclose it."

The Deepwater Horizon spill also revealed a striking lack of information to allow the spill's effects to be evaluated. To begin with, decades of research on dispersants' effects on biodegradation had resulted in mixed results, providing evidence for "enhancement, inhibition and no effect" (Johnson and Torrice, 2010).

Furthermore the ecology of the deep ocean itself is understudied and therefore little understood. Aquatic toxicologist Carys L. Mitchelmore told *Chemical and Engineering News* "Ultimately we don't even know what's down there to ascertain risks. It is just one huge black box" (Johnson and Torrice, 2010). Mitchelmore emphasized the experimental nature of the spill response: "This is one big ongoing experiment. We're going to be learning a lot from this spill, and that's wrong. We need basic information ahead of time" (Johnson and Torrice, 2010).

BP's use of chemical dispersant underwater, for example, was the first time this method was attempted. Part of the blame for this ignorance lies with government agencies. After the Exxon Valdez spill, the Oil Pollution Act allotted $19 million annually for oil spill research, but agencies have not bothered spending all this allotment: in 2008, only $7.7 million was spent (Johnson and Torrice, 2010).

The US government was also accused of prematurely declaring victory in the war on pollution in the Gulf when in early August it trumpeted a report by the National Oceanic and Atmospheric Administration (NOAA) to the effect that 75 percent of the spill had been cleaned up or naturally dispersed. Critics quickly attacked the report's conclusions as well as its failure to explain how it arrived at its calculations. John Kessler, of Texas A&M University, who led a NSF study of the spill told the *Guardian* (Goldenberg, 2010d) "Recent reports seem to say that about 75 percent of the oil is taken care of and that is just not true. The fact is that 50 percent to 75 percent of the material that came out of the well is still in the water. It's just in a dissolved or dispersed form." Two weeks later a senior NOAA scientist admitted that three-quarters of the oil was still in the Gulf. The denial by BP and ignoring by NOAA of the existence of deepwater plumes of oil was undermined by a paper in *Science* which reported a plume 22 miles long, 1.2 miles wide and 650 feet high, consisting not of pure oil but including toxic oil compounds such as benzene and xylene. (Goldenberg, 2010e).

TRANSFER OF BLAME

BP's immediate reaction was to blame the rig operator. After news of the explosion reached BP headquarters in London, Hayward brought together a crisis management team, including BP's general counsel and head of media. The first public statement by Hayward sought to transfer blame to Transocean, who owned and operated the Deepwater Horizon. It was "their people, their processes" and "not our accident." By the time he got to Houston his tune had changed somewhat, having recollected BP's role as supervisor and coordinator of the project, whose staff were involved in decisions on the rig: "[w]hen he landed in Houston Hayward said that although the blame might lie with Transocean, the oil was BP's, and it was his job to clean it up." (Crooks, 2010b, p. 20). In an interview with the BBC on May 4, Hayward was back at the old line: "This was not our accident. . .This was not our drilling rig. . . . This was Transocean's rig. Their systems. Their people. Their equipment" (Webb, 2010b).[6]

At the first Congressional hearings on May 11, BP blamed Transocean which blamed Halliburton, which blamed BP. BP America's chief executive Lamar McKay said "BP hired Transocean to drill that well. Transocean had the responsibility for ensuring the safety of that operation," blaming the spill on the BOP "that didn't work" (Goldenberg, 2010f). Steven Newman of Transocean said BP drew up the drilling plan and BP engineers were in charge when the drilling was wound up, stating "Offshore oil and gas production projects begin and end with the operator, in this case BP." He transferred the blame from the BOP to the cement casings of the well: "The most significant clue is that these events occurred after the well construction process had been completed," pointing the finger at Halliburton, whose Tim Probert said Halliburton had carried out the work to specifications laid down by BP, completing the circle of blame (Goldenberg, 2010f) On 14 May, BP claimed against Transocean's insurers, seeking up to $700 million; BP was rebuffed by insurers, who said their responsibility was only for damages to the rig.

The legal hearings under the aegis of the Coast Guard provided a foretaste of the legal battles to come, exhibiting the legal tactics the different TNCs were expected to follow. BP used the hearings to direct attention at Halliburton and Transocean. At the August hearings BP got Transocean officials to admit the BOP had not gone through the certification process required by federal regulations, copies of which BP officials then provided to the press at the hearings. While BP succeeded in getting a Transocean manager to admit Transocean was responsible for "well control," all did not go BP's way. One BP engineer admitted he never opened a report from Halliburton warning the cementing work could allow gas to enter the well, while one of BP's managers on the rig, Ronald Sepulvado, was forced to admit under repeated questioning that he was the person who made the final decisions on the rig (Casselman and Gold, 2010).

The dispute between BP and Transocean intensified with Transocean writing a seven-page letter to BP on August 18, which it copied to Administration officials and Members of Congress, accusing BP of unwillingness "if not an outright refusal" to provide documents that "only BP has" which are "critical to an honest assessment of the incident." Noting BP had not provided documents requested by Transocean since June 21, Transocean's lawyer wrote "This is troubling. . .because it appears that BP is withholding evidence in an attempt to prevent any entity other than BP from investigating the cause of the April 20th incident and the resulting spill" (Pfeifer and McNulty, 2010).

BP's report into the incident followed the corporate line, minimizing design errors and instead concentrating on the cementing job by Halliburton, problems with the BOP which it attributed to maintenance issues and actions of the rig workers in the hours before the blowout. In essence, the report focuses on immediate rather than root causes. The report follows BP's transfer-of-blame strategy, with BP taking half-responsibility for only one of the main causes of the incident (Urbina, 2010d). In what some have seen as a hardening of attitude by BP, in October its new CEO attacked media and industry critics for 'fear-mongering', saying "We fully accept our share of responsibility, but I hope . . . we can reach a balanced and informed judgement about what happened' (Groom, 2010).

LEGAL TACTICS

BP made money speedily available to settle claims by some of those immediately affected, but the legal cases are likely to be fiercely contested and long drawn out. In Texas City, BP was quick to accept moral responsibility, but not legal liability (Mac Sheoin, 2010, pp. 30-31). A previous example is provided by the conduct of Exxon in the case of the Exxon Valdez spill in 1989, which "dragg[ed] the process out for decades, outlasting its opponents, and finally settling in 2008 for a pittance compared to the destruction and suffering the corporation's spill had created" (Solnit, 2010, p. 30). A local lawyer representing fishing people suing BP drew direct comparisons: "BP is saying they will pay legitimate claims, just like Exxon said after the Valdez oil spill. Yet once publicity died down, Exxon litigated for over twenty years" (McNulty, 2010).

A more outrageous legal tactic was illustrated by allegations by two surviving Deepwater Horizon crew members that they were coerced into signing legal waivers in the immediate aftermath of their escape from the burning oil rig. Stephen Davis said he and other survivors were kept in isolation for 40 hours and not allowed to phone or radio their families, while Transocean attorneys pressured them to sign legal waivers. Furthermore, "The company was waiting for the men at Port Fourchon with portable toilets for drug tests" (Goldenberg, 2010a). Another survivor, roustabout Christopher Choy, said survivors were met in the galley of a boat by company lawyers who told them "You need to sign these. Nobody's getting off here until we get one from everybody" (Dykes, 2010).

INDUSTRY RESPONSES TO THE DEEPWATER HORIZON DISASTER

Just as Bhopal led to concerns over the safety of the chemical industry globally, so Deepwater Horizon sparked concern about safety conditions in the offshore industry globally. In Brazil in August 2010 the National Petroleum Agency shut down Petrobras' P-33 platform in the Campos basin following a leak and fire in July which led the Sindipetro union to claim the rig was unsafe. Union representatives called for four more platforms in the area to be closed, saying ten of the forty-five platforms in the Campos basin are "in a very worrying situation." However the fact that Brazil intends to use its vast deepwater reserves as a catalyst for national development and their political importance to President Lula and his successors means there is little likelihood that drilling there will not go full speed ahead. Small details such as the death of 282 Petrobras workers in the last fifteen years at the company's rigs and refineries are likely to be dismissed as necessary sacrifices on the road to national prosperity (Gall, 2010; Millard, 2010). In Mexico, union officials and experts condemned low wages, poor training, restrictions on union membership and corrupt practices in the offshore industry, with a union official saying "[w]orkers hired by private contractors do not have the training for offshore drilling. Contractors and sub-contractors are only interested in making money" (Godoy, 2010).

News from the South China Seas early in September showed that Deepwater Horizon was not preventing BP from participating in new deepwater projects in Asia. BP's new project involved Chevron buying 59.18 percent and BP the rest of block 42/05 around 250 km south of Hong Kong, with water depths varying from 198 to 1980 metres, showing other oil majors were still prepared to partner with BP and the PRC government was prepared to allow BP to participate despite Deepwater Horizon (Winning, 2010).

In Europe, Norway announced a ban on drilling in new areas until the causes of Deepwater Horizon are known. Britain has typically confined itself to initiating a parliamentary inquiry which can confidently be expected to accept such drilling is essential, though no doubt with added safeguards. While the EU energy commissioner has suggested a de facto moratorium on new drills, the lack of community control over energy policy (which is decided at national level) means this call is likely to be ignored.

Another impact of Deepwater Horizon, which paralleled that following Bhopal, was the creation of an insurance crisis for the deepwater drilling industry, with insurance costs rising 50 percent after Deepwater Horizon and expected to rise as much as 100 percent by the end of the year (News wires, 2010; for the insurance crisis following Bhopal see Jones, 1988, pp. 198-203.).

While a wide variety of critical voices have suggested the Deepwater Horizon disaster should not only lead to tougher regulation but should also put into question oil companies' plans to continue deepwater extraction and increase extraction from unconventional sources, the industry's response has been united: the drilling must go on. BP rejected suggestions from environmental groups and shareholders that the company should concentrate in future on lower risk areas, with a BP spokesperson saying:

> When the leak is plugged and we return to normal we will be carrying out an assessment of where the new BP goes from here. We will—like the rest of the industry—be working out how we can do things differently in terms of safety but not where we do them. The position is the same now as it was at the strategy update earlier in the year. We are committed to three core areas of deepwater oil, unconventional gas and enhanced recovery on super-sized fields. (Macalister, 2010b)

Indeed, in late July BP announced its intent to proceed within a few weeks with deepwater drilling off the coast of Libya, which would involve drilling 200 metres deeper than the Deepwater Horizon well, leading to calls for a moratorium in the Mediterranean until results of the Deepwater Horizon inquiries were in, with questions being raised as to whether Libya or any other Mediterranean state had the capability to deal with an oil spill of that magnitude (Dinmore and de Sabata, 2010a; Dinmore and de Sabata, 2010b; Kanter, 2010). For the US, in early September BP warned the government that if it was excluded from deepwater in the Gulf this would affect its ability to pay for the clean-up of the Gulf coast. One month later, President Obama announced the end of the moratorium on deepwater drilling.

For the industry in general the response is "drill, baby, drill," though carefully of course. This response is due to the continuing importance of deepwater sources of oil for TNCs kept from onshore sources by national oil companies. Since 2006 the number of rigs able to drill in 900 metres of water or over has increased 43 percent to 146, another sixty-five are under construction (Casselman, 2010a) and the oil industry expects to invest $167 billion on deepwater development from 2010 to 2014, a 37 percent increase on the previous five-year period (Crooks, 2010c) Shell's boss Peter Voser said "Given the rise in the population and the rise in the developing world of energy needs, we will have to develop those resources in deep waters, so my expectation is that we will go forward with it, but it will need some changes." In response to Deepwater Horizon, Voser implied that the well would not have erupted with such devastating consequences if Shell rather than BP, had been in charge: "We would not have drilled the well in the same way." (Weardon, 2010).

Voser's comments represent a change in the industry's response. The united front presented in response to the leak began to fragment early in June. Interestingly, this breaking of ranks set in when the Obama Administration announced a six-month moratorium on new deepwater drilling in the Gulf, with oil industry executives fearing global knock-on effects on deepwater drilling in Africa, Brazil and South East Asia. In an interview with the *Wall Street Journal Europe* (Casselman, 2010c), Chevron officials said the company used a safer well design and argued "If we'd have had best practices employed on this well we wouldn't have this situation that we have today," while Chevron chairperson and CEO John Watson bluntly said "This incident was preventable." In remarks to a House Energy and Commerce Subcommittee, making the case for continued new deepwater drilling in the Gulf, executives of ExxonMobil, Chevron, Shell and ConocoPhillips detailed how they would have done things differently from BP. Rex Tillerson of ExxonMobil said "It appears clear to me that a number of design standards that I would consider the industry norm were not followed. We would not have drilled the well the way they did," while James Watson of Chevron said "Not all standards we would recommend were in place. For example, the casing design and mechanical barriers appear to be different than what we would use" (Kirchgaessner, 2010b). The companies' attempts to persuade Congress that their plans would cope better with an oil spill than BP's were not convincing, with Henry Waxman, Chair of the House Energy Committee, noting their plans were largely identical and written by the same company, while the chair of the subcommittee said that, like BP, three of the companies' plans described how they would protect walruses, despite these mammals not having inhabited the Gulf for three million years (Kirchgaessner, 2010b). Some indication of the priorities behind the oil companies' disaster plans can be seen in the fact that Exxon's plan devoted forty pages to media response compared to nine pages devoted to environmental impacts (Goldenberg, 2010g).

Another part of the oil majors' response was unveiled by a joint announcement on July 21 by Chevron, ConocoPhilipps, ExxonMobil and Shell, of plans to set up a separate non-profit organisation, the Marine Well Containment Company, to design, build and operate a rapid response scheme to capture oil spilled 10,000 feet under the sea, which could be deployed within twenty-four hours. This response was similar to that after the Exxon Valdez when the industry set up the Marine Spill Rescue Corporation (MSRC) to set up and operate a response system to surface oil spills.[7] This initiative was part of the industry's attempt to persuade the Obama Administration to abandon its temporary moratorium on deepwater drilling. But if it didn't work, the industry had plans to fight on other fronts also. The API announced in mid-August plans to organize "citizen rallies" to express concerns at increased regulation and proposed legislation that might follow Deepwater Horizon (Kirchgaessner and Pfeifer, 2010).

REGULATORY RESPONSE

While it's too early yet to estimate long-term regulatory effects, the response by US regulators does not bode well for the future. In response to the woefully inadequate number of federal inspectors for offshore oil rigs, Salazar announced plans to increase this number by a mere 10 percent (Johnson and Torrice, 2010). Salazar also announced he would request an increase in Congressional appropriations for rig inspections to increase from \$23 million to \$52 million, while also requesting a tripling of the time period within which MMS processes exploration applications to ninety days. While an improvement, ninety days is still a fairly short period for reviewing environmental impacts. Even the moratorium on deepwater drilling had its problems. First, only new exploratory drilling was affected, while existing producing drills were unaffected. Second, the moratorium was only temporary. Regulators were unwilling to consider an outright ban. The moratorium affected 33 exploratory wells being drilled in the Gulf of Mexico, but allowed the continued operation of some 3,500 rigs in Gulf waters and 46 deepwater rigs. Given that rigs can service multiple wells, the number of wells allowed to continue in production is around 30,000 (Johnson and Torrice, 2010).

THE TWO DISASTERS: DIFFERENCES AND SIMILARITIES

Having devoted the majority of this chapter to tracing common elements in the Bhopal and Deepwater Horizon disasters, before drawing our conclusions we need to note the three major ways in which these disasters were dissimilar and the reasons for these. The first difference relates to the nature of disasters themselves.

The other two relate to the nature of power relations between the TNCs and the host country states.

The first difference was that Deepwater Horizon was a crisis that continued beyond the acute stage of the original explosion and sinking of the rig to the chronic oil spill stage, ensuring continued coverage, while Bhopal quickly disappeared off the radar of the international media. The Deepwater Horizon crisis continued until the well was finally plugged, while the Indian media were able to declare the Bhopal crisis quickly over, categorizing it as an acute rather than a chronic crisis. Yet despite one being an acute and the other a chronic disaster, both disasters revealed in their impacts on communities and the social fabric characteristics of what disaster researchers in the social sciences refer to as "corrosive" effects of toxic disasters (Edelstein, 1988; Freudenberg, 1997; Picou, Marshall, and Gill, 2004).

The second difference relates to the official hearings and investigations into the Deepwater Horizon disaster causes undertaken by the US government. Representatives of the culprit corporations were hauled before various committees of the US Congress and appeared at inquiries run by the Coast Guard and the MMS. Meanwhile, the prospect of litigation in the American courts also ensured continued media coverage. In India, by contrast, the investigation was not made public but placed in the hands of the CBI, while a banner of state secrecy was draped over investigations by state bodies. Thus the state in the US examined at length and in public the causes and consequences of the Deepwater Horizon disaster, while in India the state treated information on the disaster as state secrets. Furthermore, the political class in the US responded to the concerns of the affected citizens and businesses, while the Indian political elite failed to take the concerns raised by Bhopal on board. One reason for this was the geographical spread of the possible victims in the US. The spill affected a number of US states as well as two major industries—fishing and tourism. By comparison, the Bhopal victims were geographically concentrated marginal slum dwellers, mainly Muslim, with little political influence, easily abandoned and ignored not only by the national but also by the local political elite. In the US, the prospect of litigation by individuals and affected interests and organizations ensured that there was a much higher likelihood that issues of liability would eventually be determined in a court of law while in India, the state took over the litigation.

This difference also explains the profoundly different patterns of social movement mobilization following the two disasters. The acuteness of the Bhopal disaster, not to mention its high death toll, was sufficient to mobilize survivors who experienced a profound injustice. As it became clear that neither UCC nor the Indian government were acting swiftly enough to treat the harmed or to begin the process of compensating victims, survivors found much to mobilize around. As the years went by and compensation was either not forthcoming or slow to be disbursed, and as legal proceedings continually disappointed survivors with their outcomes, more and more grievances were heaped upon the injustices of the initial trauma and loss. As a result, even today, there is a strong social movement in

Bhopal. In contrast, though still early in the post-disaster phase, little social movement organizing has taken place around the Deepwater Horizon disaster. A spattering of solidarity protests took place through May and into June with small turnouts in places like New York City, New Orleans, and Tampa Bay, Florida. A "Spill into Washington, D.C." protest took place in the US capital on Labor Day, and in October an anti-dispersant rally was held. In addition, a number of virtual protests and boycotts spread around the Internet, primarily through Facebook, but none had any lasting impact or resulted in the formation of an ongoing social movement organization.

The final difference relates to the importance of the country to the TNC, both as a production site and as a market. In the Bhopal case, India was a minor market for UCC whose loss was no big deal for the US TNC. Nor did UCC have any great capital invested there, so the TNC was easily able to pull out of the country. In comparison it was not so easy for BP to pull out of the Gulf as the US was not only a major market but also a central area for production. These dissimilarities were in part due to the different forms of process industry and different investments involved. Bhopal involved straightforward pesticide production which could be undertaken anywhere. The Deepwater Horizon involved resource extraction in one of the remaining major areas accessible to oil TNCs. The USA was also a major market for BP.

THE POWERS OF CORPORATE ACTORS

Despite these differences, this chapter has demonstrated striking similarities between the disasters at Bhopal and on the Deepwater Horizon in terms of causation, corporate negligence and crisis management. We have shown both disasters' origins can be found in similar patterns of corporate negligence. The TNCs involved in both disasters share a history of previous incidents, overlooked equipment and design failures and faults, compromised safety through cost-cutting, over-extended workers and harassed union organizers, and failed to undertake adequate disaster planning and implementation. In addition, in both disasters the corporate actors were aided in minimizing the damage to their bottom lines by complicit governments. Both governments failed to regulate the respective industries sufficiently and knowingly allowed environmental, occupational and public health and safety to be compromised to ensure that the TNCs could maximize profits.

As corporate actors, BP and UCC responded to their respective disasters by framing them as unpredictable events that could not have been prevented. Both attempted to portray themselves as responsible corporate actors whose supposedly stellar safety records represented proof that the disasters could not reasonably have been predicted or avoided. Yet both corporations had plenty of evidence in their corporate histories that "accidents" had happened with great regularity. Similarly, both corporations' CEOs framed their companies as socially responsible by accepting moral responsibility. We demonstrated that the similarity of proximate causes in both disasters point to a potential pattern in highly risky industrial operations, a pattern that would suggest disasters are, in fact, quite normal. Although BP's cost-cutting measures were part of a strategy to bring an oil well online more quickly and at a lower cost, while Union Carbide's cost-cutting measures were aimed at minimizing economic losses from a pesticide plant that never turned a profit, the disastrous outcomes in both cases point to the heightened likelihood of normal accidents when cost-cutting compromises already vulnerable complex and tightly coupled systems.

In both cases, state and capital shared a common interest in limiting the crisis, though for different reasons. In Bhopal UCC's interest was to limit its liability while the state's interest was to prevent social unrest, encourage a speedy return to normality and minimize long-term costs. In the US the aims of state and capital in managing the response were similar to those in India. BP wanted to limit its liability and the US government wanted to limit and quickly end the pollution crisis. In both cases also, the industries, through their industry associations, undertook program initiatives in order to pre-empt state regulation.

Whether the proximate causes of line management failures, the use of contractors or other cost-cutting measures to complete or make a project profitable, or the over-extension and under-training of workers, or the ultimate causes of State failures to regulate industries effectively, capture of regulatory agencies by corporate interests, or the exploitation of extra-territoriality by TNCs, the commonalities across these otherwise dissimilar industrial disasters raise important questions about the ways TNCs conduct business in a global economy.

While Perrow's analysis of "normal accidents" is of use, it is limited as it remains at the technical level. It needs to be supplemented by Wright's classic argument that accidents in the North Sea were not exceptional occurrences but normal events resulting from work structures involving sub-contracting and pressures to complete projects and accelerate production (the 'speed-up') (Wright, 1986). Thus these deaths result not just from technical factors but also from the labor process. Both the technical organisation of production and its accompanying labor process point to another level of causation and explanation, that of capital and corporate decision-making, where decisions are made on the basis of profit rather than public and environmental safety.

Our analysis can push the concept of normal accidents to yet another level. In our comparison of the Bhopal and Deepwater Horizon disasters, we have highlighted how the unique positions of TNCs allow them to exploit aspects of the global economic order to prioritize production and profit over safety while utilizing the dependence of the state on economic producers and resource extractors to escape the real costs of the industrial disasters for which they are responsible. In other words, our analysis points to Bhopal and the Deepwater Horizon not just as instances of normal accidents in the context of globalizing industrial operations, but also to the responses of UCC and BP as "normal disaster responses." There is nothing inherent in the system of global capitalism that can prevent such disasters. Yet the system is self-healing, preventing responses to such disasters that might actually undermine the system's ability to go on functioning. The (im)balance of power between TNCs and states, as well as their mutual dependence on one another, ensures that no responses will be taken that compromise their respective abilities to function.

Each time a major disaster strikes an hazardous industry or a technology there is a flurry of commentary describing the crisis as something that will change the industry or technology on a basic level. Bhopal was described as the chemical industry's Hiroshima. Deepwater Horizon has been described as the industry's Three Mile Island. Yet previous experience with such game changers is that after the initial crisis has passed and the publicity has died down, the industry continues to operate as before. All the indications so far are that the situation with Deepwater Horizon will turn out the same.

NOTES

1. We use "catastrophe" and "disaster" interchangeably, which is consistent with the Oxford English Dictionary definitions of both words. We prefer a slightly narrower definition of catastrophe, one that does not characterize catastrophes as the result of "misfortune, mishap." As both catastrophes at the center of our analysis were not the result of "bad luck," we place the qualifier "corporate" in front of the word "catastrophe" in our title. For brevity we henceforth refer to both events as "catastrophes" or "disasters," with the understanding that corporate decisions, not misfortune, shaped both catastrophes.
2. This low death-toll is partly a result of luck. Captain Hung Nyugen of the joint Coast Guard/MMS inquiry noted that 115 crew members were saved because a cargo ship happened to be docked at the rig and moved quickly to assist those who abandoned the rig. (Hammer, 2010).
3. Readers should not interpret this to mean all is well in the British section of the North Sea. A recent academic study presents data showing "the degree to which victimization and intimidation of the workforce on offshore platforms acts to prevent an open and constructive approach to safety management" (Whyte, 1998), while the RMT offshore branch organizer in Aberdeen told the *Guardian* "really serious incidents are not being reported because of widespread bullying and intimidation. I cannot follow up these cases because it would expose the guys to losing their jobs." For useful background on the North Sea, see Carson (1982) and Tombs and Whyte (1998).
4. For information on flags of convenience, see Donn (2002) or DeSombre (2006).
5. Research on embedding reporters with the US military in "Operation Iraqi Freedom" (the invasion of Iraq) found "embedded reports were more favourable in overall tone towards the military and in depiction of individual troops" than nonembedded television reporting (Pfau et al., 2005), while, in the case of newspaper reporting, "compared to nonembedded reporting, embedded print coverage of 'Iraqi Freedom' was more favourable in overall tone towards the military and in depiction of individual troops, but this bias did not produce more positive overall coverage compared to recent conflicts (Pfau et al., 2004).
6. BP has a history of using this tactic. In the first supertanker oil spill, the *Torrey Canyon,* which in 1967 spilled 119,328 tonnes of crude oil off England and France, while the tanker was carrying BP's oil and was chartered by BP, BP was able to avoid

blame and evade responsibility which was placed on the tanker's crew, owners and operators (Barkham 2010).

REFERENCES

The American Lawyer (April 1985:130).

Barstow, D., Dodd, L., Glanz, J., Saul, S., and Urbina, I. (2010, June 20). Regulators failed to address risks in oil rig fail-safe device. *New York Times*. Retrieved at http://www.nytimes.com/2010/06/21/us/21blowout.html.

Barkham, P. (2010, June 24). Legacy of the Torrey Canyon. *Guardian*, p. G2: 4-7.

Bhopal People's Health and Documentation Clinic. (1998). The Bhopal Gas Tragedy, 1984-? A Report from the Sambhavna Trust. Bhopal.

Blackmon, D. A., O'Connell, V., Berzon, A., & Campoy, A. (2010, May 27). There was 'nobody in charge.' *Wall Street Journal*. Retrieved from http://online.wsj.com/article/SB10001424052748704113504575264721101985024.html.

Blowout. (2010a, June). 'Caring' for the survivors? Disgraceful! *Blowout*. Retrieved from http://www.oilc.org/oilcorg/blowout-magazine-issue-80-s/page-12-deep water-continued.html.

Blowout. (2010b, June). MMS–sex, drugs and well control! *Blowout*. Retrieved from http://www.oilc.org/oilcorg/blowout-magazine-issue-80-s/page-8-deepwater-horizon.html.

Carson, K. (1982). *The other price of Britain's oil: safety and control in the North Sea*. Oxford: Martin Robertson.

Casselman, B. (2010a, May 11). Sunken-rig owner Transocean had a rising tally of accidents. *Wall Street Journal Europe*, pp. 14-15.

Casselman, B. (2010b, July 21). Supervisor says BP rig had a flaw in hydraulics. *Wall Street Journal Europe*, pp. 3.

Casselman, B. (2010c, June 14). Chevron sets itself apart from rival BP in the Gulf. *Wall Street Journal Europe*, pp. 1,7.

Casselman, B., and Chazon, G. (2010, May 19). Deepwater oil rigs are lacking preparations for disaster. *Wall Street Journal Europe*, pp. 14-15.

Casselman, B., and Gold, R. (2010, August 29). Legal tactics emerging at oil hearings. *Wall Street Journal*. Retrieved from http://online.wsj.com/article/SB10001424052748703618504575459873056690954.html.

Chemical and Engineering News (February 11, 1985). p.37.

Chouhan, T.R. (2004). *Bhopal: The Inside Story*. Apex Press: New York.

Clemmer, D.I. and Diem, J.E. (1985) Major mishaps among mobile offshore drilling units 1955-1981: Time trends and fatalities. *International Journal of Epidemiology*, 14, 106-12.

Cookson, C. (2010, June 10). Ecologists express concern dispersants will prolong pollution far below the surface. *Financial Times*, p. 9.

Crace, J. (2010, April 30). Hell and high water. Guardian. Retrieved from http://www.guardian.co.uk/environment/2010/apr/30/deepwater-horizon-oil-workers-death.

Crooks, E. (2010a, July 2). BP: The inside story. Financial Times. Retrieved from http://www.ft.com/cms/s/0/4e228e56-84ae-11df-9cbb-00144feabdc0.html.

Crooks, E. (2010b, July 3) BP: anatomy of a disaster. *Financial Times Magazine*, pp.16-25.

Crooks, E. (2010c, May 4). Crisis exposes risks of the deep. *Financial Times*, p. 8.

Crooks, E., and Morris, H. (2010, June 16). High-level tensions behind the clean-up effort. *Financial Times*, p. 8.

DeSombre, E. R. (2006). *Flagging Standards: Globalization and Environmetnal, Safety, and Labor Regulations at Sea*. Cambridge, MA: MIT Press.

Dinmore, G., and de Sabata, E. (2010a, July 24). BP set for deep-water drilling off Libya. *Financial Times*, pp. 1,5.

Dinmore, G., and de Sabata, E. (2010b, August 2). Move to halt BP drilling off Libya. *Financial Times*, p. 6.

Donn, C. (2002). Two-tiered employment in the global economy: the world maritime industry. Working Paper WP2002-0092. Management Division, Le Moyne College Working Paper Series. Retrieved from http://www.lemoyne.edu/ library/mgmt_wp/wp2002-002.pdf.

Dykes, B. M. (2010). Deepwater Horizon survivors allege they were kept in seclusion after rig explosion, coerced into signing legal waivers. Retrieved from http://news.yahoo.com/s/ynews/20100521/sc_ynews/ynews_sc2191 acc 21/8/10.

Eckerman, I. (2005). *The Bhopal Saga: Causes and Consequences of the World's Largest Industrial Disaster*. Hyderabad, India: Universities Press.

Edelstein, M. R. 1988. *Contaminated Communities: The Social and Psychological Impacts of Residential Toxic Exposure*. Boulder, CO: Westview.

Edwards, D., and Kane, M. (2010). After blocking CBS crew, Coast Guard denies 'BP rules.' Retrieved from http://www.rawstory.com/2010/05/coast-guard-bps-rules/.

Faber, D. (2009). Poisoning American politics: the colonization of the state by the polluter-industrial complex. *Socialism and Democracy,* 23(1), 77-118.

Fifield, A. (2010, May 27). Anger at 'dysfunctional' watchdog. *Financial Times*, p. 7.

Freudenburg, W. R. (1997). Contamination, corrosion and the social order: An overview. *Current Sociology*, 45(3), 19-39.

Gall, N. (2010, August 12). Oil euphoria Lula's legacy at risk. *Financial Times*, p. 11.

Guardian (July 12, 1984:11)

Givel, M. (2007) Motivation of chemical industry social responsibility through Responsible Care. *Health Policy,* 81(1), 85-92.

Godoy, E. (2010). Oil workers at sea without a safety net. Inter Press Service News Agency. Retrieved from http://ipsnews.net/news.asp?idnews=52481.

Goldenberg, S. (2010a, May 20). Flames, fumes, a vision of hell. . .survivor of rig explosion recalls a night of horror. *Guardian* p. 25.

Goldenberg, S. (2010b, May 29). Oil is public health risk, scientists. *Guardian*, p. 6.

Goldenberg, S. (2010c, August 4). Use of chemicals to tackle BP oil spill alarmed US agency scientists. *Guardian*, p. 4.

Goldenberg, S. (2010d, August 6). House 'overly optimistic' on oil spill. *Guardian*, p. 19.

Goldenberg, S. (2010e, August 20). US scientist retracts assurances over state of BP spill. Guardian, p. 19.

Goldenberg, S. (2010f, May 12). Oil giants trade recriminations as senators begin hearing into Louisiana oil disaster. Guardian, p. 18.

Goldenberg, S. (2010g, June 16). We could not have halted gulf gusher either, oil chiefs admit. *Guardian*, p. 19.

Grayson, David. (2010). Beyond BP –version 4 of a case being developed real-time: 23/06/10—Teaching note. Retrieved from www.http://ecch.com/uploads/BeyondBPCaseV4-230610.pdf/

Groom, B. (2010, October 26). BP chief hits out at 'rush to judgement' over Gulf oil spill. *Financial Times*, p. 1.

Hannah, B., Morehouse, W., and Sarangi, S. (2004). *The Bhopal reader*. New York: Apex Press.

Hammer, D. (2010, May 12). Kenner hearing: Marshall Islands-flagged rig in Gulf oil spill was reviewed in February. *Times Picayune*. Retrieved from http://www.nola.com/news/gulf-oil-spill/index.ssf/2010/05/kenner_hearing_marshall_ island.html.

Hirschberg, S., Spiekerman, G., and Dones, R. (1998) Severe accidents in the energy sector. Villigen: Paul Scherrer Institut. PSI Bericht Nr. 98-16.

Hoyos, C. (2010, June 30). Anadarko approved key BP well design. *Financial Times*, pp.17.

IAOGP (International Association of Oil and Gas Producers). (2010). *Major accidents*. London IAOGP. Report No. 434-17.

ICFTU-ICEF. 1985. The trade union report on Bhopal. Geneva, Switzerland.

ITF (International Transport Federation). (2002). Offshore workers in the spotlight. http://www.itfseafarers.org/offshore-spotlight.cfm.

Johnson, J., and Torrice, M. (2010). BP's ever-growing oil spill. *Chemical and Engineering News*, 88(24), 15-26.

Jones, T. (1988). *Corporate killing: Bhopals will happen*. Free Associations Books.

Kanter, J. (2010, August 2). E.U. officials wary after spill in gulf. *International Herald Tribune*, p. 18.

Kirchgaessner, S. (2010a, May 7). Critics turn attention to rig watchdog. *Financial Times*, p. 10.

Kirchgaessner, S. (2010b, June 16). Blame game fails to win over Congress. *Financial Times*, p. 8.

Kirchgaessner, S., and Pfeifer, S. (2010, August 19). Oil companies ready to fight tougher regulation on drilling. *Financial Times*, p. 4.

LaPierre, D., and Moro, J. (2002). *Five past midnight in Bhopal: The epic story of the world's deadliest disaster*. New York: Warner Books.

LeBlanc, L. (1981). Tracing the causes of rig mishaps. *Offshore,* March.

Lyall, S. (2010, July 13). BP ignored plenty of warning signs. *International Herald Tribune*, pp. 1, 14.

Macalister, T. (2010a, September 6). North Sea safety risks of Gulf spill rig firm. *Guardian*, pp. 1.

Macalister, T. (2010b, July 3). BP brushes off call to keep away from ecologically risky areas. *Guardian*, p. 34.

Mac Sheoin. T. (2009). Waiting for another Bhopal: global policies to control toxic chemical incidents. *Global Social Policy,* 9(3), 408-433

Mac Sheoin. T. (2010). From Bhopal to BP Texas City. *Monthly Review,* 62(4), 21-33.

McGreal, C. (2010, June 1). BP clashes with scientists over deep sea pollution. *Guardian*, p. 14.

McNulty, S. (2010, June 8). Exxon Valdez case points to lengthy battle for litigants. *Financial Times*, p. 9.

Millard, P. (2010, August 4). Brazilian government sees prosperity in oil below seafloor. *International Herald Tribune*, p. 16.

MUA (Maritime Union of Australia). (2010). Flag of convenience behind rig disaster. Retrieved from www.mua.org.au/news/flag-of-convenience-behind-rig-disaster

News wires. (2009, June 1). *Pemex sees offshore platform deaths suit.* Upstream online.com. Retrieved from http://www.upstreamonline.com/live/article 179860.ece.

News wires. (2010). *Macondo sends insurance into stratosphere.* Upstream online.com. Retrieved from www.upstreamonline.com/live/article218598.ece.

Obama, B. 2010. "Remarks by the President in a Discussion on Jobs and the Economy in Charlotte, North Carolina," Office of the Press Secretary: The White House. Retrieved from http://www.whitehouse.gov/the-press-office /remarks-president-a-discussion-jobs-and-economy-charlotte-north-carolina.

Offshore Data Services. (1983). *Accidents: offshore mobile drilling rigs 1955 to present.* Houston: ODS.

Perrow C. 1984. *Normal accidents: Living with high-risk technologies*. New York: Basic Books.

Peters, J. W. (2010, June 9). Efforts to limit the flow of spill news. *New York Times*. Retrieved from http://www.nytimes.com/2010/06/10/us/10access.html.

Pfau, M., Haigh, M., Gettle, M., Donnelly, M., Scott, G., Warr, D., and Wittenberg, E. (2004). Embedding journalists in military combat units: impact on newspaper story frames and tone. *Journalism and Mass Communication Quarterly* 81 (1): 74-88.

Pfau, M., Wittenberg, E. M., Jackson, C. Mehringer, P. Lanier, R., Hatfield, M., and Brockma, K. (2005). Embedding journalists in military combat units: how embedding alters television news stories. *Mass Communication and Society* 8(3):179-95.

Pfeifer, S., and McNulty, S. (2010, August 20). Transocean at loggerheads with BP over rig documents. *Financial Times*, p. 5.

Picou, J. S., Marshall, B. K., and Gill, D. A. (2004). Disaster, litigation, and the corrosive community. *Social Forces,* 82(4), 1493-1522.

Pilkington, E. (2010, July 24). BP rig's alarms were switched off 'to help workers sleep.' *Guardian*, pp. 1.

Platt's Oilgram News. (2001, April 17). Critics say Petrobras growth push behind. *Platt's Oilgram News,* 79(73), 15.

Power, S. (2010, May 26). Gifts, porn use cloud drilling agency report. *Wall Street Journal Europe*, p. 3.

Priest, Tyler (2008) Wake-up call: accidents and safety provision in the Gulf of Mexico. Retrieved from http://www.gomr.boemre.gov/PI/PDFImages/ESPIS/4/4530.pdf.

Shrivastava, P. 1987. *Bhopal: Anatomy of a crisis.* Cambridge, MA: Ballinger.

Solnit, R. (2010, August 5). Diary. *London Review of Books,* 32(15), 28-31.

Stephens, J., and Flaherty, M. P. (2010, June 30). Sheer scale of oil spill stalls effort to sop it up. *Washington Post.*

Tombs, S., and Whyte, D. (1998). Capital fights back: risk, regulation and profit in the UK offshore oil industry. *Studies in Political Economy,* 57, 73-99.

Urbina, I. (2010a, August 6). Other wells worried oil rig owner. *International Herald Tribune*, pp. 4.

Urbina, I. (2010b, May 28). BP used riskier of 2 methods on well before blast. *International Herald Tribune*, pp. 6.

Urbina, I. (2010c, June 7). On oil rig a muddled chain of command. *International Herald Tribune*, pp. 7.

Urbina, I. (2010d, September 9). BP faults its contractors, taking little blame for spill. *International Herald Tribune*, pp. 1,17.

Wearden, G. (2010, June 28). Shell says deep-water drilling must go on. *Guardian*, p. 23.

Webb, T. (2010a, May 14). Gulf oil spill? It's just a drop in the ocean, says BP chief. *Guardian*, p. 1.

Webb, T. (2010b, June 2). Clumsy response threatens to make a bad situation worse. *Guardian*, p. 16

Weston, C. (1989, July 3). Offshore rig chiefs 'should be held liable for deaths.' *Guardian*, pp. 21.

Whyte, D. (1998). Risk and crisis management in the public sector: Overcoming the fear factor: workforce involvement in health and safety offshore. *Public Money and Management,* 18(4), 33-40.

Winning, D. (2010, September 8). Chevron buys stake in deep sea off China. *Wall Street Journal Europe*, p. 22.

Wright, C. (1986). Routine deaths: Fatal accidents in the oil industry. *Sociological Review,* 34(2), 265-289.

CHAPTER FIVE
THE EFFECTS OF OIL SPILLS ON ECOLOGICAL SYSTEMS

JEFFREY R. WOZNIAK

ABSTRACT

Recent events surrounding the British Petroleum (BP) Deepwater Horizon oil spill have clearly illustrated the wide range of impacts major oil spills can have on ecological systems. To date there has been limited research conducted on the effects of oil spills on natural systems and even less work on the long-term effects of oiling. Over the past decades, increased interest in environmentalism, heightened public perception, and improved government oversight have all worked in concert to drive policy and potential system interventions, focused on the restoration of ecological systems impacted by major oil spills. Low frequency, high impact disasters, such as the BP oil spill, offer the opportunity to periodically revisit our ever-changing understanding of the effects of oiling on ecological systems. To best understand these effects, it is first imperative to recognize the high level of connectivity between individual ecosystem components and that a perturbation to one component has the potential to elicit a response (positive or negative) at some given time step (short- or long-term) across the entire ecosystem. Through reviewing the existing data of the effects of oiling on natural systems and assessing the current status of the Gulf of Mexico, both scientists and managers can formulate the best adaptive management plan moving forward that seeks to identify the synchronicity between oiling events, ecological responses, and restoration efforts.

INTRODUCTION

Low frequency, high impact disasters, such as the BP oil spill, offer the unique opportunity to revisit our ever-changing understanding of the effects of oiling on the ecological systems. Following a major oil spill event, the range of ecological responses can drastically vary across different species, spatial locations and ecosystem types. Further adding to this complex situation is a high level of uncertainty originating from the sudden nature in which oil spills occur and the wide range of interventions and remediation efforts which can immediately follow the spill. Changes to external drivers which are independent of ecology, including variations in government oversight and policy, increasing levels of environmental stewardship and the globalization of the energy industry have lead to constantly changing ecological view points and adaptive management interventions.

Following the BP oil spill, there are unfortunately more questions than answers and greater uncertainty surrounding the potential ecological effects of oiling. Does oiling impact all ecosystem types in the same fashion and does the amount of oil spilled correlate with increased ecological damage? At what ecological cost is the prospect of economic or energy gain worthwhile? How do the effects of oiling vary across difference spatial and temporal scales? Do oil spills bring us closer to ecological tipping points? This chapter seeks to discuss these questions and to summarize the effects major oil spill disasters have on ecological systems. Specific scientific examples of the ecological effects of oiling on coastal ecosystems will be presented. In addition, the importance of determining and preserving the ecosystem goods and services provided by coastal ecosystems, which are often very difficult to qualify and quantify with an absolute economic value, will also be reviewed.

THE ECOLOGICAL EFFECTS OF OILING

The presence of oil in natural systems can lead to a multitude of negative effects on individual species (Smith and Proffitt, 1999), communities (Hester and Mendelssohn, 2000) and the across multiple levels (top-down and bottom-up) of coastal food webs (Valiela et al., 2004). In general, oil can effect ecological systems both physically, through the coating of foliage resulting in reduced transpiration and photosynthesis and chemically, through oiling of soil impacting ionic balance (Pezeshki et al., 2000). While much information is available on how oiling can effect individual species (mainly estuarine plant species), much less is known about the combined effects of oiling and a wide range of parameters, which are mainly abiotic. These abiotic factors include the type of oil (region where the oil originated and the degree of chemical refinement), the concentration of oil present in the ecosystem, the geographic location of the spill, regional weather patterns, and the persistence of oiling (see Alexander and Webb, 1985 and Hester and Mendelssohn, 2000). The overall ecological response to oiling in ecosystems can be greatly altered by different combinations of these factors, especially over longer time steps.

How an ecological system responds to oiling will also significantly vary based on structural attributes of the ecosystem: ecosystem type (e.g. estuarine wetland vs. sandy beach), vegetation type (different species specific effects), soil type (e.g. organic vs. mineral) and the extent of emergent plant cover (Hester and Mendelssohn, 2000). In addition, the restoration plan and the specific method employed to remove oil from the ecosystem can also have meaningful effects on ecological systems. Often the incorrect removal techniques can result in more negative (long-term) effects to the ecosystem than the initial oiling event. Methods of oil removal include physical removal, the application of dispersants (Pezeshki et al., 1998 and Fuller et al., 2004) or cleaners (Fiocco et al., 1991), burning of excess surface oil (Allen and Ferek, 1993, Lin et al., 2005a), cutting vegetation and natural remediation (also see OTA, 1990 and Lin et al., 2005b). Interestingly, in coastal wetland ecosystems which are highly susceptible to soil subsidence, the natural remediation of the ecosystem (i.e. when oil is left to naturally biodegrade) is often more advantageous to the marsh vegetation and the overall health of the ecosystem in the long-term (DeLaune et al., 2003). In coastal and estuarine wetland, this natural remediation will result in less physical disturbance (caused be the physical impacts of humans and the machinery involved in oil removal) to the marsh ecosystem and a higher level of long-term ecosystem health (Hester and Mendelssohn, 2000).

COMPARING THE ECOLOGICAL EFFECTS OF OIL SPILL EVENTS

The Deepwater Horizon spill has recently been compared to the 1979 Ixtoc oil spill which occurred in the southern Gulf of Mexico (GOM; Bay of Campeche, Mexico). Prior to Deepwater Horizon, the Ixtoc spill was the largest oil spill disaster in the region with 147 million gallons spilled (Jernelov and Linden, 1981). As both spills occurred in the GOM they shared several of the same attributes (e.g. effected similar offshore and inshore ecosystems, possessed similar abiotic and biotic mechanisms of oil breakdown, and impacted equivalent flora and fauna). However, as the Ixtoc spill occurred in Mexican waters, data on the ecological effects of this spill are limited, making both the short and long-term quantitative comparisons of the ecological impacts of these spills very difficult.

To date there have been few long-term monitoring projects centered on determining the effects of major oil spills over longer time periods (decade to multi-decadal). The one exception is the Exxon Valdez spill, where the environmental impacts have been closely monitored since the spill occurred along the Alaska coastline in 1989. In addition to Valdez being a smaller spill (~30million gallons), there were also very different ecological effects in the first months following the initial release of oil. Factors including the close proximity to the coastline and the rapid release of oil lead to much more of an acute, negative ecological response to the Valdez spill. Moreover, the geographic location of the spills (northern Pacific vs. northern GOM) can lead to significant variation in how oil is biologically processed and biodegraded (i.e. cold-temperate (slow) vs. warm-temperate (fast); Spalding et al., 2007). Based on these geographic, biological and environmental differences, it is again difficult to compare the ecological impacts of the BP spill to the Valdez spill. This highlights an important fact that when attempting to determine the ecological effects of a major oil spill, each event must be treated as an individual, stand alone oiling event. Based on the aforementioned factors, novel and adaptive response plans will need to be employed for each new spill and each spill will most likely result in different impacts to the local ecological systems.

In major spill events the cumulative quantity of oil spilled is often used as a metric to rank oil spills and to gauge their overall ecological impacts; however, the quantity of oil spilled is often not the best means to estimate the ecological impacts. As previously mention, there are several factors that must be considered when assessing the ecological impacts of any oil spill event. In addition to these factors (e.g. type of oil, persistence, weather patterns, etc.), it is of the utmost importance to acknowledge the importance of how the effects of oiling will vary across time and space. The short term effects of oiling can be very easy to identify and can include vegetation die-off, oiling of birds, and fish kills. Many of these short-term effects were reported by the media following the BP spill; but the short-term ecological effects of oiling most often extend beyond these readily visual effects to include mortality in submergent vegetation, benthic invertebrates, deepwater fish, and other ecosystem components not readily visible to the naked eye. The ecological effects of oiling can also display a great deal of spatial variation. For example, the persistence and concentration of oiling in a coastal wetland can exhibit high variation over relatively small spatial scales. Here, a given acre of wetland habitat can experience heavy, persistent oiling, while adjacent acres receive relatively little oil. Factors including multiple abiotic parameters and specific differences in how individual species respond to oiling can greatly impact the overall ecosystem response to an oil spill. When attempting to quantify and qualify theses ecological effects it is also critical to employ long-term data collection efforts (in combination with baseline data) to be able to most accurately summarize the long-term ecological response to an oil spill. It is important to note that the initial ecological response may not mirror the final outcome or the overall ecosystem response; the most significant effects of an oil spill may be made manifest on much longer time steps.

ASSESSING THE BP DEEPWATER HORIZON SPILL

Wetlands provide multiple goods, services and commodities to humans from the population, to ecosystem and global levels (Mitsch and Gosselink, 2007). Many attempts have been made to place a value on both individual ecosystem types and specific ecosystem goods and services (see Costanza et al., 1997 and Pimm, 1997);

however, the valuation (either the intrinsic value or actual economic value) of the goods and services provided by ecosystems is very difficult to calculate (Mitsch and Gosselink, 2007). Specifically, coastal marshes, including those impacted by the BP oil spill along the GOM coast, are among some of the most abundant, fertile habitats on earth and since the middle ages have been highly manipulated by humans on a grand scale (e.g. altering species composition, species distribution, and ecosystem function; Bromberg Gedan et al., 2009). These coastal wetland ecosystems provide more ecological services than any other coastal environment (UNEP, 2006 and Bromberg Gedan et al., 2009) including the export of energy-rich material to deeper waters, nursery and stock habitats, shoreline stabilization, intercept land-derived nutrients and contaminants, recreation and tourism and aesthetic and heritage value (Mitsch and Gosselink, 2007 and Valiela et al., 2004). It is clear that there are multiple, negative effects on coastal ecosystems following major oil spill events; so what are the short-term, and what will be the long-term, effects in the GOM region following the BP Deepwater Horizon oil spill?

The effects of the BP oil spill have been noted in several ecosystem types including the deepwater/openwater habitats of the GOM, more nearshore openwater habitats, coastal fringe (sandy beaches), coastal wetlands (vegetated), and along the estuarine gradient moving inland towards the terrestrial uplands. Oiling from the BP spill has impacted all levels of the aquatic food web—including primary producers (benthic community, phytoplankton, macrophytes, etc.), primary consumers (shrimp, crabs, clams, small fish, etc.), secondary consumes (larger fishes, birds, etc.), through other higher order consumers, (e.g. more charismatic fauna often seen in media reports including birds, turtles, etc.). Furthermore, the presence of oil in an estuarine food web can significantly impact the ecosystem via both top-down or bottom-up mechanisms of control. The combined effects of these mechanisms will act as a biological control to govern species assemblages and the overall health of the aquatic ecosystem (Valiela et al., 2004).

The process of quantifying and qualifying all of the ecological effects in the GOM is equally difficult to determine. While at the time of publication, actual data on the short-term effects of the BP oil spill are still being collected and analyzed, and it may be some period of time before the ecological effects (and economic effects when considering ecosystem goods and services) across the entire GOM ecosystem can be fully realized. To accurately assess the short-term effects of oiling, researchers must have high quality baseline (pre-spill) data and possess the ability to begin to collect data immediately following the initiation of oiling. Unfortunately, many researchers are unable to fulfill these two requirements, often due to permitting, research site access, funding and political issues. Equally important is the need to continue to monitor the ecosystems impacted long after the oil has stopped flowing; after all visible signs of oil have passed, and far into the future to be able to capture potential novel long-term effects of oiling that may one day present themselves well after an oil spill event. These long-term effects can linger for decades, especially in the presence of anoxic soils, but few studies have examined these long-term effects (Culbertson et al., 2008). For example, four decades after the barge *Florida* spilled 700,000 liters of No. 2 fuel oil in Buzzards Bay, MA, the effects on large scale ecosystem functions are still evident (Culbertson et al., 2008). To most effectively assess the ecological effects of oil spills and to formulate the most efficient clean-up response plans, it is critical to consider how oiling has impacted each individual ecosystem type, each individual level of those ecosystems and how those impacts have/will change across varying spatial and temporal scales. The best way to protect and preserve these delicate coastal ecosystems is most likely through an integrated approach of ecosystem-based management (Bromberg Gedan et al., 2009) that takes all of these factors into consideration on an adaptive management basis.

ECOSYSTEMS OF THE GULF OF MEXICO MOVING FORWARD

The GOM has been impacted by both anthropogenic disturbances (oil spills, urbanization, and extensive coastal wetland loss) and natural perturbations (hurricanes, natural oil seepages, etc.) for many decades. With the occurrence of each new disaster (e.g. Hurricane Katrina or the BP oil spill) the daily news cycle and the public focus shifts to the GOM region and to the potential solutions required for the future restoration and conservation of the ecological integrity in the region. Unfortunately following many major oil spills, once the flow of oil ceases, so does the majority public interest in moving forward with restoration solutions. Obviously, the need to restore Gulf Coast ecosystems does not immediately diminish, as does media reports and public interest. This paradigm illustrates a clear disconnect between the need to restore/protect and the public/government/economic resources need to successfully conduct restoration efforts. This is a critical issue that must be addressed if there is hope that the coastal ecosystems of the GOM will be able to persist in the future, especially if additional oil spill events occur.

A broader question following the BP oil spill disaster is what are the ecological lessons learned? In the event of a future oil spill, what ecological information can we take from the BP oil spill to more efficiently manage real-time mitigation efforts, to continue to monitor the future ecosystem responses to the BP spill, and to improve our ecological understanding of the GOM? A common statement following the BP oil spill is that the GOM is a "resilient ecosystem" which has endured decades of anthropogenic impacts and appears to be functioning at a high level. Indeed the GOM is a resilient ecosystem, however with each new spill or negative influence on the system a portion of the "ecological capital" that the ecosystem possessed is lost. The GOM is aging. It becomes more susceptible to decreased function with every additional disaster. A valid fear is that the "Gulf is tough" statement may actually lead to a passive management/regulatory regime which may actually increase the potential for future ecological losses. On a positive note, we know the solutions to many of the current ecological issue in the region. For example, scientists and researchers are beginning to formulate the most effective response plans for future spill events. If the energy industry takes the steps necessary to create more low risk offshore drilling operations, we should be able to extend the time between major oil spill disasters and significantly decrease the long-term anthropogenic impacts on the GOM.

To place these types of questions into an ecological context, we can use the term "tipping point" to describe potential outcomes. A tipping point is an ecological event that is used to indicate a point in time where a critical threshold is reached which results in the deterioration of a critical aspect of the ecosystem. The ecological losses associated with reaching a tipping point can include overall ecosystem function, the downward trajectory of an individual species, community or habitat type. So is it possible to determine if the BP oil spill caused an ecological tipping point for the GOM? While preliminary assessments indicate that a system-wide tipping point was not reached, this is a complicated and appropriate question to ask across all ecosystem types across the greater GOM. The "correct" answer to this question will mostly likely fluctuate based on the scope, habitat type and species/component of interest. Moving forward, the fear surrounding future spill events is that the combinations of disasters like thc BP oil spill and the other anthropogenic disturbances in the GOM (channelization of the Mississippi River, coastal marsh subsidence, nutrient loading, etc.) may lead to an unprecedented and irreversible tipping point in the region. Our ability to forecast and then avoid these tipping points is of critical importance for the long-term ecological health and success of the GOM.

ACKNOWLEDGMENTS

I would like to thank the Public Broadcasting System (PBS 8 in Houston), the program "Houston8," Laura Lucas, and Ernie Manouse for facilitating many of the ideas presented herein. This material was developed in collaboration with the Florida Coastal Everglades Long-Term Ecological Research program under National Science Foundation Grant No. DEB-9910514.

REFERENCES

Alexander, S.K., and Webb Jr., J.W. (1985). Seasonal response of *Spartina alterniflora* to oil. In *Proceedings of the 1985 Oil Spill Conference*. (355-57). Washington, D.C.: American Petroleum Institute.

Allen, A.A., and Ferek, R. (1993). Advantages and disadvantages of burning spilled oil. In *Proceedings of the 1993 International Oil Spill Conference*. (765-72). Washington, D.C.: American Petroleum Institute.

Bromberg Gedan, K., Silliman, B.R., and Bertness, M.D. (2009). Centuries of human-driven change in salt marsh ecosystem. *Annual Review of Marine Science, 1*, 117-41.

Costanza, R., d'Arge, R., de Groot, R., Farber, S., Grasso, M., Hannon, B. et al. (1997). The value of the world's ecosystem services and natural capital. *Nature, 387*, 253-60.

DeLaune, R.D., Pezeshki, S.R., Jugujinda, A., and Lindau, C.W. (2003). Sensitivity of US Gulf of Mexico coastal marsh vegetation to crude oil: comparison of greenhouse and field responses. *Aquatic Ecology, 37*, 351-60.

Fiocco, R.J., Canevari, G.P., Wilkinson, J.B., Jahns, H.O., Bock, J., Robbins, M. et al. (1991). Development of Corexit 9580-a chemical beach cleaner. In *Proceedings of the Oil Spill Conference.* (395-400). Washington, D.C.: American Petroleum Institute.

Fuller, C., Bonner, J., Page, C., Ernest, A., McDonald, T., and McDonald, S. (2004). Comparative toxicity of oil, dispersant, and oil plus dispersant to several marine species. *Environmental Toxicology and Chemistry, 23*, 2941-49.

Hester M.W., and Mendelssohn, I.A. (2000). Long-term recovery of a Louisiana brackish marsh plant community from oil-spill impact: vegetation response and mitigation effects of marsh surface elevation. *Marine Environmental Research, 49*, 223-54.

Jernelov, A., and Linden, O. (1981). Ixtoc I: a cast study of the world's largest oil spill. Ambio, *10*, 299-306.

Lin, Q., Mendelssohn, I.A., Bryner, N.P., and Walton, W.D. (2005a). In-situ burning of oil in coastal marshes. 1. Vegetation recovery and soil temperature as a function of water depth, oil type, and marsh type. *Environmental Science and Technology, 39*, 1848-54.

Lin, Q., Mendelssohn, I.A., Carney, K., Miles, S.M., Bryner, N.P., and Walton, W.D. (2005b). In-situ burning of oil in coastal marshes. 2. Oil spill cleanup efficiency as a function of oil type, marsh type and water depth. *Environmental Science and Technology, 39*, 1855-60.

Mitsch, W.J., and Gosselink, J.G. (2007). *Wetlands.* (fourth edition). Hoboken: John Wiley & Sons.

Office of Technology Assessment, U.S. Congress. (1990). *Coping with an oiled sea: an analysis of Oil Spill Response Technologies* (OTA-BP-O-63). Washington, D.C.: U.S. Government Printing Office.

Pezeshki, S.R., Jugujinda, A., and DeLaune, R.D. (1998). Responses of selected U.S. Gulf coast marsh macrophyte species to oiling and commercial cleaners. *Water, Air, and Soil Pollution, 107*, 185-95.

Pezeshki, S.R., Hester, M.W., Lin, Q., and Nyman, J.A. (2000). The effects of oil spill and clean-up on dominant US Gulf coast marsh macrophytes: a review. *Environmental Pollution, 208,* 129-39.

Pimm, S.L. (1997). The value of everything. *Nature, 387*, 231-32.

Smith, D.L., and Proffitt, C.E. (1999). The effects of crude oil and remediation burning on three clones of smooth cordgrass (*Spartina alterniflora* Loisel.). *Estuaries, 22*, 616-23.

Spalding, M.D., Fox, H.E., Allen, G.R., Davidson, N., Ferdana, Z.A., Finlayson, M. et al. (2007). Marine Ecoregions of the world: a bioregionalization of coastal and shelf areas. *Bioscience, 57*, 573-83.

United Nations Environmental Programme (2006). *Marine and coastal ecosystem and human well-being: a synthesis report based on the findings of the Millennium Ecosystem Assessment.* Nairobi: UNEP.

Valiela, I., Rutecki, D., and Fox, S. (2004). Salt marshes: biological controls of food webs in a diminishing environment. *Journal of Experimental Marine Biology and Ecology, 300*, 131-59.

CHAPTER SIX
THE GULF OIL SPILL, ECOLOGICAL DEBT, AND ENVIRONMENTAL JUSTICE IN LOUISIANA: LESSONS FROM SOCIOLOGY

TIMOTHY J. HANEY

A shorter version of this article originally appeared on Truthout.org (12 May 2010) as "The 2010 Oil Spill and Our Ecological Debt to Louisiana." The analysis presented here is expanded considerably, but identical sections are reprinted with permission of Truthout.org.

INTRODUCTION

On April 20, 2010, a drilling rig named Deepwater Horizon, operated by the oil giant BP, exploded and sank off the coast of Louisiana killing eleven workers and forcing the evacuation of the platform. For the next ninety days, between 795,000 and 3.9 million liters of crude oil leaked into the Gulf of Mexico each day (Reuters, 2010). It will take months to assess the environmental impact of the disaster and potentially years for the region to recover. But who is ultimately responsible for this catastrophe and how might our own dependence on oil irreparably damage the coastal communities of Louisiana? Furthermore, how does this event connect to Louisiana's unique history and geography, particularly to the Hurricane Katrina disaster only five years earlier? This chapter answers these questions by placing the oil spill in a historical, geographical, and economic context. Using the concept *ecological debt*, it ties the spill to larger issues of social and environmental justice that affect Louisianans on a daily basis. I argue that it is impossible to view the oil spill apart from Hurricane Katrina and other injustices that residents have suffered at the hands of the oil and gas industry.

ECOLOGICAL DEBT

Much of the debate following this ecological/technological disaster focused on BP's culpability and whether or not corporate executives knew of critical safety risks and engineering flaws on this particular platform (Leopold, 2010). Many have wondered who will pay for cleanup of the spill, with BP eventually agreeing to foot the bill (Breed and Mohr, 2010). This debate, though merited, tends to place the blame squarely on BP (framed as greedy, corporate power wielders) while failing to acknowledge the responsibility of the average citizen/consumer for creating the conditions that allow technological and ecological disasters such as this to occur. Louisiana is the place, more than any other, where technology and dirty industry meet environmentally sensitive areas. Accordingly, we must recognize that the people of Louisiana shoulder the burden of the region's oil and gas industry, while seeing disproportionately few of the benefits. Those of us living in other parts of North America enjoy being able to purchase cheap gasoline, but we experience few of the environmental impacts caused by its drilling, refining and transportation. Sociologists often use the term "ecological debt" to explain the discrepancy in the proportion of carbon released by countries of the global north (most of it) and countries of the global south (little of it), coupled with the severe environmental consequences of climate change endured by the people of the global south (Foster, 2009: 242-49). According to Foster (2009: 243), it is "the debt accumulated by Northern, industrial countries toward Third World countries on account of resource plundering, environmental damages, and the free occupation of environmental space to deposit wastes, such as greenhouse gases, from the industrial countries." In other words, one part of the globe reaps the benefits of extraction and consumption while another suffers the consequences. Over time, a moral and economic debt accrues to the region that is able to prosper from this exploitive relationship. Borrowing this term, it is evident that the people of North America owe an enormous ecological debt (and economic debt) to the people of Louisiana. This disaster affects Louisiana's natural ecosystem which, in turn, affects the livelihoods of people (shrimpers, crawfishers etc.) who depend upon these resources to survive. Some have argued that the Gulf of Mexico has become an "industrial sacrifice zone," a place where we turn a blind eye to pollution in order to keep the wheels of economic growth turning (Philpott, 2010).

The Gulf Coast (particularly Louisiana) has also become somewhat of a social sacrifice zone. Because of the interdependence of human societies and the environment, any ecological sacrifices made also affect the livelihoods of one of North America's most socially vulnerable hinterlands. Giroux (2006) argues that because the residents of southern Louisiana are largely poor and/or black, they are treated by the government and by large corporations as expendable. He argues that events such as Katrina (and the oil spill, presumably) "are the consequence of a systemic, violent form of social engineering in which those populations in the United States marginalized by race and class are now considered disposable—that is, simply collateral damage in the construction of a neoliberal order" (11). The jobs and livelihoods of fishermen in the region are far less important to capital than the profits to be made from offshore drilling. The oil spill, while a novel event, is part and parcel of a much larger historical sequence whereby raw materials are extracted from and processed in the Gulf South, the costs are contained there, the benefits are enjoyed elsewhere, and we quickly forget those who suffer in the process.

It is no secret that oil greases the wheels of our economy; no other fuel is as energy dense, capable of so much output for so little energy invested. As James Kuntsler observed, "fossil fuels provide for each person in an industrialized country the equivalent of having hundreds of slaves constantly at his or her disposal" (Kuntsler, 2006: 31). This imagery of slavery is particularly apt when talking about the Deep South, a region barely three generations removed from chattel slavery. However, I would point out that in this particular case the people of the Gulf Coast are subservient to those in other regions of the country. They bear most of the costs of our oil-dependent economy, with disproportionately few returns.

Though we can and do implement measures to decrease the probability of events such as the oil spill, disaster scholars often point out that cities, states and nations should be thinking about how to eliminate any possibility of such events. According to Clarke (2005: 41), humans tend to think in terms of what will probably happen. He argues that "we shouldn't concentrate so much on probabilities that we forget the possibilities." Planning should incorporate possibilistic, not probabilistic thinking; quite often, planners create safety standards that correspond to the most likely set of events. In this case, it was highly unlikely that such a massive blow-out of an oil rig would occur and that the world's experts would have no method for capping the resultant leak. Thus, BP and Transocean drilled no relief well. Similarly, it was very unlikely that such a powerful hurricane would pass directly over the New Orleans area. Thus, the levee system was allowed to deteriorate. With low-probability, high-impact events such as oil spills, sound planning considers what could possibly happen in a worst-case scenario. And the only way to eliminate the possibility of catastrophic oil spill events is to decrease or eliminate our use of oil. Why did only possibilistic planning take place prior to the oil spill and to Hurricane Katrina? As Clarke (p. 50) points out, probabilistic planning often protects the powerful, because it is less expensive to do. At the same time, "it often results in the nonpowerful being placed in danger." In this case, experts from BP and the federal government placed a bet on the likelihood that the rig would blow-out and create an under-sea leak, while offsetting this probability with the cost of preventing that unlikely blowout. In the end, after considering the health of the ecosystem and the well-being of the region's residents, they presumably decided that additional safety features (including a relief well) were too expensive. As a result, many Gulf Coast residents, the nonpowerful in Clarke's framework, were placed at risk once again.

In the remainder of this article, I argue that because of our dependence on oil and the failure of corporations and the government to protect the region, we owe the people of Louisiana and of the Gulf Coast both an ecological and economic debt. I also suggest several basic steps that we can and must take in order to repay this debt and to create a safer, prosperous, and more sustainable region for Louisiana's residents.

"HALF-DEAD WASTELANDS": RECONFIGURING LOUISIANA'S LANDSCAPE TO FACILITATE INDUSTRY

To understand how this ecological debt to Louisiana accumulated, we must first understand the unique geographic history of the region. Most know that southern Louisiana is a marshy, low-elevation, hurricane-prone region, as Hurricanes Katrina and Rita exposed in 2005. Fewer people recognize that this vulnerability was created and sustained by the oil, gas and petrochemical industries. From the 1910s to the 1940s, oil companies set up refineries in south Louisiana as quickly as they could; by 1951, there were 51 plants along the lower Mississippi (Colten, 2005: 127). One of the challenges that industry faced was navigating the 70 miles of winding Mississippi River that connect the Gulf to the city of New Orleans, home to much of the region's refining capacity. Industry required large shipping channels, which could easily be carved through the swampy landscape. As a result, local authorities and business leaders met with the Army Corps of Engineers to pitch the excavation of a straight waterway connecting New Orleans to the Gulf of Mexico. Completed in 1965, this canal came to be known as the Mississippi River Gulf Outlet ("MR-GO,"' in local parlance). MR-GO was never heavily trafficked, and by the 1990s, only five percent of port traffic utilized the MR-GO. Besides allowing saltwater from the Gulf to intrude upon sensitive wetlands, creating "half-dead wastelands" (Campanella, 2002: 78), it also created a convenient hurricane superhighway. Instead of traversing many miles of wetlands or a meandering river, storm surge generated by a hurricane could now travel straight to the heart of New Orleans, as it did on August 29, 2005, inundating the city. In sum, the creation of shipping channels was done to facilitate the production and refining of oil and petrochemicals, but these channels endangered both natural ecosystems and human populations alike. The Hurricane Katrina tragedy would not have been so severe if the natural landscape of southern Louisiana had not been altered to facilitate the oil and gas industry (Grunwald, 2005). The devastation wrought by Hurricane Katrina depended on the industry-driven reconfiguration of landscape, which helped multinational corporations prosper, but also helped people outside of the region fill their tanks with cheaper gasoline.

The creation of shipping channels along with the building of New Orleans' levee system also altered the way that sediments travel down the Mississippi River and deposit at the mouth. For thousands of years, the delivery of sediment at the mouth of the Mississippi ensured the replenishment of Louisiana's wetlands. Precisely because of the canal and levee system, which divert sediment from taking their nature course, the Louisiana wetlands are disappearing at an alarming rate.

During the past 50 or so years, Louisiana has lost as average of 25 square miles per year, which equates to a football field every 30 minutes (Tidwell, 2007). Since the 1950s, as much as 80 percent of the nation's wetland loss has occurred in Louisiana (Tibbetts, 2006). This means that each day, week, month and year human settlements in south Louisiana move closer and closer to the Gulf of Mexico, with less and less protection from storm surge and inundation. Today, the only things deposited at the mouth of the river are fertilizer and pesticide runoff from Midwestern farms, along with sludge from Louisiana's petrochemical industry, which combine to create an enormous dead zone in the northern Gulf. The changes in landscape were undertaken, in large part, to facilitate the needs of commerce and industry, but they also put New Orleans' predominantly African-American and poor population at risk. If Hurricane Katrina had made landfall in 1945, instead of 2005, the surge that arrived in New Orleans would have been as much as five to ten feet shallower than it was (Perrow, 2007: 23). Quite possibly, there would have been no flooding at all.

CANCER ALLEY: HEALTH EFFECTS AND ENVIRONMENTAL RACISM

It has also been well documented that the mere existence of refineries and petrochemical plants in the region make life qualitatively worse for the people of Louisiana. Though surely these facilities provide good jobs in a region with a lagging economy, they also exert untold misery on Louisiana's most vulnerable people. In his (2005) book *Diamond: A Struggle for Environmental Justice in Louisiana's Chemical Corridor*, Steve Lerner documents the struggles of the community of Norco, Louisiana (the word Norco actually derives from "New Orleans Refining Company"). Located squarely within Louisiana's "Cancer Alley," residents of Norco's African American "Diamond" neighborhood experienced chronic health problems and severely truncated life expectancies over a 30-year period. Residents told stories of breathing problems, headaches, nausea, decreased fertility, blurred vision, allergies and frequent diagnoses of hyper-aggressive cancers (Lerner, 2005: 43-53). Lerner also documents the unwillingness of the state's Department of Environmental Quality to investigate any such claims, worried instead about the economic base that oil companies provided for state coffers. Industry underestimated the total pollutants they released, while the state either failed to inspect the facilities or refused to fine responsible parties (119-21), jeopardizing the health and life expectancies of hundreds of black residents.

Residents of inner-city New Orleans also experienced such chronic health problems, due largely to the petroleum and petro-chemical industries. Before Katrina, over 50 percent (some studies place this figure at around 70 percent) of children living in the inner-city neighborhoods of New Orleans had blood lead levels above the current guideline of 10 micrograms per decileter (Mielke, 1999). Both pre-and-post Katrina, the City of New Orleans contained numerous toxic sites.

According to Bullard (2009: 155), "Many of these sites (pre-Katrina) could be considered unidentified brownfields with no plans for clean-up, representing another tacit transfer of environmental burdens onto the population without their knowledge or approval." Frickel and Elliott (2008) demonstrate that most of New Orleans' contaminated industrial sites eventually convert to other land uses (commercial or residential), subjecting new residents to unanticipated health risks. In other words, the petro-chemical industry in Louisiana (as well as other polluting industries) place residents at risk, often without their knowledge. Following Katrina, this problem was only exacerbated, as the catastrophic flooding produced 575 hazardous petroleum and chemical spills (Perrow, 2007: 181).

According to Robert Bullard's seminal work *Dumping in Dixie*, resource extraction and refining boosters convinced many local communities in the Gulf states that they must choose between jobs (at the facilities) and their health. Yet, because hazardous facilities tend to be sited near poor and vulnerable communities, those who suffer the most are also, paradoxically, those who benefit the least from these facilities (Bullard, 2000: 10-32). As Bullard pointed out, "costs are more localized, while benefits are more dispersed." Bullard argued that this phenomenon is made possible by "A colonial mentality" in the South whereby "local government and big business take advantage of people who are politically and economically powerless" (97). By participating in an oil economy that permits and depends upon this sort of environmental inequity, we are amassing an enormous ecological debt to the people of Louisiana, as they bear the consequences (through jeopardized health) and we absorb the benefits (through cheaper products and the lifestyle of convenience that oil allows).

COMMUNITY RESILIENCE

Oil spills and other environmental catastrophes also threaten the livelihoods and strong ties to community that define southern Louisiana. Even though the leak is finally plugged, many who depend upon the natural resources of the Gulf and wetlands are out of work. Already, legal teams are descending upon the region and many lawsuits have been filed on behalf of those whose incomes are threatened (Anderson and Watkins, 2010). This combination of job loss and protracted litigation may combine to erode social capital within tight-knit coastal communities.

Disaster scholars have uncovered an interesting, but paradoxical situation that arises within communities affected by disasters; though strong, community-based social networks are crucial for surviving a disaster, the event has the potential to erode these same networks. Klinenberg (2002) found that during the 1995 Chicago heat wave, residents of tight-knit neighborhoods with thriving civic cultures were less likely to perish than those whose neighborhoods were marked by social isolation. In short, people rely on one another to get through crisis situations.

At the same time, disasters have the potential to erode those community bonds and undermine the community as a source of resilience. Since the publication of Kai Erikson's *Everything in Its Path: Destruction of Community in the Buffalo*

Creek Flood, sociologists have been sensitive to the effects of environmental and technological disasters on communities (Erikson, 1976). Disasters affect individuals, but they also tear apart bonds among neighbors, coworkers and family members, effectively disabling the community as a source of resilience. Erikson argues that communities are like living organisms, whereby the cells of the body "are dependent on one another for definition, they do not have any real function or identity apart from the contribution they make to the whole organization" (p. 193). Following the 1972 flood in Buffalo Creek, West Virginia, Erikson reported that residents "do not seem to belong to anything and that there are no longer any familiar social landmarks to help fix their position in time and space" (204). The loss of possessions, jobs, landmarks and loved ones, coupled with spatial dislocation, result not only in psychological damage, but can combine to threaten a community's stock of collective social capital. More recently, a study undertaken by Elliott, Haney, and Sams-Abiodun (2010), comparing two New Orleans neighborhoods following Hurricane Katrina, documents how disasters have the potential to undermine social capital within poor communities. Following displacement, wealthier communities are better able to keep in contact with neighbors, secure necessary assistance from their networks, and reestablish their neighborhood or community, while even tight-knit poorer communities cannot use their place-based social capital as a source of resilience, ultimately making the reconstitution of the community much more difficult. The oil spill may exert just such an effect on coastal Louisiana (it will take years to know for certain), and the potential loss of both individual and collective identity are realistic outcomes.

Though affected individuals can and should seek legal action to recover their lost income, sociological research finds that legal battles only intensify this loss of community. Research into the Exxon Valdez oil spill shows that protracted litigation can erode small communities and affect the social bonds that they share, often pitting one community member against another (Picou et al. 2004). While the Gulf oil spill will surely affect the environment and, by extension, the livelihoods of those engaged in resource-based occupations, we must be cautious to work to preserve the integrity of south Louisiana communities, wherever possible. The litigation experienced by communities and families can potentially serve as a second disaster, long after the oil has been cleaned and represents one additional way in which the people of the region are affected by their proximity to intensive oil extraction.

ECONOMIC INEQUALITIES

At an economic level, through most of Louisiana's recent history, the state was victimized by contracts with the federal government that did not allow the state to profit from its oil riches. The 2006 Spike Lee film about Hurricane Katrina, "When the Levees Broke," drew popular attention to the fact that the state of Louisiana receives no revenues from the oil that is drilled off of its coast, despite the fact that other states do receive such proceeds. Yet, Louisiana produces and/or refines roughly one-fifth of all oil consumed in the United States. As a number of voices in the film echoed, if Louisiana were to secede from the union and sell their oil at fair market prices, that new nation would be exceedingly wealthy. But, as it stands, Louisiana remains one of the poorest states in the nation, largely because the state government fails to benefit from the enormous wealth created off its shores and in its refineries. In 2006, then-Gov. Kathleen Blanco threatened to block new offshore leases unless the state received its fair share of proceeds. Ultimately, this led to passing of the Domenici-Landrieu Gulf of Mexico Energy Security Act (2006), which ensured that Louisiana receives 32 percent of all of the revenues generated from offshore operations covering eight million acres in the Gulf. Although this change is positive, it is only now taking effect and it cannot retroactively mend the reality that for decades the state government received little revenue from such operations, resulting in a debt still owed by the remainder of the nation.

The above examples and analyses document that the production of oil and petroleum products is actually quite a bit more costly (particularly for people of the Gulf Coast) than the prices we pay at the pump would have us believe. According to Lester Brown, president of the Earth Policy Institute, American consumers paid about $3 per gallon for gasoline in mid-2007. At that same time, the actual cost per gallon (taking into account all short-term and long-term indirect costs) was actually about $15 per gallon (Brown, 2008: 7). The $12 difference between what consumers pay for gasoline and the actual cost of that gasoline is paid by the people in producing regions who disproportionately suffer from "natural" disasters, pollution and health complications, as well as the military costs involved in protecting supply lines from the Middle East and North Africa. The recent oil spill adds yet another market distortion to this equation; we continue to pay low prices for gasoline, while the people of the Gulf Coast must deal with the environmental, social and economic fallout from the catastrophe.

Crisis situations also present unique opportunities for profiteering, as labor and environmental regulations are either suspended or ignored. Indeed, Klein (2007) argues that disasters are often carefully engineered in order to provide opportunities for large corporations to make money. Though neither the oil spill or Hurricane Katrina were engineered, as such, Hurricane Katrina taught us that federally-allocated money is not always utilized in an equitable fashion or to hire local workers. Following Katrina, the Federal Emergency Management Agency (FEMA) awarded a $50 million no-bid contract to AshBritt for debris hauling and removal. As Bullard and Wright (2009: 175) report, this amounted to $23 per cubic yard. AshBritt, in turn, sub-contracted another firm, who subcontracted another, who sub-contracted another, who sub-contracted another, who sub-contracted another. In the end, much of the hauling was done by Les Nirdlinger, a debris hauler from New Jersey, who was paid $3 per cubic yard. Thus, almost 90 percent of the funds were spent on this sub-contracting pyramid scheme, rather than on actual debris removal. None of these contractors were based in Louisiana, and as a result, all of this federal money for debris removal left the affected area, providing no jobs for locals. Only after some time is passed will we know how many of the federal and BP-provided cleanup dollars actually reached affected individuals, families, and communities, and how much was channeled to corporate subcontractors outside of the region.

ENSURING SOCIAL AND ENVIRONMENTAL JUSTICE

The intersecting inequalities experienced by residents of Louisiana (and the rest of the Gulf Coast) during the last five years require a roadmap that acknowledges the debt owed to the people of the region. Much of this roadmap can be taken from the environmental justice movement, now more nearly forty years old, and its goal of eliminating environmental inequities. Though a broad term, encompassing many important ideas, Gosine and Teelucksingh (2008: 9) define *environmental justice* as "a state in which social conditions [are] absent of structured, racialized, and class inequalities." More broadly, it is a movement that seeks to address inequities related to natural resources and the environment by connecting

> a range of social movements, including anti-racism movements, Aboriginal rights and sovereignty movements, labour union movements, and the mainstream environmental movement. . . [It] seeks to eliminate harmful practices in housing, land use, industrial planning, and health care (11).

Much of the groundwork and grass-roots activism is being done by organizations such as the Louisiana Bucket Brigade, Lake Ponchartrain Basin Foundation, Natural Resources Defense Council, and EcoJustice. These groups, all involved in the environmental justice movement, can utilize many of the steps below to address the inequalities pointed out above.

First, to ameliorate these environmental vulnerabilities it is crucial that we fund, at the federal level, a comprehensive flood protection system for the city of New Orleans and for all other low-lying inhabited regions of the Gulf Coast. Though funding bills have passed Congress, the system requires more input before people of the region can have any assurance that they will be protected from future storm surge events. By mid-2008, the Army Corps of Engineers had spent only $4 billion of the $14 billion allocated by Congress for levee restoration and improvement. Some experts say that there are leaks in the new levees, that much of the work must be redone, and that billions more will be needed to construct a sufficient flood protection system (Burdeau, 2005). For residents of the New Orleans East neighborhood, flood maps produced in 2007 by the Army Corps of Engineers show no improvement in protection since Hurricane Katrina (Army Corps of Engineers, 2007). Few people in the New Orleans metropolitan area believe that they are safer today than they were on August 28, 2005, the day before Hurricane Katrina, and there is little engineering evidence that they are. Though federal funding has added more than $4 billion of value to New Orleans' flood protection system since Katrina, the pace has been slow and the Army Corps of

Engineers admit that many projects will not be complete for several years. According to one levee district official, "our position is the levees and floodwalls are designed to protect the property of our citizens, not their lives and they should always heed the advice to evacuate" (Grissett, 2009). Despite the snail's pace, there is surprisingly little political momentum in Washington, DC, to fund projects in Louisiana. Often, when I discuss my research with people outside of my profession, they express little sympathy for people who "shouldn't be living below sea level anyway." This view, imbued with racism and classism, explains why funding for flood protection projects is still lacking, while failing to recognize the benefits we each accrue through the consumption of Louisiana's natural resources and industrial output.

We must also fund and undertake an initiative to replenish Louisiana's wetlands. The wetlands that separate the city of New Orleans from the Gulf of Mexico are disappearing quickly, a process hastened by both Hurricane Katrina and the oil spill, placing coastal cities and towns in peril. Since the people of the Gulf Coast provide needed resources to the rest of the nation, protecting these coastal cities through wetland restoration is of paramount importance. Though ambitious coastal restoration initiatives exist, such as the Coast 2050 partnership among various levels of government, non-for-profits and private organizations, federal monies for these sorts of projects are still lacking.

In order to create healthier communities along the Gulf Coast, we must decrease our dependence on oil. Though critics would argue that such a development would cost many oil workers and refiners their jobs, the possible benefits outweigh the costs. Such an energy transition would help to maintain the environmental (and, therefore, economic) viability of the region, while also preserving the jobs of those who work in agriculture and aquaculture. The prospects of transitioning to an oil-free economy seem bleak though, with President Obama campaigning for increased offshore drilling only a few weeks before this catastrophe (Tankersley and Simon, 2010). The oil spill should make us question whether pursuing more offshore drilling opportunities, as President Obama has proposed, is the wisest energy policy decision.

We must also ensure that corporations drilling for and refining oil adhere to the strictest safety and environmental health standards. State Departments of Environmental Quality and the Occupational Safety and Health Administration (OSHA) must ensure that corporations are held accountable for the health of the environment and their employees. When drilling, transporting and refining oil, there are always risks to the environment and to human health. Even so, the federal government and the state of Louisiana must work together to ensure that refineries in Louisiana's "cancer alley" region, as well as offshore drilling operators, are following federal emission guidelines, OSHA standards, and safety protocols. In the days and weeks following the Gulf debacle, other jurisdictions reaffirmed their commitment to preventing such a catastrophe. For example, the government of Newfoundland and Labrador, in Canada, promised to reassess the technology and oversight procedures used in their own offshore drilling operations (CBC, 2010).

If steps like this result in concrete actions to ensure environmental responsibility, they should be applauded. However, they still do little to decrease our dependence on oil.

In order to ensure the well-being of affected communities, we must compensate those whose livelihoods depend upon the health of the wetlands and Gulf waters. Not only must we compensate them out of compassion, but because each and every one of us must accept some responsibility for the loss of income experienced by the shrimpers, crabbers, fishermen of the Gulf Coast. Several Louisiana shrimpers recently filed a lawsuit, accusing the operators of the rig of negligence and seeking millions in damages. If their suit is granted "class action" status, they may represent any or all Louisianans who lost income as a result of the disaster (Agence France-Presse, 2010). BP and its rig operator, Transocean Ltd. face many other lawsuits, as well (Brubaker et al., 2010). Despite the deleterious effects of litigation on communities, which are well documented, successful litigation, combined with humanitarian aid, is the best available option for out-of-work fisherpersons and shrimpers should the region's stocks of seafood perish from this catastrophe. We must also ensure that the proper mental health resources are available for residents who experience psychological trauma. Following Hurricane Katrina only five years earlier, only one-third of those with evidence of mental health disorders were able to access mental health services (Wang et al., 2007).

In order to reinvigorate local economies and employ displaced workers, we must ensure that that every effort is made to employ local people in the continued clean-up efforts. Evidence is clear that much of the cleanup and aid money allocated after Hurricane Katrina was lost through layers of bureaucratic sub-contracting, or ultimately used to pay non-locals at criminally-high hourly rates (see Bullard and Wright, 2009: 169-182). As cleanup continues on the Gulf Coast, the federal government (as well as BP) must ensure that cleanup funds are used responsibly and spent locally.

Finally, Louisiana, as the largest domestic producer of oil, must continue to receive revenues from the oil that is drilled offshore and processed within the state. Though the state recently began collecting these proceeds, in financial terms, it only begins to make up for the decades in which the state of Louisiana was not allowed by the federal government to take a share of oil and gas revenues, as other states did. This point is important, as Louisiana suffers from chronic poverty, high unemployment and other social ills, yet all of these problems could be solved with increased tax revenue from oil and gas production.

CONCLUSION

When a disastrous event like the oil spill affects the people of an already-marginalized region, it is crucial that we remember our own role in it. Though, surely, corporations have a responsibility to protect the surrounding ecosystems and communities, errors do happen. When they do, it is important to remember that such disasters occur during the provision of a commodity that we all use every day. In North America we are dependent upon oil for food production (petroleum-based fertilizers and pesticides), food transportation, road construction, building construction, home heating and cooling and for our daily transportation needs. In this sense, decreasing our dependence on petroleum products (not simply imported oil) is the only sure-fire way to mitigate the risks associated with the recovery of oil.

The policy prescriptions outlined above add up to a massive reinvestment in the Gulf South region and in Louisiana in particular. North America extracts resources from Louisiana, but has done little to shore up the infrastructure and shield the livelihoods of people in this region. We treat it as a resource-rich hinterland and its people as an easily exploitable labor force. If we hope to mitigate the effects of industrial/ ecological disasters like the oil spill, we must first and foremost invest resources into the people of the region. As Simms (2009: 12) reminds us "The narrative running along the spine of all debts dictates that they should be paid back."

Because our dependence on oil surely will not end tomorrow, each and every one of us who utilize oil for fueling our vehicles, fertilizing our food, or making our plastic products (which includes all of us), should acknowledge the ecological and economic debt that we owe to the people of Louisiana. Although I live far from the Gulf of Mexico, I purchase the very same product that fouled their waters and challenged their livelihoods, yet I will likely see no economic or environmental repercussions where I live (except, perhaps, gasoline that is marginally more expensive). Each of us must acknowledge that our convenience comes at a cost.

Renowned disaster scholar Charles Perrow wrote, "Hurricane Katrina was not even a wake-up call. We hit the snooze alarm instead" (Perrow, 2007: 40). Hopefully the 2010 Gulf of Mexico oil spill will finally prompt us to wake up and challenge the environmental and social inequalities that have made the people of Louisiana and the Gulf Coast second-class citizens.

REFERENCES

Agence France-Presse. (2010). Louisiana shrimpers file lawsuit over U.S. oil spill. (29 April).

Anderson, C. and Watkins, T. (2010, May 1). Lawyers flock to Gulf Coast for oil spill lawsuits. Associated Press.

Army Corps of Engineers Performance Evaluation Task Force. (2007). Risk and reliability report." Available at: http://nolarisk.usace.army.mil.

Breed, A. and Mohr, H. (2010, May 3). British Petroleum says it will pay for Gulf oil spill's cleanup. Associated Press.

Brown, L.R. (2008). *Plan B 3.0: Mobilizing to save civilization*. New York: W.W. Norton.

Brubaker Calkins, L. and Fisk, M.C. (2010, May 1). BP, Transocean lawsuits surge as oil spill spreads in Gulf." *Bloomberg Businessweek*.

Bullard, R. (2000). *Dumping in Dixie: Race, class and environmental quality.* Boulder, CO: Westview Press.
Bullard, R. and Wright, B. (2009). *Race, Place, and Environmental Justice After Hurricane Katrina.* Boulder, CO: Westview Press.
Campanella, R. (2002). *Time and place in New Orleans: Past geographies in the present day.* Gretna, LA: Pelican Publishing.
CBC News. (2010, May 3). Low risk for Newfoundland oil disaster: Williams.
Clarke, L.. (2005). *Worst cases: Terror and catastrophe in the popular imagination.* Chicago: University of Chicago Press.
Colten, C. (2005). *Unnatural metropolis: Wresting New Orleans from nature.* Baton Rouge: Louisiana State University Press.
Elliott, J., Haney, T., and Sams-Abiodun, P. (2010). Limits to social capital: Comparing network activation in two New Orleans neighborhoods devastated by Hurricane Katrina. *The Sociological Quarterly* 51(4): 624-648.
Erikson, K. (1976). *Everything in its path: Destruction of community in the Buffalo Creek flood.* New York: Simon and Schuster.
Foster, J.B. (2009). *The ecological revolution: Making peace with the planet.* New York: Monthly Review Press.
Frickel, S. and Elliott, J.R. (2008). Tracking industrial land use conversions: A new approach for studying relict waste and urban development. *Organization and Environment* 21(2): 128-47.
Giroux, H. (2006). *Stormy Weather: Katrina and the Politics of Disposability.* Boulder, CO: Paradigm Publishers.
Gosine, A., and Teelucksingh, C. (2008). *Environmental Justice and Racism in Canada: An Introduction.* Toronto: Emond Montgomery Publications.
Grissett, S. (2009, May 30). New Orleans region's levee system making strides. New Orleans *Times-Picayune.*
Grunwald, M. (2005, September 14). Canal may have worsened city's flooding: Disputed project was a 'funnel' for surge, some say. *Washington Post.*
Klein, N. (2007). *The Shock Doctrine: The Rise of Disaster Capitalism.* New York: Picador.
Klinenberg, E. (2002). *Heat Wave: A Social Autopsy of Disaster in Chicago.* Chicago: University of Chicago Press.
Kuntsler, J.H. (2006). *The Long emergency: Surviving the end of oil, climate change, and other converging catastrophes in the twenty-first century.* New York: Grove Press.
Leopold, J. (2010, April 30). Whistleblower: BP risks more massive catastrophes in Gulf. Truthout.org.
Lerner, S. (2005). *Diamond: A struggle for environmental justice in Louisiana's chemical corridor.* Cambridge, MA: MIT Press.
Mielke, H. (1999). Lead in the inner cities: Policies to reduce children's exposure to lead may be overlooking a major source of lead in the environment." *American Scientist* 87(1).

Perrow, C. (2007). *The next catastrophe: Reducing our vulnerabilities to natural, industrial, and terrorist disasters*. Princeton NJ: Princeton University Press.

Philpott, T. (2010, April 29). Gulf of Mexico: From magnificent resource to industrial sacrifice zone." Grist.org.

Picou, J.S., Marshall, B.K., and Gill, D.A. (2004). Disaster, litigation, and the corrosive community. *Social Forces* 82(4): 1497-1526.

Reuters News Service. (2010, May 2). "Black Tide."

Simms, Andrew. 2009. *Ecological Debt: Global Warming and the Wealth of Nations* (second edition). New York: Pluto Press.

Tankersley, J. and Simon, R. (2010, March 31). Obama's offshore drilling plan seen as political olive branch. *Los Angeles Times*.

Tibbetts, J. (2006). Louisiana's wetlands: A lesson in nature appreciation. *Environmental Health Perspective* 114 (January): A40-A43.

Tidwell, M. (2007). *The ravaging tide: Strange weather, future Katrinas, and the coming death of America's coastal cities*. New York: Free Press.

Wang, P.S., Gruber, M.J., Powers, R.E., Schoenbaum, M., Speier, A.H., Wells, K.B., and Kessler, R.C. (2007). Mental health service use among Hurricane Katrina survivors in the eight months after the disaster. *Psychiatric Services* 58(11): 1403-11.

CHAPTER SEVEN
A NEW GEOGRAPHY OF TROUBLE

DAINA CHEYENNE HARVEY

ABSTRACT

In the early nineties Kai Erikson wrote about *a new species of trouble*. Erikson noted that past disasters were temporally marked by a beginning, middle, and end. This new species of disaster, however, was unmarked. It was in Erikson's terms "unbounded." The new form of trouble produced a special kind of fear or dread, a new form of uncertainty. I suggest that the Deepwater Horizon catastrophe is responsible for bringing to light what I call *a new geography of trouble*. In this geography, trouble is constant, producing a set of unique attitudes among residents towards the environment, the community, and the future, ultimately affecting residents' worldview. This chapter looks at the BP spill through the lens of the residents of the Lower Ninth ward in New Orleans. I introduce the concept of a geography of trouble and then examine two key features of it, environmental uncertainty and a culture of suffering.

Keywords: Geography of trouble, BP spill, Hurricane Katrina, disasters, suffering, environmental uncertainty, cognition, Lower Ninth Ward

INTRODUCTION

As oil began to flow out of the ocean floor, plans were underway throughout New Orleans to commemorate the fifth anniversary of Hurricane Katrina and the failure of the levees. Despite obvious differences there are similarities between the two events. Both events were treated as worst-cases. As worst-cases they were therefore possible, but perceived to be improbable and thus unthinkable (Cerulo, 2006; Clarke, 2006). They simply were not part of our cognitive framework. Both events were consequently met with inadequate responses. Thousands of people were stranded in New Orleans in the aftermath of Katrina, many for five days on rooftops without food or water. As police and other emergency personal dealt with looters and alleged potential terrorists, many of those stranded were rescued by neighbors. Five years later, as the nation witnessed, there was no viable Plan B for responding to the Deepwater Horizon explosion and oil spill. There were, as Lee Clarke notes, simply fantasy documents (1999). Perhaps most importantly, as worst-case events, they are treated as discrete phenomena. One is recorded as a natural disaster (though many of us are at pains to demonstrate that Katrina is a man-made disaster) and the other an industrial accident. They are seen by most as separate and distinct events. A one-two punch, if you will, that hit Southeast Louisiana.

This chapter looks at the BP spill through the lens of the residents of the Lower Ninth ward in New Orleans. For residents of the Lower Ninth ward the Deepwater Horizon explosion was part of a larger process. It was a continuation of past tragedies and toxic experiences, but it also represents future peril. The explosion and subsequent spill were not perceived to be separate from previous events, but part of a way of life for people in this community. As one resident of the neighborhood noted, the spill "is just another link, just one more thing." I call this larger phenomenon a *geography of trouble*. While these geographies are not new, they are becoming more prevalent and have yet to be conceptualized. In this chapter I conceptualize this geography and I focus on two key attributes of it, environmental uncertainty and a culture of suffering. While these two attributes do not represent an exhaustive examination of the topic, they are among the most resonant in this geography.

Unlike other contributions to this volume, this chapter does not directly deal with the people most affected by the BP spill nor the events that caused the spill, but rather treats the spill as part of a generalizable phenomenon. The BP spill highlights the uncertainty and confusion inherent to the relationship that residents of Southern Louisiana have with their environment. For many, while not the last straw, the spill is the event that has caused them to realize that they live in an environment that is under continual assault.

A GEOGRAPHY OF TROUBLE

Beginning in the early 1990s Kai Erikson noted that certain disasters produce a special kind of dread or fear (1994). From decades of field research and comparing a number of sites of industrial accidents, catastrophes, and other instances of social disruption, Erikson noticed that people were especially worried about toxic experiences. Whereas traditional disasters proceeded in a recognizable temporal order, with an identifiable beginning and end, toxic experiences did not have a clearly defined origin and those who experienced the disaster could not project to a time where they would not be affected by their toxic experience. The experience was boundless, producing an endless contamination. For Erikson these new experiences represented a "new species of trouble" (Erikson, 1994).

The sites that Erikson studied, however, were essentially one-offs. The toxic experiences were produced by a single experience. To be sure, the residents' experiences consumed and defined their life. Many were afraid to go outside, some were paralyzed with indecision about moving or leaving the community, they worried incessantly about their toxic futures, family members were ill or had died from their experiences, and they felt trapped in their toxic lives. Erikson's troubled sites are very similar to what Mitchell at about the same time was calling surprises (1996). A surprise differs from previous disasters in that those who experience it have nothing to compare it to. There is no cognitive anchoring—the tendency to use a piece of past information to help one get through the current experience—because the surprise is unique. It is a disaster or catastrophe that was treated as possible, but not probable and is therefore completely unexpected.

In addition to focusing on one of a kind events, most of which are treated as unthinkable, both Erikson and Mitchell looked at communities where the trouble, while all encompassing, came from a single source. There are communities, however, where trouble comes from multiple sources and is reoccurring. Toxic events which would come as a surprise in any other community here become a way of life. Trouble becomes routinized, a way of understanding the world. I call these places geographies of trouble. In these places, communities are under continual assault and trouble becomes a constant.

Louisiana is one of the most polluted states in the nation. Scholars of environmental racism and environmental justice tell us that its' poverty and numerous towns and neighborhoods where the ethnic make-up is heavily minority provide two of the main determinants for where pollution ends up (Bullard, 2000; Pellow, 2004; Sze, 2007). Those who are socially marginalized, however, not only often suffer from living in contaminated places, but are also exposed to hazards other than pollution (Grineski, et al., 2010). For example, in addition for being infamous for "Cancer Alley," a stretch of land between Baton Rouge and New Orleans lined with petrochemical plants (Louisiana is the one of worst states in the nation for water and air releases of recognized carcinogens and toxins), Louisiana has one of the busiest ports and hence the risk of industrial shipping accidents (or perhaps a terrorist attack). Additionally, it has over 40,000 miles of pipelines and close to 200 oil refinery and chemical processing plants—which are always at risk of releasing contaminants or worse exploding. Parts of the State are continually at risk for major hurricanes. And of course the world now knows the risk of living near levees in New Orleans. Those areas that are more heavily socially marginalized suffer more exposure to hazards.

In Southeast Louisiana there were more than two hundred hazardous waste repository and handling sites that were at serious risk of being breached during Hurricane Katrina. There were 575 reported toxic spills during Katrina, including an amount of oil equal to three-quarters of that spilled during the Exxon Valdez disaster. After that hurricane season, there was over 100 million cubic yards of debris scattered throughout and around New Orleans. Imagine if you will debris covering a thousand football fields, with each field having a six-story pile of trash (Godsil, Huang, and Solomon, 2009:115). The danger in these places is increased by the gradual accumulation of trouble. It is not just one incident that residents are forced to suffer, but a series of experiences over a period of time that plagues the community. The Lower Ninth ward is one of those communities where trouble is a constant.

The Lower Ninth ward was ground zero for Hurricane Katrina. The Environmental Protection Agency declared every single home uninhabitable. There were over 20,000 residents displaced there after Katrina. The majority of the almost nineteen hundred deaths during Katrina were residents of the Lower Ninth. Five years later less than twenty percent of the residents have returned. There are no grocery stores, no hospital, no fire station, no police station, and no high school in the neighborhood. It is today the most blighted neighborhood in the country. In addition to suffering through previous Hurricanes and being flooded during Hurricane Rita—almost immediately after Katrina's flood waters had receded—the community is surrounded by thirty-one hazardous waste sites, including the Agriculture Street Landfill Superfund Site. The Lower Ninth borders the Mississippi River, the Industrial Canal, and the Mississippi River Gulf Outlet (MRGO). These two water projects funnel flood surges, especially during Hurricanes over the levees and into the community. Toxic materials have been dumped from the Canal into the community and residents often have to fight against the storage of toxic materials from the widening of the Canal or MRGO in their community. The neighborhood also has two illegal dump sites. Recently an ExxonMobil refinery in Chalmette blanketed the Lower Ninth ward with nineteen tons of catalyst and over 100,000 pounds of sulfur dioxide. The plant has had over four hundred toxic incidents in the last five years, releasing 6,221,685 pounds and 11,679,706 gallons of pollutants. Additionally the per-capita murder rate is ten times the national average and even before Katrina the poverty rate was three times the national average. Then of course there is the BP oil spill.

The Lower Ninth ward was not affected by the spill in the same way that St. Bernard, Plaquemines and Terrebonne parishes have suffered. There were a handful of residents who directly lost jobs and many more that are marginally employed who worry about job security. Because they are further removed from the site of the spill than those in other parishes, and hence not as involved with various agencies or institutions, the residents of the Lower Ninth do not know they are at risk for coming into contact with the Corexit that was sprayed on the oil and now comes down in the Lower Ninth during rain nor are they told about other dangers stemming from the spill. Nonetheless, the BP spill has affected how they see the world. For those who had hoped something had been learned from Katrina the spill was the last straw.

> New Orleans is certainly unto itself, but it is surrounded by, it's got another layer of insulation from the United States in that it's surrounded by Southeast

> Louisiana. And Southeast Louisiana, if it weren't for New Orleans, Southeast Louisiana would be the most unique culture in the United States . . . so why, people who have lived here for 200 years are going to have to leave, *wow,* everything has changed, *wow*. . . the oil spill was profoundly dissettling. . . and there again the response, just seem, seems, they went home, we still got nowhere. And there was this hope. . . what does Rahm Emanuel say, "Don't let a crisis go to waste"; we just keep shooting ourselves in the foot. It just can't continue. (*Sandy, White female, 50-60 years old*)

Sandy had hoped that residents of the Lower Ninth ward would be able to gain some political and economic traction from Katrina. She sees the failure to use the BP spill in a similar way as perhaps the last opportunity to draw attention to community issues such as wetland restoration, lack of jobs, and the vast economic and political inequality that residents are forced to endure.

For others it was a bitter reminder that neither they nor the community were over Katrina. Several residents noted that the spill hurt more than any incident since Katrina and that they now had to face the fact that they would not recover from Katrina. They noted that the size of the spill and its' aftermath was like Katrina and that people in neighboring parishes would not recover from this, just like they were not going to recover from Katrina. As Rick (*Black male, 20-30 years old*) said after the well was capped, "They saying it's over, just like New Orleans is rebuilt. It ain't over. It ain't gonna be over for a long, long time. They just walking away from it. Again."

But for many the BP spill has forced them to realize or come to terms with how their community is in constant peril. Mirroring Sandy's comments above, another resident, Paul (*Black male, 40-50 years old*), suggested that the spill was not as bad as it seems, and certainly not as bad as Katrina. Sitting on his front porch he points to several houses, jabbing each one with his finger. "And that one, and that one, and that one, look, look at that, each one done that way by Katrina." For Paul the spill only affected a handful of people in the neighborhood. He points to some unknown house in the distance, his finger unsure of itself, "so maybe that guy lost his job." But later talking about the release from the Chalmette refinery, Paul mentions how he was sick and his neighbor was not. Then perhaps realizing his previous comment about the spill says, "that's how it is here, always something happening to someone. Constantly I mean. Ain't nowhere else like that. It didn't hurt me, but I'm sure that spill caused trouble for this place." Likewise Tim (*White male, 50-60 years old)* notes that the community has fought all three levels of government against projects such as building the Industrial Canal, the MRGO, dumping, plans to turn the neighborhood into green space post-Katrina, the locks on the canal, the list is endless. "This has been a long time coming," says Tim. "We keep telling them that this, x or y, is going to happen. Then it does. It is not easy living here. It is a constant fight with somebody."

For some the geography of trouble is a continual struggle. Others, while living in the geography do not engage in the struggle, but are subject to the trouble nonetheless. A geography of trouble suggests that what may come is as worrisome as what has already happened. It also hints at a vagueness, uncertainty, or confusion—that we don't always know what the source or effect of the trouble is or will be. We just know it exists. But the trouble nonetheless has the potential to affect us in remarkable and unthinkable ways. Trouble in these geographies becomes normalized in much the same way that accidents are normalized or come to be accepted in high-risk technologies (Perrow, 1999; Vaughan, 1996). These are simply high risk places.

Kasperson et al. (1988) note that some accidents act as signals, creating "ripple effects," which as Slovic writes, extend beyond those directly affected, eventually encompassing ever widening pools of victims (2002). The BP spill was this kind of accident and many communities have felt its' ripple effects. The rippling effect is exactly how I would describe the cognitive effects that disasters like the BP spill have on communities in geographies of trouble. These communities are exposed to real risk and have suffered through a number of disasters, but they are also susceptible to cognitive ripples. Disasters like the BP spill alters the way people in these communities see the world, the environment, their future, and even how they interact with others. The disaster has had a rippling effect through different cognitive processes, ultimately producing a cognitive bias so that residents expect trouble.

While some argue that our cognitive systems depend upon stability and thus hide potential risk from us (Langadec, 1990), others have realized that certain cognitive biases normalize potential risk (see especially Cerulo, 2006; Clarke, 2006; Vaughan, 1996). Below I touch on some of the cognitive biases within two conditions that highlight the geography of trouble, environmental uncertainty and a culture of suffering. While not exhaustive, environmental uncertainty and a culture of suffering, especially in the words of residents of the Lower Ninth, reveal the links between the BP spill and this kind of geography.

ENVIRONMENTAL UNCERTAINTY

In their ethnography of toxic experience in Flammable, a shantytown in Argentina, Auyero and Swistun note that toxic uncertainty becomes a way to understand everyday life (2009). The toxic experience in Flammable is not a reaction to a single, large scale event, but rather a routine struggle within the community that has lasted for decades. At the heart of this struggle is the confusion and uncertainty surrounding the possible source(s) of the contamination, competing and contradictory explanations about why some residents are sick while others are healthy, and also whether the government or the companies themselves will pay for contaminated homes and for relocation. Perhaps most importantly the toxic uncertainty is *omnitemporal*. It is part of the past, present, and the future. In this section I build upon Auyero and Swistun's use of toxic uncertainty, noting how environmental uncertainty contributes to a geography of trouble.

The reaction to the BP spill in the Lower Ninth has revealed an uncertainty similar to that in Flammable. I call this type of uncertainty environmental uncertainty because rather than confined solely to toxic experiences it centers on broader environmental concerns, including both the actual and projected environment, and because of environmental degradation the continuity and survival of the community. As some residents have been told they are sick and that the chemicals sprayed on the oil might cause them more problems, the BP spill has reminded residents of the Lower Ninth that they are continually subject to environmental trouble. Even healthy residents have begun to worry about the potential affects of possibly storing toxic materials from the clean up in the community. These worries lead to an uncertainty that is palpable to anyone visiting or working in the community. Below I examine five features of this environmental uncertainty and how the BP spill has highlighted this uncertainty in the Lower Ninth.

ANYTHING IS POSSIBLE HERE

Because of the history of the community which involves constantly fighting different levels of government over proposed projects and usually the effects of those projects once completed, being left out of the planning for and then the actual rebuilding after Katrina, the mistrust of both government and big business, and the inundation of other kinds of trouble, several residents when asked about the BP spill and its perceived effects on the community commented that anything is possible here. One resident, summing up the last five years and what the future might hold, said, "I lost my home in Katrina, my job in the BP spill, now I'm getting into a class action over the Exxon-Mobil in Chalmette." When asked if he is optimistic about the future, he says he can't be, "no one knows what is going to happen here." (*Black male, 30-40 years old*)

In a city that prides itself on the unexpected, unconventional, and unusual, residents find it difficult to understand what is going on in the community. At a recent neighborhood association meeting, after discussing several major zoning ordinances-which may or may not be implemented in the near future, residents were briefed on the oil spill, told of an incident involving a handful of juveniles that the media were labeling a gang, and then invited to participate in a class action lawsuit regarding the Exxon-Mobil refinery in Chalmette. One attendee, leaving the meeting exasperated, noted that there is no other community like this. Turning to me as left she said, "who else has to deal with all this." (*Black female, 30-40 years old*).

CONFUSION

The authors of the National Commission on the BP Deepwater Horizon Oil Spill and Offshore Drilling's working paper number three, titled "The Amount and Fate of the Oil," write:

> By initially underestimating the amount of oil flow and then, at the end of the summer, appearing to underestimate the amount of oil remaining in the Gulf, the federal government created the impression that it was either not fully competent to handle the spill or not fully candid with the American people about the scope of the problem. (p. 1)

Because of their experiences with Hurricane Katrina many residents thought that the government would in the words of one respondent "get this one right." There is a wide gulf, however, in what they are told by those in power (whether government or corporations) and their own experiences. This causes a great deal of confusion in the neighborhood as some believe what they hear in the media and others are distrustful of the media or company spokespersons.

How could residents not be confused. Since before the well was officially capped they have been told that the effects of the spill would be minimal and that the majority of the oil has evaporated or been cleaned up. Residents are told the beaches are safe and then read that it sits just inches below the surface. After being told there is little to no surface oil, fishermen and boats working the BP spill find mile long plumes of weathered oil which takes them hours to clean off their boats. Less than a week later, however, the Louisiana Department of Wildlife and Fisheries tells them it is only algae (Marshall, 2010). In an interview with NPR, Darryl Malek-Wiley, the local organizer for the Sierra Club's Environmental Justice Program, described how confusing it is for local residents that used booms and other items from the spill are not considered hazardous waste by the EPA. He followed up in a personal conversation to note that he feels above all else the spill is simply confusing for the community because there are competing explanations for how the spill will affect the Lower Ninth. The government, the media, and BP tells them not to worry, but environmental activists and some community members tell them they should be very worried. Residents state that one day they are told not to worry about the effects and the next that they shouldn't eat seafood from the Gulf or that they should refrain from going outside when it rains.

Other instances where the spill has created confusion and uncertainty: (1) loss of identity—the Lower Ninth is suppose to be one, if not the, greenest neighborhood in the country—How will the spill affect this identity?; (2) employment is more precarious—many service related jobs are tied to tourism and the seafood industry—Will people in the community lose their jobs?; (3) loss of help—much of the rebuilding in the Lower Ninth is through volunteers and non-profits—Will volunteers leave the community and go to neighboring parishes to help out with the spill?

DENIAL

As in other contaminated communities many residents are in denial about the danger (Auyero and Swistun, 2009; Edelstein, 1988). When asked if the BP spill made her think about moving, one resident said it's just like any other place.

"California has earthquakes, Texas has hurricanes, and New York has terrorists. I'm staying here. There is nothing wrong with this place." (*Black female, 40-50 years old*). While many of us live in places that are at significant risk for being affected by some disaster or accident, this woman lost her home in Katrina and is trying to join a class action lawsuit against the Exxon-Mobil refinery in Chalmette because she complains of having several symptoms common to the toxins released in that accident. Her home is in the process of being rebuilt, but she has lost lots of money and time to fraudulent contractors and on two occasions someone has broken in and stolen wiring and hardware, furthering complicating her situation. She, however, is unfortunately not alone. Several residents I spoke with denied that the BP spill was that bad. A couple residents noted that this was the media's way to malign President Obama. Some said that they personally had not seen any oil on fishing trips, so the water must be clean.

WHAT'S IN THE ENVIRONMENT?

Some residents do not worry about their environment being contaminated. Some believe that it is contaminated and do not give it much thought while others believe that someone would tell them if it was dangerous to live there. When asked if she knows that Corexit, the dispersant sprayed on the spill, might be in the community, Janice seems unsure, then upset, "no one tells us anything" she says. Mike (*Black male, 30-40 years old*) says that the release from the refinery probably wouldn't have been as upsetting for the community if it had not been for the BP spill. "You go outside all summer and it smells like gas, it's unpleasant, but I can't do nothing about that. Now I go outside and I wonder if I am being poisoned. It has doubled my risk, it added one with the other." Mike discounted what 'smelling' gas might mean for the environment, but that experience coupled with a more recent one has caused him to worry about becoming sick. The actual questioning, however, rarely comes up in conversations or interviews. Residents simply do not want to know the answer to the question.

THE FUTURE OF THE NEIGHBORHOOD

Residents worry that the BP spill could prove to be too much for some members of the neighborhood. If people stop eating seafood and the drilling moratorium continues, people are going to leave. Sandy said that friends of hers who have worked on the spill have told her it might be time to leave. "A friend of mine, an activist, said maybe its time to think through exit strategies, and I thought whoa, and then I thought about the last five years and yeah, maybe it is time. We've reached that point." Again, it is not just one event, but a series of events that BP helped to put an exclamation point on. Mary echoes Sandy's comments. "[With Katrina] it was just a little water." But then we learned that we had the black mold [in our homes], that the poisons were in the ground, then we find out that they want to put all the Katrina stuff in the Ninth—back into the ground. Now we worry about them putting oil stuff here. It's too much. Why would I want to stay here?" For many residents the BP spill forced them to realize that there is an exponential nature to the threat of living in such a place. Janice says as much when asked what the next five to ten years are going to be like for her, "[I'm going to] get my truck right, get my house right, then I get right, but who knows, soon as I get that house fixed, bam! Something gonna come along. Almost makes it not worth the trouble." (*Black female, 30-40 years old*).

The BP spill highlighted for many the precarious nature of living in this particular neighborhood. Many residents have come back to abandoned houses and empty streets. Some have rebuilt where they are the only ones living on the block. If the spill makes it impractical to stay or for others to return, they wonder how the neighborhood will survive.

A CULTURE OF SUFFERING

In his work on environmental suffering, Auyero reminds us that the Latin root of the word "patience" is to endure, to suffer (2010). To be patient is to suffer. In a geography of trouble people endure, they suffer and this marks the community.

> And I looked around that room and Cecil B. DeMille, Frank Capra, it was like a Frank Capra thing, Lisa Jackson is in the front of the room with some other folk, there were people from the neighborhood, people from the North Shore, people from the Parish, people from further over . . . people from Houma Nation, there were people in, umm, work clothes, people who were clearly, uhh, I don't know like for example if they were from Lafourche or Thibodaux, or New Orleans East, but there were representatives of the Vietnamese community, and you looked around that room and people that were gathered together, and they were like having a pretty civil conversation...and it was like yeah, we can do this, we've done this before. Everyone in this room is a survivor.

Sandy's observation of the first community meeting to discuss the impact of the oil spill in the neighborhood is important because it defines what it means to live in this community. She sees the neighborhood as a community of survivors. Her description of the scene is also useful in that one's survivorhood marks membership in the community. While earlier in our conversation she defined the community in terms of geographical markers, as do most others in the neighborhood, she feels a bond with other survivors in the region. Her inclusion of people from nearby parishes or ethnic groups shows that trouble extends membership in the community. This is noteworthy because these communities are often at odds over resources in an endless game of survivorhood (i.e. which community is worse off and thus more deserving). Just as Edelstein (1988) found in his examination of several "contaminated communities" where people come to identify with victims in other toxic communities, residents in the Lower Ninth see others who are forced to suffer as community members. The BP spill extended the culture of suffering and therefore the boundaries of the community. It also strengthened the bonds of membership in the community by giving residents the opportunity to practice normalizing the trouble.

Rather than a culture of silence (Beamish, 2000) where accidents are treated as open secrets, involving a style of mental focusing that relegates trouble to the background where it is simply not noticed or taken for granted (Zerubavel, 1997, 2008), in a geography of trouble the trouble becomes part of the social fabric of the community. Trouble becomes a way of interpreting and framing events. In a culture of silence members of a community are co-conspirators. They agree that what they notice as individuals will not be noticed by the group. As Erikson observed in his study of the Buffalo Creek flood, communities involve conspiracies to make a world filled with danger seem safe (1976: 240). For residents of the Lower Ninth, however, these conspiracies highlight danger, at times even exaggerating it. The conspiracy becomes one of normalizing trouble.

> We know a lot about risks, hazards, and disasters, it's our life. (*White male, 50-60 years old)*

> That's just another thing. We use to it. (*Black female, 60-70 years old).*

These two statements reflect residents' orientation to the BP spill. Rather than being surprised by the spill, they interpret it as an everyday, routine occurrence. To be sure, both spoke of the spill as being a 'worst', but both also simultaneously saw it as being similar to other events. The first speaker for instance described his long battle against the widening of the industrial canal and for the closing of the MRGO. These two struggles have lasted decades and the majority of residents who opposed these two water projects did so because they knew the trouble they could bring, such as destruction of the wetlands and the funneling of flood waters into the community, not to mention possible levee breaks. The second speaker, who may sound callous to some readers, spoke of the spill as a national tragedy, a disaster which will affect the region for many years, but in surmising how she thinks about the spill, after a lengthy response, she ends up normalizing it. Because it is usually backgrounded, discussion of trouble in most communities is reduces social interaction. In the Lower Ninth, discussion of trouble brings people together here. As William said:

> Yeah, the spill was bad. But it was like Katrina. It got me going to meetings again and talking to friends and neighbors. People I hadn't seen in two to three years, since finishing up my house, the repairs on my house, start coming around again. (*White Male, 40-50 years old*)

Karen Cerulo writes that our culture cognitively biases us towards positive asymmetry (2006). We are much better at imagining best case scenarios than worst cases. Whereas most of us can easily imagine what we would do if we won the lottery or imagine meeting Mr. or Ms. Right, few of us daydream of being bankrupt or meeting Mr. or Ms. Wrong (which is actually more likely). She notes that there are exceptions to this asymmetry and that some communities practice negative asymmetry. Medical practitioners and computer information technicians, for example, are two communities she cites that practice negative asymmetry. Both are trained to see trouble. While not pathologically pessimistic, living in a geography of trouble causes people to practice negative asymmetry. This also highlights the suffering inherent to the culture of these places.

Explaining a phone call she received during a conversation at her house, Mary's (*Black female, 50-60 years old*) comment typifies the negative asymmetry that accompanies this culture of suffering.

> I know a lot of people who took during Katrina. People who lived on the third floor who got $27,000. They didn't get no water. I could have got something. FEMA giving money away. You suppose to get either $2,000 direct deposit in your account or a check. Some peoples got both. I got this friend who got both. She told me go home and check the balance, check your account, cause I got the check first. There it was and I was like uh, oh. Somebody made a mistake. Now I ain't that kind of person. I ain't gonna take what's not mine. So I called FEMA and told them they double gave. They be double giving to everyone. It took me, oh I don't know, maybe, it was over an hour, to explain to that lady what was going on. She asked if I had still had the check. I said yeah, because I didn't want to get into trouble. She gave me an address to mail it back. I did. But it came back and I had to call again and they said no honey she gave you the wrong address and I mailed it right that time. That second lady she told me a computer did it. It been happening for awhile. So now FEMA telling people to call in and get an extra $5000. Why you think they doing that? [I mention a recent lawsuit regarding racial housing discrimination, but she interrupts] I know why. They gonna catch people who double took. I been getting phone calls all day from friends, telling me about the money. They greedy. That's what it is. It is just a way for the government to get their money back. They double gave and they want that money back. This is just a trick. A lure. They luring people into a trap with that new money. You'll see. They gonna get back that other money, millions and people going to jail. BP the same way. They going use this as a way to get greedy people. BP gonna give away all this money and look good on tv, then they gonna spend years getting it back from people who took. I got friends calling me cause of my business, telling me I can get BP money. These the same friends who took during Katrina. They all going to jail.

CONCLUSION

At a recent forum on the BP spill several speakers noted how the spill and the consequent suffering compares to Katrina, with some saying the spill would be far worse and others conceding that Katrina was probably a larger disaster. In the question and answer period several attendees commented that there was no "hierarchy of suffering" and that we needed to see these events as "somehow connected."

In this chapter I have maintained that one way to see the connection between these events and others is to recognize that they happen in specific geographies. For many the BP spill highlighted problems with living in this area and for others caused them to relive Katrina. Geographies of trouble have both direct and indirect effects. People in the Lower Ninth ward have lost jobs, some are ill, but even more devastating have been the cognitive ripples. The BP spill has changed the way many residents think about the future, their environment, and their neighborhood. As yet one more 'worst' case, it has made many in the community realize that they live in dangerous place.

We need to recognize these geographies for several reasons. Freudenberg et al. (2009) notes that we must begin to question the assumption that environmental damage is somehow good for the economy. There is no short term versus long term trade off that makes environmental degradation worth it. This assumption that Freudenberg recognizes usually pertains to very distinct places. This assumption is only made in impoverished places where people have weak ties to political power. Often it is only made in places where the assumption has been made before. This assumption represents a type of durable inequality that ties the most marginalized to the worst places. Essentially then, we need to recognize these geographies because corporations and politicians already have.

Second, more and more of us live in these troubled places. Flood plains continue to be developed. We pay little attention to our infrastructure. More of us are moving to urban centers or regions where terrorists can easily strike. One in four Americans lives in the immediate vicinity of a superfund site. Nine of every ten Americans live in places at risk for a major disaster (Ripley, 2008). We live in dangerous places and rarely recognize it. Worse we rarely plan for it. Clarke (2006) notes that as a society we often focus on the probabilities of events rather than the event's possibility. We thus rarely expect the worst to happen. But worst cases happen all the time. This is especially true in geographies of trouble. Here the possibilities far surpass the probabilities. Residents of Southeastern Louisiana should not statistically have been victims of two worst-cases in five years. They were. It was always possible. People in geographies of trouble should especially heed possibilities and learn to ignore probabilities. We all should.

Finally, these geographies and the people who live there have a lot to teach us. Failing to recognize these geographies keeps us from having to think about them and the people who live there. It also keeps us from having to think about the relationship between ourselves and our environment. When we can pass off negative externalities to other communities we can ignore the moral and political dilemmas of the role we play in environmental degradation. To recognize geographies of trouble is to acknowledge that, as Freudenberg et al. (2009) write, these disasters are usually not instances of nature hitting us, but us hitting nature. These hits, as Auyero (2009) mentions, leave indelible marks on those who endure them, but increasingly it seems that it is society itself that is being marked.

REFERENCES

Auyero, J., and Swistun, D.A. (2009). *Flammable: Environmental suffering in an Argentine shantytown*. Oxford: University of Oxford Press.

Auyero, J. (2010). Chuck and Pierre at the Welfare Office. *Sociological Forum* 25: 337-52.

Beamish, T. (2000). Accumulating trouble: Complex organization, a culture of silence, and a secret spill. *Social Problems 47*: 473-98.

Bullard, R. (2000). *Dumping in Dixie: Race, class, and environmental quality*. Boulder, CO: Westview Press.

Cerulo, K. (2006). *Never saw it coming: Cultural challenges to envisioning the worst*. Chicago, IL: University of Chicago Press.

Clarke, L. (1999). *Mission impossible: Using fantasy documents to tame disaster*. Chicago, IL: University of Chicago Press.

Clarke, L. (2006). *Worst cases: Terror and catastrophe in the popular imagination*. Chicago, IL: University of Chicago Press.

Edelstein, M. (1988). *Contaminated communities: The social and psychological impacts of residential toxic exposure*. Boulder, CO: Westview Press.

Erikson, K. (1994). *A new species of trouble: The human experience of modern disasters*. New York, NY: W.W. Norton & Company.

Freudenburg, W., Gramling, R., Laska, S., and Erikson, K. (2009). *Catastrophe in the making: The engineering of Katrina and the disasters of tomorrow*. Washington, D.C.: Island Press.

Godsil, R., Huang, A., and Solomon, R. (2009). Contaminants in the Air and Soil in New Orleans After the Flood: Opportunities and Limitations for Community Empowerment. In R. Bullard and B. Wright (Eds.), *Race, Place, and Environmental Justice After Hurricane Katrina: Struggles to Reclaim, Rebuild, and Revitalize New Orleans and the Gulf Coast*. Boulder, CO: Westview Press.

Grineski, S., Collins, T., de Lourdes Romo Aguilar, M., and Aldouri, R. (2010). No safe place: Environmental hazards and injustice along Mexico's northern border. *Social Forces 88*: 2241-65.

Kasperson, R., Renn, O., Slovic, P., Brown, H., Emel, J., Goble, R., Kasperson, J., and Ratick, S., (1988). The social amplification of risk: A conceptual framework. *Risk Analysis* 8: 177-87.

Langadec, P. (1990). *States of emergency: Technological failures and social destabilization*. London: Buttterworth-Heinemann.

Marshall, B. (October, 27, 2010). Definitely algae, not oil, scientists conclude of Gulf substance. *Times Picyaune*. Retrieved from http://www.nola.com/news/gulf-oil-spill/index.ssf/.2010/10/definitely_algae_not_oil_scien.html.

Mitchell, J. K. (1996). Improving community responses to industrial disasters. In J.K. Mitchell (Ed.), *The long road to recovery: Community responses to industrial disaster*. Tokyo, Japan: The United Nations University Press.

National Commission on the BP Deepwater Horizon Oil Spill and Offshore Drilling (2010). *The Amount and Fate of the Oil*. Staff Working Paper #3.

Pellow, D. (2004). *Garbarge wars: The struggle for environmental justice in Chicago*. Cambridge, MA: Massachusetts Institute of Technology.
Perrow, C. (1999). *Normal accidents: Living with high-risk technologies*. Princeton, NJ: Princeton University Press.
Ripley, A. (2008). *The unthinkable: Who survives when disaster strikes—and why*. New York, NY: Crown.
Slovic, P. (2002). Terrorism as hazard: A new species of trouble. *Risk Analysis 22*: 425-26.
Sze, J. (2007). *Noxious New York: The racial politics of urban health and environmental justice*. Cambridge, MA: Massachusetts Institute of Technology.
Vaughan, D. (1996). *The challenger launch decision: Risky technology, culture, and deviance at NASA*. Chicago, IL: University of Chicago Press.
Zerubavel, E. (1997). *Social mindscapes: An invitation to cognitive sociology*. Cambridge, MA: Harvard University Press.
Zerubavel, E. (2008). *The elephant in the room: Silence and denial in everyday life*. New York, NY: Oxford University Press.

This research has been generously funded by The Natural Hazards Center at the University of Colorado at Boulder and the Public Entity Risk Institute with support from the National Science Foundation and Swiss Re and The Horowitz Foundation for Social Policy.

CHAPTER EIGHT
ECOLOGICAL IDENTITY AND DISASTER RECOVERY IN AN OIL-STAINED LANDSCAPE: CURRENT AND FUTURE POLICY IMPLICATIONS

DEMOND SHONDELL MILLER, JASON DAVID RIVERA and BRANDON ERIC FLEMING

> This spill has affected everyone along the Gulf Coast. We all value the natural beauty of our coast and the connection with nature. We're moved by the site of brown pelicans. We laugh with joy when we see the dolphins playing off the boat in the Mississippi Sound. There is something so special about our connection to our gulf that links everything our livelihoods, our sense of connectedness, our spiritual awareness and this disaster has just taken away the sureness that everyone would wake up with and it would be there. And that uncertainty and experience has been across the board. That will go with us whatever the trajectory is for the rest of our lives. (Dr. McIntosh quoted by Jamil, 2010, Global Research, CA)

INTRODUCTION

This chapter discusses the role of ecological identity and place attachment in the aftermath of the Deepwater Horizon oil spill and its potential implications for future policy. We maintain that the physical environment presents natural characteristics used by members of a society to sustain themselves. The physical characteristics of a place initially define the interactional potential of the society living in that space, which allows specific interaction patterns to manifest in distinct ways that become part of the localized cultural knowledge and contributes to the location's interactional past.[1] Members of a society develop a sense of place and a place attachment unique to individuals and their specific geographical location. Over time, the individual's interactional past, sense of place, and place attachment contribute to the society's overall connection with the environment and helps to shape interaction patterns with the environment. How society views its interactional past with the environment influences its ability to redevelop and rebuild society in the aftermath of catastrophic events. Natural resources indicative to a place, in addition to the society's entrepreneurial spirit, have led to the development of a place's social and economic landscapes.

Local understanding of one's physical landscape is paramount if societies and cities are to become less vulnerable; however, the general public must alter their perception of responsibility for this to come to fruition. According to Pelling (2003), citizens tend to view environmental problems and their vulnerability to natural hazards as the responsibility of the government, and they therefore argue that disaster preparedness should not be a priority in their daily lives. By adhering to this idea, citizens rely on the fiduciary role of the government for disaster mitigation and the limiting of vulnerability to environmental hazards to higher echelons of government (i.e. policy makers). A failure to mitigate environmental risks from a technological disaster result in a disaster landscape that is incompatible with the pre-disaster landscape causing a shift in one's ecological identity (from a pre-disaster sense of relative safety and security to a post-disaster sense of uncertainty and anxiety about one's health, economic status or social interactions). Moreover, a regional ecological identity shift occurs that is characterized by a shift from a sense of relative regional environmental safety to extreme uncertainty. In order to illustrate how this process has been exhibited within the Gulf in the past, a brief examination of the Ixtoc I oil spill is presented. This is observed through individuals' and communities' reactions to oil and toxins constantly washing ashore rendering the coastal ecosystem *unsafe* and thereby detrimentally affecting regional industries and general day-to-day existence.

The chapter will conclude by addressing the myriad of current[2] and future human, social, economic, and environmental consequences of the Deepwater Horizon oil spill. The loss of lives is a tragedy. Social, economic and environmental consequences are yet to be fully estimated, but there is little doubt that they will be significant. Even though the impacted physical environment is expected to recover, we must bear in mind that this region is home to a variety of vulnerable species and habitats. Oil spills are visible incidents that trigger policy action at various administrative and geographical levels, receive widespread media attention and cause civil society alarm. Without long-term social and political attention towards risk, only minor policy interventions will develop.

ECOLOGICAL CATASTROPHE AND THE DISASTER LANDSCAPE

After an ecological catastrophe, the abrupt alterations of the physical landscape sets in motion a series of societal disruptions. Fried (1963) characterizes these disruptions as the result of altered perceptions of a place by survivors in the aftermath of events similar to the *Exxon Valdez* oil spill. Fried maintains:

> Any severe loss may produce a disruption in one's relationship to the past, to the present, and the future. Losses generally bring about fragmentation of routines, of relationships, and of expectations, and generally imply an alteration of the world of physically available objects and spatially oriented action. It is a disruption on the sense of continuity which is an ordinarily taken-for-granted framework for functioning in a universe which has temporal, social, and spatial dimensions . . . *the loss of an important place represents a change in a potentially significant component of the experience of continuity* [emphasis added]. (Fried, 1963:153)

During times of natural or technological disasters, one's sense of identity, namely one's ecological identity, place and space are disrupted. Read (1996) asserts that "the way in which [humans] actually and symbolically create landscape within the cultural community probably reflects other organizing principles of that society and its world view" (p. 2), and, in many ways, the physical landscape affects our life patterns and the way we interpret the world. By understanding the changes in the physical landscape of an area after an ecological tragedy, people are capable of appreciating the workings of nature (Miller, 2006a; Saito, 1998; Miller and Rivera, 2008).

It is the alteration of a population's perception of the world around them that influences their perceptions about themselves, their community, and society. Hamerton (1885, as cited in Riley, 1992) contended that people's moods and personalities correspond with specific, real-world, mood-inducing landscapes. Places have the ability to influence our understanding of the world and indirectly guide us in the establishment of specific moral views about the world around us. Therefore, when a landscape is abruptly changed by oil and tar balls washing ashore in an area that stretches from the Texas coast to the panhandle of Florida, there is an abrupt change in the way society relates to the landscape. Human relationships with a landscape change as their experiences and memories attained in that place are altered, which may have positive or negative results (Riley, 1992). In short when a place with a strong tourist identity associated with its natural setting, as a relatively safe family-friendly location suddenly becomes a toxic landscape, complete with a host of known volatile chemicals and carcinogenic dispersants, the "place," as it is phenomenologically conceptualized by a new place identity for the impacted area forms and the images of the once safe Gulf Coast becomes associated with the disaster landscape; the positive qualities of the area are diminished as a hyper-reactive media presents images consistent with the risks of harm.

The news media is a main arena in which ''various social groups, institutions, and ideologies struggle over the definition and construction of social reality'' (Gurevitch and Levy, 1985; Brinkland and Lawrence, 2002; p 18), usually through the use of verbal or visual symbols of the reality they seek to construct. In constructing reality, groups illustrate ''problems'' for which groups seek "solutions." Perceptions of the problem serve as a catalyst for group action in pursuance of an agenda-setting process that is brought to the attention of the larger public and subsequently policy makers (Schattschneider, 1960/1975; also see Brinkland and Lawrence, 2002). Hence, in the days following the Deepwater Horizon oil spill, media coverage was transfixed on the "slowness" or inadequacy of government and private sector responses, which set the stage for a contested environment of blame and responsibility "accepting," or lack accepting responsibility. Moreover, governmental responses and litigation began, based on promises made by BP, state and federal governments, in addition to local industry leaders. It is in this milieu that "places disappear and lose their identity when they are no longer distinguished from their context, when they vanish in their context" (Huigen and Meijering, 2005: p. 29). Places become part of a larger socio-political whole—as an oiled landscape filled with uncertainty about the future prospects for a sense of social normalcy and a pre-disaster natural environment. Hence, when topographic features are indistinguishable, humans have a difficult time connecting to place.

ECOLOGICAL IDENTITY AND PLACE AND A SENSE OF PLACE

Thomashow maintains that ecological identity, ". . . refers to all the different ways people construe themselves in relationship to the earth as manifested in personality, values, actions, and a sense of self. Nature becomes an object of identification" (1995: p. 3). Applying this to the Deepwater Horizon oil spill, residents of the Gulf of Mexico region experience a disconnection from their natural environment and the ecological identity formed prior to the oil spill. Because the oil spill destroys a sense of comfort and security associated with the routine patterns and social networks in the region, the disaster or oil-stained landscape defy their ecological experiences prior to the oil spill, simultaneously forcing many to reconsider their cognitive understanding and risks associated with the *new normal*—a toxic future of uncertainty. Dissecting an individual's ecological identity and its impact varies among individuals; however, multiple similarities persist because, ". . . identity is a very complex notion, referring to all of the different ways people construe themselves in social relationships as manifested in personality, values, actions and sense of self" (Thomasahow, 1995: p. 3).

Daily life after the oil spill will be impacted as ". . . notions of what it is to be expected in the normalcy of everyday activities are challenged" (Miller and Rivera, 2008: p. 10). As massive amounts of oil wash ashore, these events exceed our ability to readily comprehend the long-lasting effects on society as a whole, thus making the experience all the more devastating. Individuals working in the oil, fishing, tourism and hospitality, shipping and other related industries will suffer the impact for generations to come. Breaks from cognitive structures, once hallmarks of the Gulf Coast, become more commonplace as many become afraid to eat the seafood and enjoy the sands of the beaches; therefore they seek other social and economic opportunities that are deemed "safer" and less likely to be affected by the possibility of contamination or negative media coverage.

LOSING PLACE

The loss of place, sense of place, and place attachment are some of the cultural mechanisms that govern, and are in direct relation to disasters, disaster recovery and future policies. Eyles (1989) defines place as,

> a centre of felt value, incarnating the experience and aspirations of people. Thus it is not only an arena for ever day life. . .[it also] provides meaning to that life. To be attached to a place is seen as a fundamental of ourselves and our identities. Places are thus conceived as profound centres of human experiences. As such, they can provide not only a sense of well-being but also one of entrapment and drudgery. To be tied to one place may well enmesh a person in the familiar and routine (as cited in Brown and Perkins, 1992: p. 281).

The notion of place is associated with a connection or a "self being" with one's surroundings. If an individual is disconnected from the natural setting in which he or she lives by a disaster or catastrophic event, they may feel as if they no longer belong to that particular area, which subsequently disrupts the ability to identify with that place.

Sense of place and place attachment are two concepts that intertwine, but have one specific dissimilarity: "[p]lace attachment is used to examine the relationships between people and both neutral and built environments" (Trebtleman, 2009: p. 3). The word attachment implies that there is a positive meaning or resemblance from the word because of the past experience in scholarly literature, sense of place does not (Trentleman, 2009). Hummon (1992) states, "sense of place involves a personal orientation toward place, in which one's understandings of place and one's feelings about place become fused in the context of environmental meanings" (262). Places are deemed significant by society's use of a place. Trentleman addresses this notion by stating, ." . . place is socially constructed, by whom, and with what interests, as well as with how constructions are adopted and how various conceptualizations of places affect different groups" (Stokowski, 2002; Williams, 2000; Trentleman, 2009: p. 196). For the individuals who are affected by the oil stained land, the importance that society has put upon the places is now destroyed. People identify with what they know and understand from society. After the spill in the Gulf coast region, people who identified with those lands and waterways no longer had a bond with that particular place because of the tarnished ecology of the Gulf area.

INTERACTIONAL PAST AND POTENTIAL OF OIL IN THE GULF: THE IXTOC I OIL SPILL

The interactional past of the Gulf of Mexico in reference to oil spills stretches back more than thirty years. The interactional potential of the area serves as an instrumental resource for local domestic communities and subsistent populations; however, the natural resources of the Gulf also contribute to the development of the socio-cultural, socio-economic, and political landscapes of several nations throughout the Western Hemisphere. What is interesting to note in this regard is that although the Gulf of Mexico has such significant influences on a broad range of communities' and nations' sense of place and place attachment, the interactional past of the region, in reference to oil spill experiences, has seemed to have vanished from the general public's memory. In most of the coverage that has taken place in reference to the Deepwater Horizon oil spill, domestic popular media outlets have acknowledged/mentioned the occurrence of the Ixtoc I oil spill that also took place in the Gulf more than thirty years earlier in 1979.

A blowout at the Ixtoc I oil platform occurred off the coast of Mexico in the Bay of Campeche, which is situated in the southern part of the Gulf of Mexico. According to Pemex, the state oil company of Mexico that owned and operated the facility, about half a million tons of oil was released over more than nine months. The amount of oil released, in addition to the length of time that the oil was released into the Gulf, made this oil spill the world's largest peacetime oil spill to occur to date (Jernelöv and Lindén, 1981). The only other oil spill that compares to the size and scope of the Ixtoc I spill was an oil spill that occurred during the first Gulf War, in Kuwait, when Iraqi soldiers emptied all the oil-storage tanks along the coast as a military strategy against the US military. In reference to the Ixtoc I spill, crews on the drilling platform neglected warning signals that indicated well pressures were extremely high, some safety procedures were skipped to save time and money, and the blowout resulted in an explosion and fire that sank the platform, subsequently damaging the well casing that caused oil to be released into the Gulf (Jernelöv 2010).

According to Jernelöv (2010), the environmental effects of the Ixtoc I oil spill were still visible five years after the event; however, due to cleanup efforts in the immediate aftermath of the disaster the environment was recovering.

> Ecologically, one had to look close to see any remaining damage, and fisheries were, if anything, doing better than before the spill. The reduction of fishing pressure after the spill obviously helped the oil-affected populations to a more rapid recovery, and better equipment helped the local fishermen to catch them (p. 365).

Although the restoration of natural resources did occur in reference to this specific case, the physical landscape of the Mexican coast played a significant role in the extent of damage inflicted by the spill. The part of the Mexican coast that was affected by the oil was mostly sandy beaches that acted as protective barriers for the lagoons and other waterways behind them (Jernelöv, 2010). In comparison to the Deepwater Horizon disaster and the physical landscapes that the disaster effects, sensitive wetlands will and have been affected by the oil spill, which will have different rates of recovery and subsequently impact human industries that rely on these resources further into the future than in the case of the Ixtoc I oil spill.

The Ixtoc I oil spill also affected the socio-economic landscapes of the state of Campeche, Mexico and southeastern Texas. In reference to Texas, in the first month after the oil spill commercial fishermen had not reported any changes in their fishing grounds. Additionally, state and federal inspectors were not able to find any contaminated catches, mainly due to the fact that the majority of the oil from the spill had not drifted toward Texas (Hill, 1979). Although this boded well for the fishing industry of southeastern Texas, the local tourism industry reported losses totaling $155 million (in 1979 US dollars) due to a lack of visitors in reaction to oil residue washing up on the beaches (Hill, 1979). In reference to the state of Campeche, Mexico, where the oil spill had taken place a few miles off the coast, subsistence and commercial ways of life were indefinitely altered. In recalling the event and its effects on the fishing industry, Paul Bonastre, a fisherman in the fishing village of Champotón, Campeche said, "We never captured the same amount of product" (as cited in Miglierini, 2010). Moreover, Miglierini (2010) explains that in the months following the disaster many fishermen and other people associated with the industry left Champotón to seek other work, some of whom never returned.

In addition to these socio-economic impacts, the disaster also affected the political landscape at the federal level in the United States and internationally. Domestically, Congress took steps in order to raise the oil spill contingency fund to $6 million (in 1979 U.S. dollars) in order to aid businesses and individuals that were affected in southeastern Texas as a result of the disaster (Hill, 1979). In reference to U.S. actions taken against the responsible party liable for damages from the spill, little was done. According to Hill (1979), due to foreign trade negotiations that were taking place at the time between the United States and Mexico that were taking place at the time in which the United States was attempting to procure oil and gas from Mexico, the U.S. Department of State was hesitant to hold the responsible party liable for damages. International discussion on the issue arose when Texan politicians representing the interests of the tourism industry that had incurred "indefinite losses" pressured the State Department to do something. In reaction to a response to a message sent by the State Department to Mexico's Foreign Ministry, the Mexican government responded that there was not international law justification for the government or the government owned and operated oil company to be held liable for the damages caused by the spill. Not only did the Mexican government refuse to compensate affected communities based on a lack of legal justification, Mexican President José Lopez Portillo said that Mexico had no intention of paying for any spill damage because the United States had not compensated Mexico for a cross-border contamination of saline water in the Colorado River (Hill, 1979). These political and legal maneuverings on the behalf of responsible parties, and the federal government's lack of initiative to push for some manner of compensatory agreement, left affected communities in the void; thereby contributing to a lengthening of community and industry recovery processes.

THE EVENTS OF THE DEEPWATER HORIZON OIL SPILL: SETTING THE STAGE FOR AN OIL-STAINED LEGACY

In order to understand the social, cultural and political ramifications of the Deepwater Horizon oil spill, one must understand the dynamics of the actual event. What immediately follows is a chronological account of the incident, which sets the trajectory of the region's social, cultural and political responses that will influence community development far into the future.

On April 20, 2010 an explosion erupted on the Deepwater Horizon, a British Petroleum (BP) licensed drilling rig in the Gulf of Mexico. Eleven crewmembers were reported missing[3] and approximately seventeen were injured. BP had installed a blowout, which was intended to prevent a release of crude oil in an event such as this one, but it failed to activate (Guardian Research, 2010; Associated Press, 2010a). On April 22, the Deepwater Horizon sank, and an oil slick, approximately a mile long, covered the Gulf of Mexico. On April 23, the U.S. Coast guard suspended the search for the missing workers, who at that point were presumed to be dead (Guardian Research, 2010; Goldenberg, 2010a).

By April 25, underwater robots were used in an attempt to stop the leak of an estimated 1,000 barrels per day of crude oil being dispersed into the Gulf of Mexico as reported by the U.S. Coast Guard's underwater camera (Guardian Research, 2010: Goldenberg, 2010b). Moreover, on April 25th the Coast Guard began "controlled burning" of the oil as BP made its first attempts to repair the hydraulic leak in the blowout preventer. They were ultimately unsuccessful. Four days later, Louisiana declared a state of emergency as the oil slick approached the coast and threatened the natural resources of the state and Gulf Coast. In reaction to these events, President Obama deemed BP responsible for the cleanup efforts. On April 30, President Obama banded drilling in new areas of the Gulf until the oil spill's cause had been identified and addressed. BP chairman Tony Hayward states that they will take full responsibility for the spill. On May 6, toxic oil washes ashore onto the Chandueler Islands and it is estimated that BP would lose 15 billion Euros (Guardian Research, 2010: McVeigh, 2010).

On May 8, as ball-like tar washed ashore upon Dauphin Island, Alabama, BP attempted operation "junk shot" pumped tires and golf balls into the well in an attempt to clog the valve. May 10th, BP announced that they would try to place a containment dome, known as "top hat," which would funnel the oil to the surface (Guardian Research, 2010). On May 12, BP tried to place the dome to cover the leak. The same day, a congressional hearing blamed BP for ignoring safety warnings just hours before the explosion. On the 13th of May, it was estimated that the well was leaking oil at a rate of 70,000 barrels per day (Guardian Research, 2010; Goldenberg, 2010c). In a report by the *New York Times*, Urbaina (2010) noted that ". . . a half-dozen current and former agency scientists" for the Minerals Management Service (MMS) claim they were "regularly pressured by agency officials to change the findings of their internal studies if they predicted that an accident was likely to occur or if wildlife might be harmed;" furthermore, the article went on to state that "the MMS had approved hundreds of drilling projects without obtaining legally required permits" (Urbina, 2010). On May 14, BP planned to insert a 4-inch tube into the ruptured 21-inch riser pipe to propel the water to the surface. This plan was not intended to cease the leak, but to decelerate it. On May 18, the no fishing zone is doubled in size, increasing it to 19 percent of the Gulf's region (Guardian Research, 2010; Associated Press, 2010b; Webb, 2010a).

On May 20, "BP reports that the amount of oil drawn off by the tube has increased to 5,000 barrels per day. Given that BP had previously estimated the entire leak at 5,000 bpd and the tube only draws off a fraction of the spill, the leak must be much larger than previously estimated. Experts that testified at the congressional hearing estimated "20,000-100,000 barrels per day" (Guardian Research, 2010). On May 26, BP pumped thousands of barrels of mud into the well to stop the leak but operation "top kill" was unsuccessful (Guardian Research, 2010; Clark, 2010).

On June 4, BP placed a cap onto the valves that were engineered to prevent the rig from leaking (Guardian Research, Clark 2010), this measure proved successful. By the 6th of June, BP's new containment cap is capturing an estimated 10,000 barrels of oil per day (Guardian Research, 2010, Goldenberg, 2010c). On June 12, the estimate of oil leaking from the well was doubled from 20,000 bpd (barrels per day) to 40,000 bpd (Guardian Research, 2010; Webb, 2010b). On June 21, a Deepwater Horizon worker claimed that the oil rig was leaking before it exploded (Guardian Research, 2010; Wearden, 2010).

On July 11, BP tried to seal the leak by removing the cap from the well so they could install another containment system on July 13th (Guardian Research, 2010; Goldenberg, 2010d; Meikle, 2010). After they put the tighter fitted cap onto the leak, a pressurized test was done to determine if the leak had been seized. On the 15th, BP confirmed that the leak had been stopped for the first time in 87 days (Guardian Research, 2010; Goldenberg, 2010d). On the 1st and the 2nd of August, BP offered one-lump sum payments to claimants who waived their right to sue for the spill and would attempt operation "static kill." This operation would pump heavy mud and cement into the well to stop the leak (Guardian Research, 2010;

Webb, 2010c, 2010d). On the 4th, BP stated that "static kill" had stopped the leak and the U.S. government announced that the majority of the oil had been cleaned up by relief efforts (Guardian Research, 2010; Gabbatt, 2010). On Sept 19, five months after the first explosion, BP completely sealed the well (Guardian Research, 2010; McGreal, 2010).

THE CATASTROPHE HITS HOME: THE HUMAN PERSPECTIVE

When technological disasters occur, contamination of the environment produces a series of events that continually impact victims and their families into the future (Erikson 1994; Freudenburg, 1997; Gill and Picou 1991; Picou and Gill 2000; Picou, Brunsma and Overfelt, 2010). Similar to the *Exxon Valdez* oil spill, contamination and subsequent resource losses caused by the Deepwater Horizon oil spill have the potential of threatening people's personal safety and security, thereby resulting in social disruption, psychological stress, and loss of institutional trust (Edelstein, [1988] 2004; Erikson, 1994; Freudenburg, 1997; Kroll-Smith and Couch, 1993; Ritchie and Gill, 2010).[4] In the advent of a massively destructive natural disaster, the physical landscape changes, literally, overnight, which, in turn, abruptly changes or displaces the daily routine of the inhabitants of the disaster landscape (Kyvig and Marty, 2000; Miller and Rivera, 2006a). Miller and Rivera (2007a) contend that:

> The disaster landscapes exist not only as a structure but also as a parallel to the "normal structure" (predisaster) as individuals constantly refer back to it as a reflexive notion of the way reality ought to be represented. Critical to the understanding of human agency in the aftermath of a disaster is the existence of the disaster social structures, such as the socio-cultural, socioeconomic, and the political landscapes. (p. 144)

These social impacts that affect both individuals and entire communities, have the potential of having far reaching affects on the political landscape as well, resulting in changes in public policy or executive enactments that seek to aid the recovery and well-being of citizens within a specific place.

THE SOCIO-CULTURAL LANDSCAPE

Empirical research shows clear and compelling evidence that illustrates the "debilitating consequences" of technological disasters at the community and societal levels as chronic psychological stress that collectively undermines the community's social structure (Marshal et. al., 2004; Picou et. al., 1992; 1997). Furthermore, after twenty years of legal maneuvers the *Exxon Valdez* oil spill case of compensation was heard before the US Supreme Court. In an *Amici Curiae* brief filed by Sociologists, Psychologists, and Law and Economics Scholars in Support of Respondents the long term socio-cultural impacts were well documented.

> [t]hese affects are particularly acute when technological disasters strike renewable resource communities-communities within a defined area whose residents' 'primary cultural, social and economic existences are based on the harvest and use of renewable natural resources' [(Picou et al., 2004; p. 1496)]. In such cases, technological disasters can create a domino affect: they are likely to result in contamination if the environment and natural resources, Picou et al. [(2004; 1496)], which then disrupts subsistence or commercial harvests of those natural resources, Picou et al. (1992 :239), which in turn creates perceptions of 'uncertainty, ambiguity and continuing disruption [that] generate patterns of long-term community stress,' And even after initial stress levels subside for the community as a whole, some groups may continue to experience chronically elevated stress levels based on their special relationship to the disaster-for example, because of a physical proximity to the contamination or 'a unique relationship to the resources that have been contaminated, i.e., subsistence and/or commercial harvests' [(Picouet al., 1992 : p. 239)].

It is these general patterns that tend to exacerbate the prolonged recovery period. For natural resource dependent communities, the ambiguity and uncertainty that occurs in the aftermath of technological disasters furthers the negative social impacts in a region heavily reliant on commercial fishing (fishing, processing, boat supplies, and related tourism and fishing activities) and subsistence fishing (harvesting, trading or stocking one's own family supplies of wild catch or supplying neighbors) all are a part of how many in the coasted Gulf Coast region, especially Louisiana, Mississippi, and Alabama, define themselves (see Picou et al., 1992, 1997; Gill and Picou 1997; Arata et al., 2000) and how they relate to the natural environment around them. Immediately following the oil spill, the oil began washing ashore Grande Isle, Louisiana; locals began to discuss the amount of social disruption taking place during the height of the tourist season. This distress soon spread throughout the Gulf region. Several news stories in local papers commented on the impact of the spill. A headline from the *Times-Picayune* read, "I'll Never Be the Same." This headline underscores what sociologist Anthony Ladd pointed to in an interview:

> You don't know when the BP check is going to show up in the mail, if ever. You don't know when the feds and the state are going to do their thing, toward recovery. It's a chronic spiral of people into often deeper and deeper levels of anxiety, and research shows that one of the major sources of anxiety is the litigation process itself. So on top of everything else the disaster throws at you, then you have the decade long experience of trying to litigate your way back to your economic livelihood or trying to get some kind of economic compensation for what you've lost and of course it never comes" (Ladd cited in Jamail 2010, Global Research).

At the time of writing this chapter, two months after the spill was contained, there are no comprehensive mental health or social scientific studies of the communities along the Gulf Coast; however, many are presently underway. Suffice it to say that in the aftermath of such a traumatic event for those who reside, work, and own businesses in the region, the lack of subsistence activities from the Gulf's waters has brought with it a wave of unemployment, substance abuse, and changes in livelihoods, all which increase the level of stress and decrease the quality of life. When compared to Hurricane Katrina,

> Something like this stresses families, kids, relationships. After Katrina we saw a spike in domestic violence and divorces or break ups. Traumatized kids saw their parents fighting. Substance abuse increases, which leads to fueling domestic violence and crime. It's a ripple effect and the longer it goes on, with unemployment or inability to return to their way of life, the worse things will get. We already have a high stress level now this on top of it. We saw a rise in suicides after Katrina, so people see their families fall apart and unemployment continues, these are real concerns, along with the psychological effects" (Johnson as Quoted in Jamil, 2010, Global Research).

In order to avoid these types of social reactions, proactive and methodical steps need to be taken in order to return affected populations to normalcy, or something as close to normalcy, as soon as possible.

THE SOCIO-ECONOMIC LANDSCAPE

To date, the economic impact of the oil spill and how daily life has been changed has yet to be calculated. However, there are indicators of the impact it continues to have on the local, state (particularly Louisiana) and regional economies. While comprehensive data is still being analyzed regarding the economic impact on the Deepwater Horizon Oil Spill Disaster, the Federal Deepwater Drilling Moratoria and fishery closures have altered the economic landscape, at the very least in the short-run. A Greater New Orleans Inc. study of the initial economic impact revealed that some fishermen were able to benefit from BP payments and other BP related clean-up activities while other fishermen and fishing related industries would be likely to suffer (Greater New Orleans Inc., 2010: p. 7). Moreover, the study continued to cite that this once thriving industry virtually has come to a halt. With the establishment of BP's Vessels of Opportunity[5] program, some vessel owners were able to sustain themselves while other life-long fisherman, not in the program, face grim employment options outside the industry. In data obtained from a focus group of 75 Louisiana fishermen, ". . . BP employment has been lucrative and compelling, provided one was able to obtain a contract. Once these contracts are completed, few fisherman plan to leave the industry." (Greater New Orleans Inc., 2010: p. 7). Thus setting the stage for an environment of "Deep Horizon Winners, "those able to secure lucrative oil skimming contracts, and "Deep Horizon Losers," those who were not able to secure contracts or ultimately collapsed under the economic pressure in an openly hostile economic landscape of uncertainty. For these "Deep Horizon Losers," this could ultimately mean the demise of family businesses, traditional substance life styles and the collapse of hundreds of small businesses along the Gulf Coast. As stated eloquently by Brian B. O'Neil in his testimony before the Senate Committee on the Judiciary while comparing the *Exxon Valdez* oil spill to the Deepwater Horizon spill he stated:

> Fishermen run significant small businesses. Fishing boats cost from $60 thousand to well over a million dollars. In areas of the country where limited permit entries exist, those permits can cost from $5 thousand to $300 thousand. Boats and permits tie up significant capital often financed through bank loans. The capital also provides the fisherman's retirement fund.
>
> An oil spill may prevent a fisherman from catching fish. The fish may hurt the exvessel price of fish as most buyers do not want fish from an oiled fishery. And a spill may drastically devalue boats and permits.
>
> Because a fisher's capital is liquid an interruption in fishing can drive him into bankruptcy in a year. A fisherman needs immediate cash after his fishery is oiled, to live on and to make payments on his boat and permit. Because the affects of oil on a fishers' business may be for years, any financial accounting may have to wait for years. The uncertainties created by a spill can destroy a commercial fishery.

It is within this context that the economy is vulnerable. Yes, the global consumption and dependence on oil is staggering and the world views the 24-7 media coverage of the saga as it unfolded, complete with deep-sea cameras; however, when the media attention has calmed, the real social and economic issues persist for years to come. What was reported during the summer of 2010 regarding the commercial, recreational, tourism and ecological damage to the area has the potential to destroy the long term social and economic fabric of the interlocking sub-economic systems that makes up the patchwork of the disaster landscape across the Gulf Coast.

THE POLITICAL LANDSCAPE

The political landscape of a specific place is not only the byproduct of the top-down politics of the federal government structure, but also a culmination of the unique political experiences that have developed over time in a specific geographic area. Public policy that is created in a specific location or that has implications on a specific location directly affects the lived experiences and sense of place of individuals and social groups that inhabit a place (Miller and Rivera, 2008). Moreover, Miller and Rivera (2008) maintain that the political landscape is both shaped by and shapes the socio-economic and socio-cultural landscapes indicative to the area. When disaster occurs, there are bound to be changes to the political landscape; however, the way in which the political landscape changes is difficult to predict.

Within the political landscape long-run impacts traditionally manifest themselves in the form of federal political debate and litigation. According to Hirch (1997), litigation generally results in the deferment of any payments to victims well into the future. This lengthy process prolongs the "reality disjuncture" that disrupts a community's routines and social network (Kroll-Smith and Couch, 1993). Similar to the political landscape that emerged in the aftermath of Hurricane Katrina, a blame game has manifested itself in reference to the Deepwater Horizon oil spill. Dissimilar to what occurred in the political landscape after Katrina, there is a specific entity that is viewed as the cause of the disaster by all levels of the United States' government, British Petroleum (BP) Global. Because individual victims and the United States government, as a whole, have been able to reach consensus on who is the perpetrator of the disaster, a political landscape has emerged that actively attempts to hold those responsible liable for direct and indirect effects of the disaster. Along these lines, Congress is giving a great deal of attention to the compensatory liability provisions of the Oil Pollution Act, the Jones Act, and the Death on the High Seas Act in order to aid individuals and communities recover from the oil spill. Specifically, the Oil Pollution Act outlines the liability of an oil spill's responsible parties in reference to costs inflicted on others by the spill—cleanup costs and economic damages (i.e. damages to natural resources, destruction of private property, and loss of profits or impairment of earning capacities) (Meltz, 2010). Moreover, according to Meltz (2010), Congress is also investigating various federal statutory provisions provided in different Congressional acts that could be applied to impose civil and criminal penalties against the responsible parties to ensure that similar disasters are avoided in the future.[6]

In the interim of victims receiving compensation by the responsible party, steps are being taken by more locally affected governments to aid their populations' physical, psychological, environmental and economic needs. The House of Representatives passed H.R. 3534 in July, which proposes the formation of a Gulf of Mexico Restoration Program. This program creates a taskforce composed of the governors of the Gulf states and heads of appropriate agencies in order to develop strategies for restoring the natural resources of the Gulf and issue reports every five years (Alexander, 2010). Moreover, the bill requires the Secretary of the Interior to organize baseline studies of the Gulf region in order to gauge natural resource restoration progress. Although this bill specifically attempts to restore natural environmental structures, the bill subsequently affects the livelihoods of those communities and individuals that profit and sometimes subsist in the region off the natural resources, such as fishing and shrimping communities. Moreover, renewable resource communities (Ritchie and Gill, 2010) that inhabit the areas affected by the oil spill have the potential to continue their traditional lifestyles that might otherwise end if natural resources and the ecosystem as a whole are not actively restored.

According to Administration (2010a), prior to the company transferring the operations of handling personal and business claims to the Gulf Coast Claims Facility, BP had reported that it had paid out a total of $368 million. Since the handling of claims was transferred to the Gulf Coast Claims Facility, which was setup to independently process personal and business claims in reference to the oil spill, the facility has reported that as of September 28, it has resolved 14,443 claims totaling $213.3 million (Administration, 2010b). Kenneth Feinberg, who previously administered the September 11th Victim Compensation Fund, was appointed by President Obama to administer the independent Gulf Coast Claims Facility Fund. According to the claims schedule that Mr. Feinberg disclosed to the public, $5 billion from the $20 billion fund would be paid to victimized parties over the second half of 2010 (Administration, 2010c); however, from the numbers disclosed above, it seems that a great deal more progress must be made to meet that goal. Therefore, Congress is investigating other financial methods of stabilizing individuals and businesses in the region.

Sherlock et al. (2010) suggests tax policy is an option that Congress may want to consider in reference to providing relief to victims; however, this option is problematic. In the past, the federal government has responded to natural disasters by providing tax relief measures in order to aid victims with property losses, job displacement, and lost taxpayer records. In reference to the Deepwater Horizon oil spill, however, the disaster is not considered a natural disaster and the nature of the damages inflicted by the oil spill are also different from natural disasters (i.e. damages are in the form of lost business income and employment rather than direct property loss) (Sherlock et al., 2010). In natural disaster situations, there are some Internal Revenue Code (IRC) provisions that are triggered by a Presidential disaster declaration under the Stafford Act. Some of these provisions exclude qualified disaster relief payments from a recipient's gross income when filing taxes. Moreover, other provisions triggered by a Presidential disaster declaration include the postponement of tax deadlines, adjustments in computing alternative minimum taxable incomes, and taking net operating losses into consideration, which all work to aid in facilitating the recovery process of individuals and businesses in disaster affected areas. On April 29, 2010, Homeland Security Secretary Janet Napolitano declared the Gulf oil spill a Spill of National Significance (SONS); however, this does not result in differential tax treatment for payments or losses that arise from the event. The implementation of these tax provisions that could aid in the recovery of the Gulf Coast region in the aftermath of the oil spill, cannot be enacted without a Presidential declaration of disaster (Sherlock et al., 2010), which has not been made by the Obama Administration during the writing of this article. This lack of action to declare the oil spill an official disaster by the Office of the President raises questions in reference to the federal government's commitment to the Gulf Coast's recovery, which also seems to be a reoccurring theme within the political landscape of the Gulf Coast in reference to recovering from natural disasters.

One significant change to the political landscape of the region, and possibly the nation, has been the President's choice to freeze plans to open more offshore oil drilling facilities. As a reaction to high energy prices, Congress had lifted the offshore drilling ban in 2008 in an effort to create jobs and make the United States more independent of international oil supplies. This change in policy by the Obama Administration not only cuts the creation of jobs and oil supplies in the short-term, but also jeopardizes the potential of taking advantage of domestic oil supplies in the long-term. According to White (2010), although this policy decision was made ambivalently in light of the environmental and political consequences of the Deepwater Horizon oil spill, by halting the development of offshore drilling the president has opened opportunities for Cuba to take advantage of oil deposits in the Florida Straits. This would mean the loss of important energy resources and the subsequent economic advantages that come along with them. A silver lining to the decision is the potential for international cooperation and dialogue exchange. According to French (2010), the Deepwater Horizon oil spill offers the opportunity for the United States and Cuba to come together for dialogue, planning and technological exchange resulting in oil development strategies that are more responsible and would, hopefully, avoid the mistakes made by BP. Additionally, this international exchange has the potential of ushering in a new era of foreign relations with Cuba, which has been historically troubled.

FUTURE DIRECTIONS

Since the socio-cultural, socio-economic, and political landscapes interact with one another on a continual basis, it is important to remember that changes in one unquestionably forces changes in others. Because of these landscapes' interaction, no one landscape is more important than the rest, however, it is theoretically easier to make changes to the political landscape than in the others due the occurrence of top-down social change. With this in mind, making policy changes that strengthen oil spill victims' chances of recovering from disasters are needed. As stated above, Congress is taking steps in an effort to dissuade companies from practicing drilling techniques that can have severe consequences for the environment and society. By financially holding those parties responsible for the contamination that occurred, companies will be less likely to engage in risky practices. Moreover, if the presidents of companies that contaminate the environment also face the possibility of being served criminal penalties, as was mentioned above, decision making individuals may enact organizational change that would result in business policies that are more environmentally aware and disaster resistant.

Although this is unlikely to occur due to legal maneuvering on the behalf of company leaders, a more practical consideration may be for Congress to encourage or enforce the global offshore energy insurance market to participate in the oil spill financial responsibility program. Participation in such a program could lead to higher limits of liability on responsible parties for oil spills, but the corresponding insurance requirement could lead to the domination of drilling activities by major oil companies because many smaller oil firms and investors would be unable or unwilling to expose themselves to such liabilities (King, 2010). Program participation along these lines would be very difficult for relatively smaller businesses to accept; however, it would mean that those companies that do engage in drilling practices could be held more directly responsible for technological failures and subsequently liable for consequences associated with technological disasters.

Complications within the political landscape have the potential of contributing to the development of corrosive communities. Although BP Global seems to be facilitating methods of compensating individuals, businesses, and communities that have been affected by the oil spill, actions not taken by the President have the potential of contributing to social dynamics that result in non-therapeutic community situations. The President's decision not to make a presidential disaster declaration has resulted in a lack of economic relief options for individual and business victims. Consideration should be taken in reference to the notion of the president sharing the ability to make disaster declarations with the Secretary of Homeland Security or even with the Congress in order to avoid the influence of politics on a president's choice to issue a disaster declaration (Sylves and Búzás, 2007). Moreover, it has the potential to create situations of "haves and have nots" in reference to economic resilience from disasters that disproportionately benefits more affluent individuals and communities. Obama's actions in reference to the disaster, although seemingly proactive in holding BP responsible for the disaster, can be viewed as almost dismissive in reference to recovery when viewed through the perspective of those most socially vulnerable to the affects of the Deepwater Horizon disaster. These sentiments were illustrated in a poll administered to 492 Louisiana voters between June 12th and 13th, 2010 by Public Policy Polling. At the state level of government, 65 percent of voters reported that they approved of the way Governor Jindal had handled the aftermath of the disaster up to that point; however, when asked the same question in reference to President Obama, 62 percent reported their disapproval (Public Policy Polling, 2010).

NOTES

1. This research has been generously funded by The Natural Hazards Center at the University of Colorado at Boulder and the Public Entity Risk Institute with support from the National Science Foundation and Swiss Re and The Horowitz Foundation for Social Policy.
2. At the time of the writing of this chapter, the direct oil spill has been contained; however, oil continues to wash ashore in the coastal communities so that the economic, social and cultural impacts are beginning to surface more clearly and are being reported in local, national and international newspapers.
3. Ultimately, the nation learned that a natural gas explosion of April 20th resulted in eleven deaths (Gordon Jones, Roy Wyatt Kemp, Karl Kleppinger, Jr., Keith Blair Manuel, Shane Roshto, Adam Weise, Aaron Dale Burkeen, Jason Anderson, Donald Clark, Dewey Revette, and Stephen Ray Curtis).
4. For a more detailed synopsis of the two decades-long struggle to regain normalcy in Alaska villages and coastal communities in the aftermath of the *Exxon Valdez* oil spill,

see *Brief Amici Curiae of Sociologists, Psychologists, and Law and Economics Scholars in Support of Respondents at Exxon Shipping Co. et al. v. Baker et al., (No. 07-219), 2008.*

5. BP has established the Vessels of Opportunity program to obtain additional resources to support the oil spill response. Vessels were required to be commercial or charter fishing boats, registered in the state in they would work as a commercial vessel prior to March 31, 2010. (see http://www.flseagrant.org/index.php?option=com_content&view=article&id=224&Itemid=103).
6. Some statutes being investigated to impose civil and criminal penalties on responsible parties include those written in the Clean Water Act the Endangered Species Act, the Marine Mammal Protection Act, the Migratory Bird Treaty Act, the Oil Pollution Act, the Outer Continental Shelf Lands Act, and the Occupational Safety and Health Act.

REFERENCES

Administration. (2010a). "Panama City Hires Attorney for BP Claims Assistance." BP Claims Facility Help. September 29. http://www.bpclaimshelp.net/. Retrieved 10/26/10.

——— (2010b). "New Gulf Coast Claims Facility Opens Its Doors." BP Claims Facility Help. August 23. http://www.bpclaimshelp.net/. Retrieved 10/26/10.

——— (2010c). "About the $20 Billion Dollar Fund for Oil Spill Victims." BP Fund. July 21. http://www.thebpfund.com/. Retrieved 10/26/10.

Alexander, K. (2010). *The 2010 Oil Spill: Natural Resource Damage Assessment Under the Oil Pollution Act.* CRS Report Prepared for Members and Committees of Congress. Washington, D.C.: Congressional Review Service. September 8.

Arata, C.M., Picou, J.S., Johnson, G.D. and McNally, T.S. (2000). "Coping with Technological Disaster: An Application of the Conservation of Resources Model to the *Exxon Valdez* Oil Spill," *Journal of Traumatic Stress* 13:23-39.

Associated Press. (2010a). "BP using undersea robots to try to plug Gulf oil leak." The Guardian. 14 May., available: http://www.guardian.co.uk/environment/2010/may/14/bp-robots-gulf-oil-leak. Retrieved 10/26/10.

Associated Press. (2010b). "Deepwater Horizon oil rig fire leaves 11 missing." The Guardian., 21 April, available at: http://www.guardian.co.uk/world/2010/apr/21/deepwater-horizon-oil-rig-fire. Retrieved 10/26/10.

Birkland, T.A. and Lawrence, R.G. (2002). The Social and Political Meaning of the *Exxon Valdez* Oil Spill. *Spill Science & Technology Bulletin*, 7 (1–2) 17-22.

Brown, B.B. and Perkins, D.D. (1992). Disruptions in place attachment. In I. Altman and S.M. Low (Eds), *Place attachment.* New York Plenum Press, pp. 279-304.

Brief *Amici Curiae* of Sociologists, Psychologists, and Law and Economics Scholars in Support of Respondents at *Exxon Shipping Co. et al. v. Baker et al.*, (No. 07-219), (2008).

Clark, A. (2010). "Gulf oil spill: 'Top kill' mission to halt gush under way." The Guardian 27 May., available at: http://www.guardian.co.uk/environment/2010/may/27/gulf-oil-spill-top-kill. Retrieved 11/2/10.

Edelstein, M. (1988 [2004]). *Contaminated communities: The social and psychological impacts of residential toxic exposure*. Boulder, CO: Westview Press.

Eyles, J. (1989). The geography of everyday life. In D. Gregory and R. Walford (Eds.), *Horizons in human geography*. Totowa, NJ: Barnes & Noble, pp. 102-17).

Erikson, K.T. (1994). A new species of trouble. New York, NY: Norton.

Fried, M. (1963). Grieving for a lost home. In L.J. Duhl (Ed.), *The urban condition*. New York: Basic Books, pp. 151-71.

Fried, M. (1966). Grieving for a lost home: Psychological costs of relocation. In J. Q. Wilson (Ed.), *Urban renewal the record and the controversy*. Cambridge, MA: MIT Press, pp. 359-79).

French, A.L. (2010). *U.S.-Cuba Engagement in the Gulf: Lessons from the Deepwater Horizon Oil Spill*. Discussion held by the New America Foundation. May 26. http://www.newamerica.net/events/2010/us_cuba_engagement_in_the _gulf. Retrieved 10/26/10.

Freudenburg, W.R. (1997). "Contamination, Corrosion and the Social Order: An Overview." *Current Sociology*, 45(3), pp. 19-39.

Gabbatt, A. (2010). "BP oil spill mostly cleaned up, says US." The Guardian 4 August., available at: http://www.guardian.co.uk/environment/2010/aug/04/bp-static-kill-successful. Retrieved 11/26/10.

Guardian Research. (10/28/2010) BP oil spill timeline. http://www.guardian.co.uk/environment/2010/jun/29/bp-oil-spill-timeline-deepwater-horizon. Retrieved 10/26/10.

Gill, D.A. and Picou, J.S. (1991). "The Social Psychological Impacts of a Technological Accident: Collective Stress and Perceived Health Risks." *Journal of Hazardous Materials*, 27, 77-89.

Gill, D.A. and Picou, J.S.. (1997). "The Day the Water Died: Cultural Impacts of the Exxon Valdez Oil Spill." In J S. Picou, D.A. Gill and M.J. Cohen. (Eds.) *The Exxon Valdez Disaster: Readings on a Modern Social Problem*. pp. 167-192. Dubuque, IA: Kendall-Hunt Publishing Company.

Goldenberg, Suzanne. (2010d). "BP begins new operation to seal off leaking Deepwater Horizon well." *The Guardian* 11 July., available at: http://www.guardian.co.uk/environment/2010/jul/11/bp-deepwater-horizon-well-oil. Retrieved 11/4/10.

Goldenberg, Suzanne.(2010a). "Deepwater Horizon clean-up workers fight to prevent disaster." The Guardian, 23 April, available at: http://www.guardian.co. uk/world/2010/apr/23/deepwater-horizon-oil-rig-gulf. Retrieved 11/1/10.

Goldenberg, Suzanne (2010b). "Deepwater Horizon oil spill: Underwater robots trying to seal well." The Guardian April 26., available at: http://www.guardian.

co.uk/environment/2010/apr/26/deepwater-horizon-spill-underwater-robots. Retrieved 1/1/10.

Goldenberg, Suzanne. (2010c). "Gulf oil spill: firms ignored warning signs before blast, inquiry hears." The Guardian, 13 May., available at: http://www.guardian.co.uk/environment/2010/may/12/deepwater-gulf-oil-spill-hearing. Retrieved 10/26/10.

Greater New Orleans, Inc. (2010). A study of the economic impact of the Deepwater Horizon oil spill. available at: http.//www.gnoinc.org/.../Economic%20Impact% 20Study,%20Part%20I%20-%20Executive%20Summary.pdf. Retrieved 11/6/10.

Gurevitch, M., Levy, M.R., 1985. Preface. In M. Gurevitch and M.R. Levy (Eds.), *Mass Communication Review Yearbook*. Beverly Hills: Sage.

Hamerton, P.G. (1885). *Landscape*. Boston, MA: Roberts Bros.

Hill, G. (1979). "Ixtoc's Oil has a Silver Lining." *Audubon*, 18(6), pp. 150-59.

Hirch, W.B. (1997). "Justice delayed: Seven years and no end in sight." In J.S. Picou, D.A. Gill, & M.J. Cohen (Eds.), *The Exxon Valdez Disaster: Readings on a Modern Social Problem*. (pp. 271-307). Dubuque, IA: Kendall/Hunt.

Huigen, P.P., and Meijering, L. (2005). Making places: A story of DeVenen. In G. J. Ashworth & Graham, B. (Eds.), *Senses of place: Senses of time* (pp. 19-30). Burlington, VT: Ashgate Publishing Co.

Hummon, D.M. (1992). Community Attachment: Local sentiment and sense of Place. In I. Altman and S. M. Low (Eds.), *Place attachment*, 253-278. New York: Premium Press.

Jamail, Dahr. (2010). "Gulf residents likely face decaded of psychological impacts from BP's oil disaster. *Global Research*. August 6, 2010. http://www.globalresearch.ca/ index.php?context=va&aid=20504. Accessed November 2, 2010.

Jernelöv, A. (2010). "The threats from oil spill: Now, then, and in the future." *Ambio*. 39, pp. 353-66.

Jernelöv, A. and Lindén, O. (1981). "Ixtoc I: A case study of the world's largest oil spill." *Ambio*, 10, pp. 299-306.

Keeble, J. (1991). *Out of the channel: The Exxon Valdez oil spill in Prince William Sound*. New York, New York: Harper Collins Publishers.

King, R.O. (2010). *Deepwater Horizon Oil Spill disaster: Risk, recovery, and insurance implications*. CRS Report Prepared for Members and Committees of Congress. Washington, D.C.: Congressional Review Service. July 12.

Kroll-Smith J.S. and Couch, S.R. (1993). "Technological Hazards: Social Responses as Traumatic Stressors." In J.P. Wilson and B. Raphael (Eds.). *International Handbook of Traumatic Stress Syndromes*. New York, NY: Plenum Press, pp. 79-91.

Kyvig, D.E., and Marty, M.A. (2000). *Nearby history: Exploring the past around you* (second edition). Lanham, MD: Rowman & Littlefield Publishers, Inc.

Maag, C. (2008). "Supreme court decision on *Exxon Valdez* damages a blow to Alaskans." 2 July. Available at: http://www.nytimes.com/2008/06/26/world/americas/26iht-alaska.4.14027236.html. Retrieved 11/6/10.

McGreal, C. (2010). "Five months after BP oil rig disaster, US government declares well 'dead'." *The Guardian*. 20 September. Available at: http://www.guardian.co.uk/environment/2010/sep/20/us-oil-spill-fixed. Retrieved 11/4/10.
McVeigh, K.(2010)."Containment dome reaches Gulf oil spill scene" *The Guardian*. 6 May, available at: http://www.guardian.co.uk/environment/2010/may/06/gulf-oil-spill-containment-dome. Retrieved 11/2/10.
Meikle, J. (2010). "BP oil spill: new cap 'successfully installed' on leaking well." The Guardian, 13 July., available at: http://www.guardian.co.uk/environment/2010/jul/13/bp-oil-spill-containment-cap. Retrieved 10/29/10.
Meltz, R. (2010). *Federal civil and Criminal Penalties Possible Application to Parties Responsible for the Gulf of Mexico Oil Spill*. CRS Report Prepared for Members and Committees of Congress. Washington, D.C.: Congressional Review Service. August 16.
Marshall, B.K., Picou, J.S. and Schlichtmann, J.R. (2004). Technological disasters, litigation stress, and the use of alternative dispute resolution mechanisms. *Law & Policy*. 26 (2), pp. 289–307.
Miglierini, J. (2010). "Derrame: las lecciones de México." BBC Mundo. Junio 16. http://www.bbc.co.uk/mundo/internacional/2010/06/100615_derrame_especial_campeche.shtml. Retrieved 11/3/10.
Miller, D.S. (2006a). *The aesthetic value of landscapes and place orientation after a natural disaster*. A paper presentation at the Pacific Sociological Association, Universal City, Los Angeles, CA.
Miller, D. (1997). *The Construction of Competing Disaster Narratives: Media Coverage of the Distribution of Risk Following a Technological Disaster*. A Doctorial Dissertation. Mississippi State, Mississippi.
Miller, D.S. and Rivera, J.D. (2007). "Landscapes of Disaster and Place Orientation in the Aftermath of Hurricane Katrina." In *The Sociology of Katrina: Perspectives on a Modern Catastrophe*. Eds. D. L. Brunsma, D. Overfelt, and J. S. Picou. (Eds). Rowman & Littlefield Publishers, Inc. Lanham, MD., pp. 141-54.
Miller, D. S. and Rivera, J. D. (2008). *Hurricane Katrina and the Redefinition of Landscape*. Lanham, MD: Lexington Books.
Milligan, M.J. (1998). "Interactional past and potential: The social construction of place attachment." *Symbolic Interaction*. 21(1). 1-33.
O'Neil, B.B. (2101). Hearing of the Senate Committee on the Judiciary: "Exxon Valdez to Deepwater Horizon: protecting Victims of Major Oil Spills." Tuesday, July 27, 2010 at 2:30 p.m. Room 226 of the Dirksen Senate Office Building. Washington, D.C.
Pelling, M. (2003). *The vulnerability of cities: Natural disasters and social resilience*. Sterling, VA: Earthscan Publications Ltd.
Picou, J.S. and Gill, D.A. (2000). "The Exxon Valdez Disaster as Localized Environmental Catastrophe: Dissimilarities to Risk Society Theory." In M. Cohen (Ed.). *Risk in the Modern World*. London, UK: Macmillan, pp. 143-70.

Picou, J. S., Brunsma, D. L. and Overfelt, D. (2010). "Katrina as Paradigm Shift: Reflections on Disaster Research in the Twenty-First Century." In D. L. Brunsma, D. Overfelt and J. S. Picou (Eds.). *The Sociology of Katrina: Perspectives on a Modern Catastrophe*. (pp. 1-21) New York, NY: Rowman & Littlefield Publishers, Inc.

Picou, J.S., and Gill, D.A. (1996). "The *Exxon Valdez* Oil Spill and Chronic Psychological Stress." In *Proceedings of the Exxon Valdez Oil Spill Symposium*, edited by S. D. Rice, R. B. Spies, D. A. Wolfe and B. A. Wright. Bethesda, M.D.: American Fisheries Society.

Picou, J.S., Gill, D.A. and Cohen, M.J. (Eds.) (1997). *The Exxon Valdez Disaster: Readings on a Modern Social Problem*. Dubuque, Iowa: Kendall-Hunt Publishing Co.

Picou, J.S., Gill, D.A., Dyer, C.L. and Curry, E.W. (1992). "Stress and Disruption in an Alaskan Fishing Community: Initial and Continuing Impacts of the *Exxon Valdez* Oil Spill," *Industrial Crisis Quarterly*, 6: 235–57.

Picou, J.S., Marshall, B.K. and Gill, D.A. (2004). "Disaster, Litigation and the Corrosive Community," *Social Forces*, 82, 1497–526.

Public Policy Polling. (2010). Louisianans Support Off-Shore Drilling. Press Release of the Public Policy Polling. http://www.publicpolicypolling.com/pdf/PPP_Release_LA_615.pdf. Retrieved 10/26/10.

Riley, R.B. (1992). Attachment to the ordinary landscape. In I. Altman and S.M. Low (Eds.), *Place attachment*. New York: Plenum Press, pp. 13-36..

Ritchie, L.A. and Gill, D.A. (2010). "Fostering Resiliency in Renewable Resource Communities: Subsistence Lifescapes and Social Capital." In J.D. Rivera and D.S. Miller (Eds.). *How Ethnically Marginalized Americans Cope with Catastrophic Disasters: Studies in Suffering and Resiliency*. Lewiston, NY: The Edwin Mellen Press, pp. 51-81.

Read, P. (1996). *Returning to nothing: The meaning of lost places*. New York: Cambridge University Press.

Schattschneider, E.E. (1960/1975). In: *The semisovereign people*. The Dryden Press, Hinsdale, Ill,

Sherlock, M.F., Lunder, E.K., Liu, E.C. and Klein, H.S. (2010). *Tax Issues and the Gulf of Mexico Oil Spill: Legal Analysis of Payments and Tax Relief Policy Options*. CRS Report Prepared for Members and Committees of Congress. Washington, D.C.: Congressional Review Service. July 15.

Saito, Y. (1998). Appreciating nature on its own terms. *Environmental Ethics, 20,* 135-49.

Stokowski, P.A. 2002. Languages of place and discourses of power: Constructing new senses of place. *Journal of Leisure Research,* 43(4), 368-382.

Sylves, R. and Búzás, Z. (2007). "Presidential Disaster Declaration Decisions, 1953-2003: What Influences Odds of Approval?" *State & Local Government Review*, 39(1), 3-15.

Thomashow, M. (1995). *Ecological Identity*. Massachusetts: Massachusetts Institute of Technology.

Trenrelman, C.K. (2009). Place attachment and community attachment: A premier grounded in the lived experience of a community sociologist. *Society & Natural Resources*. 22(3), 191-210

Urbina, I. (2010). "U.S. Said to Allow Drilling Without Needed Permits." The New York Times. 13 May, Available at: http://www.nytimes.com/2010/05/14 /us/14agency.html?_r=1.\. Retrieved 10/29/10.

Wearden, G. 2010. "Deepwater Horizon worker claims oil rig leaking weeks before explosion." *The Guardian*. 21 June. Available at: http://www.guardian.co.uk/ environment/2010/jun/21/bp-oil-spill-deepwater-horizon-leak Retrieved 10/29/10.

Webb T. (2010a). "BP boss admits job on the line over Gulf oil spill." *The Guardian* 14 May, available at: http://www.guardian.co.uk/business/ 2010/may/13/bp-boss-admits-mistakes-gulf-oil-spill. Retrieved 11/2/10.

Webb T (2010b). "Scale of BP oil leak revised up to 40,000 barrels a day." *The Guardian* 12 June, available at: http://www.guardian.co.uk/environment/ 2010/jun/12/bp-oil-spill-gulf-mexico. Retrieved 10/26/10.

Webb, T. (2010c). "BP offers one-off payouts to stem Gulf oil spill lawsuits." *The Guardian*.1 August, Available at: http://www.guardian.co.uk/environment/ 2010/ aug/01/bp-gulf-oil-spill-compensation. Retrieved 11/2/10.

Webb, T. (2010d). "BP 'static kill' due in next 24 hours." The Guardian, 2 August, available at: http://www.guardian.co.uk/environment/2010/aug/02/bp-static-kill-set-for-next-24-hours. Retrieved 10/26/10.

White, J.P. (2010). "Drilling in Ecologically and Environmentally Troubled Waters: Law and Policy Concerns Surrounding Development of Oil Resources in the Florida Straights." Colorado Journal of International *Environmental Law & Policy*, 21, pp. 557-95.

Williams, D.R. (2000). Personal and social meanings of wilderness: Constructing and contesting places in a global village. In Personal, societal and ecological values of wilderness: Sixth World Wilderness Congress Proceedings on Research, Management and Allocation, 11, 24-29, October 1998, Bangalore, India, 77-82. Proceedings RMRS-P-14. Ogden, UT: USDA Forest Service, Rocky Mountain Research Station.

Winston and Strawn. (2008). "Supreme Court's Landmark Decision in the *Exxon Valdez* Case." 27 June. Available at : http://www.winston.com/siteFiles/ publications/Supreme_Court_Decision_Exxon_Valdez.pdf. Retrieved 10/26/10.

Wray, R. (2010). "BP shares top risers as engineers assess latest oil spill Operation." The Guardian 4 June. Available at: http://www.guardian.co. uk/business/2010/ jun/04/bp-shares-rise-oil-spill-cap. Retrieved 10/26/10.

CHAPTER NINE

THE CRUDE AWAKENING: GULF COAST RESIDENTS REFLECT ON THE BP OIL SPILL AND THE 2010 HURRICANE SEASON

MICHELLE MEYER LUECK AND LORI PEEK

ABSTRACT

June 1, 2010 marked the official opening of the annual Atlantic hurricane season. Atmospheric scientists predicted that the season would be "above average" in terms of hurricane activity, with a forecast of over 10 named storms throughout 2010. How would residents cope with this latest crisis? As speculation continued on the potential effects of a hurricane with the oil spill, coastal residents began preparations for a hurricane season complicated by a seemingly unstoppable spill.

Just three weeks after the start of the hurricane season, in mid-June of 2010, our research team launched the first wave of a three-year panel survey on risk perception and hurricane evacuation decision-making processes and behaviors. The survey was mailed to a spatially random sample of individuals living within 10 miles of the Gulf or Atlantic Coasts, stretching from Texas to North Carolina. We dedicated half of the final page of the survey to an open-ended question eliciting written comments from participants regarding their opinions about the interaction of hurricanes and the oil spill. Of the 629 total survey respondents, 270 offered their thoughts on the oil spill and the potential for a "na-tech" disaster if a hurricane were to occur.

This chapter presents the views and perspectives of individuals who were experiencing an unfolding technological disaster while facing an impending natural disaster. To our knowledge, this is the first study of a potential, but as yet unrealized, na-tech disaster. We compare and contrast the respondents' thoughts on uncertainty, lack of control, blame, and their emotional responses for the oil spill and hurricanes, and then discuss their concerns of a potential na-tech disaster if a hurricane would have interacted with the spill.

This research was supported by a grant from the National Science Foundation and the National Oceanic and Atmospheric Agency (Grant# CMMI-0856227). The authors would like to thank Craig Trumbo, Brian McNoldy, Alyssa Dawson, Emily McCormick, and Holly Marlatt.

INTRODUCTION

> It turns out that a big powerful hurricane ironically is probably less damaging with respect to the oil spill because it just disperses everything and the oil breaks up and degrades more quickly. It's those tropical storms and tides that would just wash stuff into the marshes that would be an ecological disaster. President Barack Obama, June 3, 2010.

Disaster researchers often categorize disasters as "natural" (e.g., hurricanes, floods, tornados, and tsunamis) or "technological" (e.g., oil spills, chemical releases, pollution, and explosions) (Fischer, 2008: 14). Natural disasters, with atmospheric or geophysical origins, are perceived by the general public as "acts of God" and thus are viewed as largely beyond human control. On the other hand, technological disasters clearly result from human action or, in some cases, inaction. Because technological disasters typically involve human error and/or the breakdown of technological or industrial systems, they are presumed preventable with the appropriate regulations and safeguards (Picou and Marshall, 2007).

The official opening of the Atlantic hurricane season on June 1, 2010 created an especially challenging circumstance for individuals along the Gulf Coast—they faced an unfolding technological disaster as a result of the BP Deepwater Horizon oil spill *and* the potential threat of a natural disaster. To make matters more uncertain, atmospheric scientists predicted that the hurricane season would be "above average" in terms of hurricane activity, with a forecast of over 10 named storms throughout 2010. The arrival of the hurricane season and the dire predictions raised a plethora of concerns and intensified the already urgent need to stop the gushing oil spill. As speculation continued, individuals along the coast began preparations for a hurricane season complicated by the seemingly unstoppable spill.

In this chapter, we analyze the ways that coastal residents framed the threat of "natural" versus "technological" disasters as they were preparing for the 2010 hurricane season. The chapter is organized in four sections. First, we briefly review the literature detailing the relationship between natural, technological, and na-tech disasters. Second, we describe our methodological approach and sample characteristics. Third, we present the results of our analysis of 270 responses to an open-ended survey question regarding the BP oil spill and potential hurricane risk. Fourth, we discuss the findings and consider the methodological and policy implications of this research.

LITERATURE REVIEW

Natural and technological disasters differ in their perceived underlying cause. More importantly, however, they also differ in terms of the types of damage they cause, the time frames in which they unfold, and the individual and community impacts they engender (Picou, Gill, and Cohen, 1997). Most natural disasters cause destruction to the built and natural environments; are confined in time and space, with a relatively clear beginning and end; and generate societal consensus that a disaster happened, that there are "legitimate," identifiable victims, and that rescue, restoration, and recovery should be automatically supported by government and community organizations (Picou and Marshall, 2007: 4). Commonly viewed as short-term disruptions, natural disasters encourage a "therapeutic" or "altruistic" community response, which fosters timely recovery and an "amplified rebound" to normalcy (Barton, 1969; Fritz, 1961).

Research on the *Exxon Valdez* oil spill (Alaska, 1989), the Three-Mile Island nuclear accident (Pennsylvania, 1979), contamination of Love Canal (New York, 1978), the Buffalo Creek dam burst (West Virginia, 1972), and other events suggests that technological disasters, in contrast to natural disasters, generate extreme and prolonged psychological problems for affected individuals. In a comprehensive review of disaster mental health research, Norris et al. (2002) found that those affected by technological disasters experience more severe and extended social, physical, and mental health issues than natural disaster survivors. Individuals affected by technological disaster often exhibit more anger, blame, chronic stress, and loss of trust in societal institutions (Erikson, 1994). This psychological stress is comparable to loss of a loved one and technological disaster victims' avoidance behaviors can mimic the psychological responses of rape victims (Gill and Picou, 1998).

Unlike the general natural disaster timeline of warning, impact, response, and recovery, technological disasters "trap portions of the population in the warning, threat, and impact stages" (Kroll-Smith and Couch 1993: 81). For technological disaster victims, there is no clear "beginning" and "end" to the disaster. This throws into question whether recovery and restoration will ever occur for populations affected by technological disasters (Picou and Marshall, 2007), thus further intensifying the negative psychological effects of such events.

Gill and Picou (1998) describe four characteristics of technological disasters that contribute to chronic individual and community stress: (1) responsibility and loss of control, (2) uncertainty, (3) construction of "victimhood," and (4) the corrosive community effect. First, as a consequence of human error or industrial failings, technological disasters "involve a loss of control over processes perceived to be controllable" (Gill and Picou, 1998: 796). Unlike natural disasters, which are generally perceived as uncontrollable "acts of God" or caprices of nature, technological disasters are *believed* to be controllable and preventable. Therefore, when technological disasters occur, the resultant loss of control leads to cycles of blame and, often, protracted and costly litigation (Picou, et al., 1997).

Second, those who have endured a technological disaster often report high levels of uncertainty regarding the short- and long-term effects of the event (Foster, Branovan, and Ukrainsky, 2003). In many cases, the invisibility of the contamination or pollution induces extreme fear of exposure for affected populations (Erikson, 1994). In contrast to natural disasters, technological disasters often spare the built environment but contaminate the biosphere and create health concerns. Damages from technological disasters are more difficult to see and to assess than the human and property tolls from a natural disaster. For example, no homes or lives were lost in the 1985 underground petroleum leak in Fort Collins, Colorado, but the levels of dread and anxiety reported by the population were so severe that many members of the affected community sought to sell their homes and vacate the area (Erikson, 1994).

Third, the uncertainty surrounding the effects of technological disasters prevents individuals from being perceived as "worthy victims" in the same way as those affected by natural disasters. The uncertainty, lack of clear or measurable impacts, and disputes over responsibility limits the ability of the community as a whole to construct a single definition of the situation. Thus, technological disasters become "contested disasters," characterized by a lack of consensus about who is a victim and how much individuals and communities have actually been affected (Picou and Marshall, 2007).

Fourth, researchers have labeled the response following technological disasters the "corrosive community" to contrast this effect with the therapeutic community that follows natural disasters (Kroll-Smith and Couch, 1990). The corrosive community is characterized by conflict and a breakdown of social relationships while the therapeutic community involves altruistic and other-focused behavior. Debate about impacts, compensation, and litigation causes friction between community members that can continue for years and decades following the event.

NA-TECH DISASTERS

When the annual Atlantic hurricane season began, the BP Deepwater Horizon oil well had poured nearly two million barrels of oil into the Gulf of Mexico. By that time, the oil spill had already earned the dubious distinction of being the largest in United States history and had eclipsed the amount released in the *Exxon Valdez* disaster by nearly eight times. The interaction of a hurricane and the oil spill created the potential for what Showalter and Myers (1994) first termed natural-technological or "na-tech" disasters. This paradigm recognizes that many disasters are a hybrid of nature and technology, with a natural catalyst (e.g., Hurricane

Katrina) directly or indirectly causing technological failures (e.g., the levee failures in New Orleans and the oil and chemical releases along the Gulf Coast).

The likelihood of na-tech disasters has grown significantly as a consequence of four interactive factors: population growth, infrastructure development in hazardous areas, increased reliance on technology, and the rising occurrence of natural disasters. Na-tech disasters produce contestation over whether the "real" cause of the event was an act of God, as in natural disasters, or human-induced, as in technological disasters, and thus whether there is someone specific to blame. Na-tech disasters also induce similar psychological impacts as "classic" technological disasters with elevated stress over the uncertainty of impacts, increased dread and despair, and potential litigation processes (Picou and Marshall, 2007). Na-tech disasters often create new conflicts as well, such as those related to insurance coverage. As observed following Hurricane Katrina, some insurance companies would only cover impacts from the hurricane (wind), not the flooding (water) caused by the failure of the levee system surrounding New Orleans.

Given their growing prevalence and theoretical significance, more research has been dedicated to na-tech disasters over the past decade, and many researchers maintain that this type of event will represent a large proportion of future disasters (Erikson, 1994; Freudenburg, Gramling, Laska, and Erikson, 2009; Picou and Marshall, 2007). Yet, most available na-tech disaster studies have been conducted in the disaster aftermath. This chapter, which draws on open-ended survey responses from coastal residents, represents the views and perspectives of individuals who were experiencing an unfolding technological disaster while faced with an impending natural disaster. To our knowledge, this represents the first study of persons positioned to be affected by a potential, but unrealized, na-tech disaster. Understanding these perspectives represents an important contribution to the literature, as it sheds light on how coastal residents comprehend and interpret risk to separate, but potentially interactive, hazards events.

Although, to date, little scholarly attention has focused on how coastal residents perceived their risk to the BP oil spill and hurricanes, the National Oceanic and Atmospheric Administration (NOAA) (2010) dedicated a special report to the possible interaction between the two hazards agents. According to NOAA, while the oil would not affect the track, speed, or magnitude of a hurricane, high winds could have spread the oil over a larger area of land dependent upon the track of the hurricane. Storm surge was also predicted to bring oil further inland, and debris from a hurricane could have become contaminated with oil. Furthermore, NOAA underscored that the rain and wind accompanying any given hurricane would not contain oil, as the water vapor in a storm is drawn from a large area and produced with the clouds in a storm.

METHODS AND SAMPLE

Just three weeks after the start of the hurricane season, in mid-June of 2010, our research team launched the first wave of a three-year panel survey on risk perception and hurricane evacuation decision-making processes and behaviors. The survey was mailed to a spatially random sample of individuals living within 10 miles of the Gulf or Atlantic Coasts, stretching from Brownsville, Texas to Wilmington, North Carolina. The 8-page survey included predominantly close-ended questions designed to gather information on, for example, household characteristics, general levels of optimism, emotional responses to hurricanes, prior evacuation behavior, lifetime hurricane experience, and perceptions of community resilience. In addition to the close-ended questions, nearly half of the final page of the survey was dedicated to an open-ended question designed to elicit written comments from participants regarding their opinions about the interaction of hurricanes and the oil spill. The question appeared in a large text box and read as follows:

> *BP Gulf Oil Spill.* This study was planned prior to the current events in the Gulf of Mexico. There has been some discussion of how the oil spill may interact with any Gulf hurricanes this season. If you have any thoughts you would like to share on this topic, please write them here. Feel free to include an additional sheet if you like.

The entire survey was completed and returned by 629 respondents before the well was sealed in September 2010, with 270 individuals (43 percent) providing hand-written comments about the oil spill. Given the focus of this edited volume and our interest in residents' perceptions of the interaction between natural and technological disasters, the present chapter draws on the open-ended responses to the oil spill question.

Of those 270 individuals who answered the oil spill question, 66 percent lived along the Gulf Coast (see Figure 9.1). Almost half lived in Florida, which is not surprising given the length of Florida's coastline (see Table 9.1). The gender distribution was nearly balanced: 56 percent were male and 44 percent were female. The age of respondents ranged from twenty-one to ninety-one years, with a mean of sixty years. Over half of the respondents were married (62 percent) and a majority was White (91 percent). For education, over a quarter of respondents (26 percent) reported completing graduate or professional school with approximately 20 percent reporting having a high school degree or less. The mode for education was some college or technical school. Household income was measured on a nine-point scale with the lowest bin less than $10,000 and the highest bin greater than $80,000. A majority of the respondents reported household incomes less than $60,000 annually. Forty-four percent of the respondents were employed full-time, while a sizeable minority (33 percent) was retired. Almost nine out of ten (88 percent) of the respondents owned their residence rather than renting. On average, the respondents had lived within 20 miles of their current residence for twenty-four years and had experienced five hurricane landfalls.

Figure 9.1: Location of Respondents to the Open-ended Survey Question

The 270 open-ended responses to the oil spill question ranged in length from a few words to several paragraphs. Most respondents offered three to four sentences of text. All hand-written responses, including words and punctuation, were re-typed verbatim in a word processing program; typographical errors, capitalization, underlining, and other features of the responses were maintained in the transcripts. Once all of the data were inputted into computer format, the transcripts were uploaded into Atlas.ti, which is a qualitative data analysis software program, and coded simultaneously by both authors.

We drew on Saldaña's (2009) first cycle and second cycle coding method to analyze the qualitative data. Specifically, during the first cycle, or initial stage, of coding, we read each response and assigned a code to all words/short phrases. After completing the first cycle of coding, we compared and refined our codes, and then moved into the second cycle of coding, where we developed more general categories to capture the emergent themes and patterns in the data.

Table 9.1: Background Characteristics of Respondents

	Oil Spill Question n = 270
percent Gulf Coast	65.9
percent State	
Texas	11.1
Louisiana	16.3
Mississippi	4.8
Alabama	4.1
Georgia	3.0
Florida	45.2
South Carolina	8.9
North Carolina	5.6
percent Male	55.9
percent White	90.7
Age (mean)	59.5 [13.3]
percent Married	61.9
percent Education completed	
High school diploma or less	19.6
Some college/Technical school	33.0
Undergraduate degree	21.1
Graduate or professional degree	26.3
percent Employment	
Full-time	44.4
Part-time	5.2
Retired	33.0
Not employed	10.4
percent Annual household income	
Less than $20,000	13.4
$20,000 - $39,999	17.1
$40,000 - $59,999	27.8
$60,000 - $79,999	16.7
$80,000 or greater	25.2
percent Homeowners	87.8
Coastal tenure (mean years)	24.4 [18.8]
Hurricane landfalls experienced (mean)	4.8 [4.8]

*Note: Standard deviations appear in brackets.

RESULTS

The survey respondents, who were accustomed to the annual threat of hurricanes, expressed a variety of feelings and opinions about the BP oil spill. Their responses highlight key perceived differences between natural and technological disasters in general, as well as between hurricanes and the BP oil spill in particular. Below, we discuss how respondents framed the oil spill as compared to hurricanes by addressing their feelings of control, the uncertainty of impacts, the long-term effects of the hazards, placement of blame, and emotional responses. Then, we detail the respondents' concerns about a potential na-tech disaster, specifically their predictions of general and specific impacts.

NATURAL VERSUS TECHNOLOGICAL DISASTERS

Respondents generally cast hurricanes and the oil spill in dichotomous and oppositional terms. This theme was especially apparent in how respondents portrayed the source, or agent, driving the disaster. Specifically, hurricanes were depicted as "natural," "regular" occurrences. "Hurricanes are part of nature—we learn to live with them—survive them. . ." The oil spill, on the other hand, was "man's doing" in the Gulf." . . . "The hurricanes we experience are natural, inevitable, and evident. This spill was man-made, unnccessary, and caused by greed."

Respondents had a sense of history, understanding, and resultant sense of control in relation to hurricanes: most had lived in coastal areas for decades and had prior experience with hurricanes. In contrast, individuals expressed a lack of knowledge and control regarding the impacts of the oil spill. As the following quote illustrates, coastal residents felt more prepared to deal with hurricane impacts, especially in the aftermath of the active 2004 and 2005 hurricane seasons, than oil spill impacts:

> I feel I could deal with just a hurricane (lost all during Katrina). I know what to do, know about mold and the dangers of it. I have no idea about oil and chemicals being used in the gulf. I find myself thinking every day about what to take if I have to leave this year knowing what I leave could be destroyed or oil soaked.

The lack of control was also associated with the magnitude of the oil spill. Respondents clearly felt ineffectual in the face of a catastrophe of such significant magnitude: "I try *very* hard *not* to worry about it, since I cannot help!"

Respondents could envision, with a high degree of certainty, what a hurricane would look and feel like; they understood the types of loss and damage that a hurricane could cause. The oil spill represented an unknown and little understood threat. The uncertainty regarding the scope and magnitude of current and possible future impacts from the oil spill was a focus of several responses, as illustrated in the following quotes:

> I have been thru numerous hurricanes in my lifetime. I have no clue what or how the added scenario with the oil in the gulf will impact our area. I want to know ASAP.

> IT IS UNKNOWN HOW IT [the oil spill] WILL EFFECT US BUT THE UNKNOWN IS SCARY.

Some survey participants phrased their responses in the form of questions, further highlighting the uncertainty surrounding the spill: *"Are we in even more danger because of the oil spill & if so what more precautions should we take?"* The uncertainty of impacts from the spill complicated the preparation for hurricane season and left respondents feeling unsure, even helpless, about what to do. The following quotes portray the uncertainty and loss of control that respondents felt when reflecting on the oil spill:

> It will be [an incomprehensible] amount of destruction. Oil and debris will be everywhere! I don't think any one will or can be prepared. What have we done? What is the future of the once Beautiful Gulf Coast? What will the future generations be missing out on?

> It is a disaster that far outreaches any hurricane. We *know* how to deal with hurricanes. We can't deal with the *death* of our gulf; emotionally, financially or any other way. . .

It is of note that of our 270 respondents, only two indicated that they were less concerned about the oil spill than a potential hurricane.

Survey respondents also depicted the oil spill as more of a long-term threat to the environment and human livelihoods than a hurricane. Respondents wondered when normalcy would return following the oil spill, as portrayed in the following quote:

> This oil spill worries me more than just a hurricane. We had our previous home destroyed by Hurricane Dennis & we re-built. In the 18 mos. we were in a much safer home and community. With the oil spill I fear it could be decades before normalcy returns.

Their ability to cope with and understand hurricane impacts caused many respondents to similarly assert that recovery from a hurricane would be a more straightforward and linear process. Recovery from the oil spill, on the other hand, was viewed as an uncertain and lengthy process that would perhaps span decades or even generations. "The BP Gulf Oil Spill—*without* hurricane interaction—is a catastrophic event for the immediate, intermediate, and the long term (generations)." Other respondents calculated the timeframe for recovery from the oil spill in generations, as did this grandmother: "The BP oil spill will impact my life forever. Not only my life but the lives of my grandgirls and their children's children." As noted by another respondent, the slow recovery from the *Exxon Valdez* spill provided little hope for people of the Gulf Coast: "Some effect will not be seen for a long time (see Exxon-Valdez Alaska oil spill—permanent changes + still going)."

BLAME ASSIGNMENT

It is well documented that technological disasters, as human failings, regularly generate widespread blame for the cause of and response to the disaster. It is not surprising, then, that blame and contestation over responsibility erupted soon after the BP oil spill began. The survey responses provided an early and uncensored look into how coastal residents were conceptualizing responsibility surrounding the oil spill. *"Who is going to jail?"* asked one respondent—providing a succinct way of describing of the issue of blame assignment. Our analysis revealed that respondents placed blame on several parties, including BP, the federal government, environmentalists, corporate interests, and the American public for their dependence on oil.

Respondents most frequently placed blame on British Petroleum for *causing* the spill. BP's "greed," "incompetence," and "inability to construct a proper well," were at the center of many accusations leveled by the respondents. Moreover, BP was blamed for their lack of preparedness and planning for a disaster of this magnitude, which contributed to their inept response. The following quotes exemplify the blame BP received from the survey respondents:

> BP should have had the ability to stop a well if they are going to drill that deep.
>
> How can a company operate without safety plans to eliminate trouble or COST OF LIFE. You should have plans, and back-up plans, and more back-up plans.
>
> And the lack of urgency at the beginning of this blowout makes me sick. BP should + will pay, but the rest of us should not suffer + lose jobs because of this.

In addition to BP, other respondents blamed the corporate oil industry more generally for placing profit above human safety and environmental protections: "The natural resource of oil belongs to all of us; oil companies in this country are unpatriotic and every citizen on the coast should be reimbursed . . ."

The federal government and federal officials, including specific references to President Obama and the U.S. Congress, were also blamed for the spill and, most frequently, for the "inept" response. The blame the federal government received varied more than the specific failing of BP—the government was criticized for a slow response and, conversely, for interfering in the response; for a lack of oversight and, conversely, for too much oversight and the moratorium on off-shore drilling. Many respondents felt that the government waited too long before responding to the situation in the Gulf, as noted in the following quote:

> I am very disappointed that the government has not been more proactive in effectively stopping the spill. It's a disgrace to know that not only do we have resources; but been offered resources from other countries, and not utilized them. It makes me question my government's priorities.

Similarly, another respondent asserted that the slow government response made the disaster worse: "Where is our Federal government when we need help so bad. It should not have gotten this bad if our gov. had acted sooner and with proven help instead of appointing committees."

Respondents associated the perceived slow response time with the creation of circumstances leading to a potential na-tech disaster: "The slow response brought the spill closer to hurricane season. Very poor industry/gov't. cooperation." As noted in the following quote, the perception of a slow and inept response from the government generated worry about government response to a na-tech disaster:

> It [the oil spill] has raised anxiety levels; created more distrust in the federal government's ability to handle coastal issues (or exert positive influence over local citizenry). Essentially; the Government's lack of leadership lowers our expectations of them successfully assisting us IF a major storm hit.

The respondents were divided over whether the government was "too slow" and should have been more actively involved, or if the government was actually "getting in the way" and hindering the clean-up process. "LET THE OIL Co. Do Their Thing," one respondent said. He continued, "They know more about it than anyone else. Government & people stay out of it." Speaking from personal experience, another respondent felt that the problem would have been solved more quickly had the government not "interfered": "If the politicians would let the professionals take care of the problem and not interfere it would be solved. . . I know; I worked 40 years in the oil field; including killing weed wells!!!"

The government's role as regulator of industry was also questioned by respondents. Some blamed the federal government for failing in their responsibility to regulate and oversee corporations, as the following respondent noted:

> If the federal government regulatory agencies were doing their job and following their responsibilities as an independent investigative oversight agency then this calamity would not have occurred. Will continue to have 'mistakes' because of the feds lax, let the corporations make their billions.

Another respondent was moved to encourage the government to further regulate corporations: "I feel strongly motivated to contact Law Makers to not allow companies such as BP to operate unless they can prove they can prevent and or fix the disaster they create 200 percent."

Again illustrating the "no-win" position of the federal government, another group of respondents blamed the government for actually overstepping its role by placing the moratorium on deep water oil drilling. These respondents believed safety was a "management issue" for oil companies—not an area for government interference. Thus from this perspective, the moratorium was perceived as further impairing the damaged economies of coastal communities and as a response that did not address the current problem: "If a hurricane makes things worse the gov. will not help by stopping the drilling."

A small number of respondents blamed environmentalists as a special interest that has compelled oil exploration in the deep waters of the Gulf of Mexico rather than other areas of the U.S. For example, one respondent felt that the environmentalists had forced oil companies into the deep water by lobbying against drilling near the coast. Another respondent asserted that environmentalists had forced drilling from other less treacherous areas as well: "We wouldn't be in this situation if the tree huggers would let us drill in Alaska + the rest of the USA."

Taking the opposite perspective, others discussed the U.S. dependence on oil as the underlying reason for the drilling in the Gulf: "We are *too* dependent on oil." Others were hopeful that the oil spill might motivate change in our national resource base: "35 years ago we (USA) citizens had an opportunity to change our dependency on oil usage + we failed. Perhaps; with this recent attention to the gulf oil spill we can end our dependency on our oil needs." Some went even further and called for a complete end to off-shore oil drilling.

EMOTIONAL RESPONSES

Thoughts of a hurricane and the oil spill also generated many strong emotions, including, most commonly, fear and anxiety, anger, and sadness. Although respondients were clearly concerned regarding the interaction between a hurricane and an oil spill, the overwhelming majority of emotion-based responses were focused specifically on the oil spill itself. For those respondents still recovering from the devastating 2004 and 2005 hurricane seasons, which destroyed millions of homes and caused billions of dollars of damage in Florida, Alabama, Mississippi, and Louisiana, the possibility of coping with the oil spill was absolutely overwhelming. As one respondent near New Orleans, LA indicated, "The oil spill may be the proverbial straw. We ARE physically, financially, and emotionally exhausted."

Respondents were facing a double dose of uncertainty. On the one hand, they were unsure of the level of damage that the oil spill would cause. On the other hand, they were equally uncertain, and even more concerned, about how a hurricane would interact with the oil spill and what sorts of damage would result. The high levels of uncertainty generated fear and anxiety for many of our survey respondents. The following quote is illustrative: "We are extremely anxious about the effects of oil on the beaches of the Gulf Coast."

Worst case scenarios were the topic of some of the written responses. For example, one resident of Florida admitted he was "Extremely tense and afraid about the situations concerning the gas pressure underneath the oil spill and hurricanes entering the gulf. very; very; frightening." Others expressed fear of potential evacuations and displacement: "Fear of long-term evacuation and perhaps permanent displacement: If we have to evacuate this year I feel I may be saying goodbye to my home for a long time if not forever." As succinctly noted by one respondent, the fear and anxiety of oil spill impacts contrasted sharply with the impacts of a potential hurricane alone, "It [the oil spill] has already caused more fear than a hurricane would."

The survey respondents also expressed a great deal of anger over the oil spill. Anger because it happened in the first place. "Someone Goofed, Big Time. We will pay for it!" Anger because the gushing oil well had not been capped. "It depresses me and makes me angry that it has not been fixed. I can't even watch it on TV." And anger because there was no clear end in sight. "BP doesn't have enough money to make it right. It will never be the same."

The apparent anger in the survey responses commonly accompanied the placement blame for the disaster or for the ineffective response and "inept" clean-up: *"Angry at BP and government for not doing something soon enough."* Many respondents called for BP and/or government officials to be fired or replaced, and their anger is clear in their responses through the use of expletives, exclamation points, underlining, and capitalization, as the following quotes demonstrate:

> I drilled more than 1,000 oil wells. These People are Assholes + Deserve to be put out of business.
>
> Replace all BP executives with truthful people. Hang em high.
>
> BP has lost all rights to drill for oil within our coast—BP should no longer exist as a company.
>
> Light a fire under Obama's ass and get him to work for us & not his campaign financiers.
>
> Washington leaders are a joke and should resign!
>
> I don't think BP gets it. I think they would get it IF this oil was in the English Channel!
>
> Respondents whose replies were short expressed their anger by simply writing "DISGUSTED" or other angry comments in large letters.

Respondents also articulated how sad and depressed they were as a result of the oil spill. Again wondering how the spill impacts would affect the people, the wildlife, and the landscape of coastal areas, sadness accompanied the worry about potential, but still uncertain impacts from a na-tech disaster: "I'm worried & sad when a storm spreads out the oil spill & speeds up the Destruction of Eco Systems." Another respondent expressed a similar perspective: "It is sad to think, but absolutely true, that hurricanes could spread the oil further than anticipated." One respondent who recalled time spent with family along a now oil-soaked coast, discussed her grief about this loss: "I sit crying as I think of the impact the oil will have. If it's not bad enough, add a hurricane to the picture."

AN OIL SPILL IN HURRICANE TERRITORY: THE POTENTIAL FOR A NA-TECH DISASTER

> I pray to the higher powers that I will never know another Hurricane Katrina here with this combined disaster!
>
> "Horror" seems to be the appropriate word of choice at present.

The above quotes epitomize the views of survey respondents who were well aware of a possible na-tech disaster if a hurricane were to enter the Gulf amidst the BP oil spill crisis. Respondents were resolute that a hurricane of any magnitude would amplify the impacts of the oil spill: "A hurricane at this time would impact the Mississippi gulf coast with as much damage caused by hurricane Katrina." Respondents frequently expressed their predictions with phrases such as "catastrophic," "unimaginable," "devastating," "destruction," and "greatly worsen."

Respondents saw the damage already caused by the oil spill as horrifying enough without the consideration of a hurricane interaction: "[The oil spill] already has ruinously affected the gulf coast for FL, AL, MS, & LA. A hurricane will multiply the ruin beyond imagination." Some respondents recalled previous hurricanes to envision the potential destruction: "If another hurricane like Ivan (Sept 16, 2004) hits our areas this year, the devastation will be catastrophic."

The most common specific predictions were first, that a hurricane would drive the oil further inland, devastating the environment. "If a hurricane picks up oil and brings it on land; it will be devastating." And, second that a hurricane would spread the oil to currently unaffected areas. "Concerned that a hurricane through the Gulf would speed the rate at which the oil reaches our area, and may in fact make the amount of oil that gets here exponentially worse." The winds, storm surge, and waves that accompany hurricanes were all perceived as ways a hurricane could drive oil inland: "The storm surge and waves will move a large amount of oil inland and cause a large impact on marshes and farmland." Some respondents predicted walls of water and oil crashing onto shore: "40 foot wall of *water + oil* coming onto land." One respondent used Hurricane Alex, the first hurricane of the 2010 season, to predict the potential impacts: "Hurricane Alex just passed and it seems to have pushed the oil closer to land."

The ability of a hurricane to spread the damages to other areas was a concern for populations along both the Gulf and Atlantic Coasts. This was apparent in our data set, as while 66 percent of the respondents lived within miles of the Gulf Coast and were ostensibly at direct risk, the remaining 34 percent of respondents lived along the Atlantic Coast. Yet these respondents, like their Gulf Coast counterparts, expressed high levels of concern that the oil would be picked up and widely dispersed, should a hurricane strike. As one Florida resident indicated, the spread of oil beyond the Gulf Coast was certain if a hurricane occurred: "The damage from the oil spill hasn't showed up in our area *yet*. We know it's coming; but it will take only **1** hurricane to bring devastation (or additional devastation) to wherever it may carry the oil to." Respondents living on the Atlantic Coast believed that a hurricane would increase the likelihood of oil spreading to their shores or entering into the North Atlantic Current, as noted in the following quote.

> I would be very concerned if (we) lived along the west side of Florida. We are a bit worried that a hurricane or tropical storm could push the oil into the currents that would bring it to the Atlantic, making the tragedy global and also greatly affecting us personally.

Others also discussed the expansion of oil damage to the Eastern seaboard: "Hurricanes + Tropical storms could potentially intensify the magnitude of this problem; especially in the Gulf but possible on the East Coast as well."

Only five of the 270 respondents felt that a hurricane would improve the oil spill situation by dispersing or breaking down the oil or have no effect, and sixteen respondents were unsure how a hurricane would interact with the oil spill. As is apparent in this response, participants were looking for expert advice: "So far 'the experts' have been more confused than anyone could imagine as to the expected outcomes of the oil spill. Therefore one can safely assume that we may all be surprised at the effect a hurricane *might* have on areas close to the spill."

SPECIFIC NA-TECH CONCERNS

The grave predictions of a na-tech disaster from the respondents were likely to be focused on, in descending order, effects on the wildlife, environment, the economy, health and safety, and property. Moreover, many respondents noted that the impacts would be interactive and multiplicative: "A hurricane will multiply the ruin beyond imagination, not just ecological, but financial. Entire lifestyles and livelihoods have been ruined." Because of the uncertainty of the oil spill impacts, and the uncertainty surrounding the interaction between a hurricane and the spill, respondents' specific concerns included both the Gulf Coast areas that were already impacted by the spill and our extended study area from Texas to North Carolina.

The oil was already "ruinous" to ecologically fragile areas along the Louisiana coast, and the potential na-tech disaster was expected to spread and amplify this destruction of ecosystems: "I believe that any hurricane that goes into the gulf will cause more devastation to the environment!" Respondents also noted the complex effects of oil on the environment: "Its effects are, and will be, both explicit and hidden and complex. Its effects [are] both profoundly negative for both the species and related ecosystems." By pushing oil inland, a hurricane was predicted to worsen the impacts on marshes and wetlands along the coast: "If a hurricane comes anywhere near the Gulf, it will bring the oil further into the marshland and totally destroy the marsh and wildlife for many, many years." As noted in the following quote, respondents provided specific examples of the fragile wildlife that would be harmed.

> Our county is known for breeding grounds for sea turtles. If a hurricane increases the amount of oil sucked into the current when eggs erupt, the young turtles head for the seaweed in the current, and they will be poisoned by the tar balls. Also our estuary, Sebastian Inlet, is a large breeding grounds for many species of fish. Again, We are Concerned about contamination of our intercoastal waters.

With an active hurricane season predicted, respondents concern about the environment was further intensified:

> I feel sick about the oil spill. It is the first time I have really felt so strongly about an environmental disaster. I am really worried about what will happen if a hurricane causes the oil to do further damage to the environment. I am hoping that the forecasts for an active hurricane season turn out to be wrong or the spill is stabilized before one hits.

Our respondents also offered a variety of specific concerns about the human impacts of a potential na-tech disaster, which we categorized into three groups, namely health and safety, jobs and livelihoods, and houses and property. The oil itself and the dispersants used during the clean-up process generated worries among our respondents as to their health effects. If a hurricane carried oil inland, respondents felt that water and food sources would be contaminated as the following quotes show.

> Will the ground and water supply be safe after going under water + oil? Chemicals being used in the oil spill are they safe for people to get on them? If our property has damage will it be safe to come back in area to try to take care of it before more damage is done?

> I fear the oil will be dispersed onto every tree, bush, blade of grass, fruits, veggies, cars, home, and this will become a bio-hazardous area and we will all eventually perish as a result of strange cancers and lung issues, if a hurricane were to enter the Gulf at this time.

> Concerned about flood waters from hurricanes affecting our food plots! How disease may impact from the oil on land.

Another concern for respondents was the spill's effect on the 2010 fishing season, which was to be the first fully-recovered season since Hurricane Katrina five years previous. The spill had already drastically impacted the livelihoods of Gulf Coast residents dependent upon fishing, thus a hurricane was seen as worsening this effect: "Increased destruction, loss of jobs + revenue hardship in general." A few of our respondents discussed the personal effects of lost livelihoods:

> We have a small crawfish farm will be impacted thru contaminated water. How long before our seafood is good to eat?

> No oil is near us and our clam crop. Our clam farmers are very worried as to the impact to include the long term?

Besides the fishing industry, respondents discussed the potential impacts of a na-tech disaster on the already bruised tourist areas: "We are very concerned because our little community is dependent on Tourism." Again, the worry of na-tech impacts expanded beyond the already affected areas of the Gulf, as noted in the response from a Texas resident: "We are very fearful that the spill will reach our south Texas coast + destroy our tourism based economy." Overall, respondents felt that a na-tech disaster would worsen the "economic disaster" for the coastal communities.

Finally, respondents expressed concern for homes and property if oil and contaminates were carried inland by a hurricane. A respondent captured this theme when she expressed: "I would hate to think about oil being picked up by a hurricane and slammed into our homes." Respondents discussed the destruction of homes and contamination of land, the unknown effects of oil and water on homes, the ability to return after evacuation, and whether insurance would cover damages. Again, we see that although this population was accustomed to how hurricanes damage property, they were uncertain about the combination of oil and hurricanes: "When a hurricane hits it will be a major disaster, with oil going into people's homes. With just storm surges you have to fight with mold, mud, mildew now it also will be the oil." When speaking of the potential devastation, respondents would personalize the impacts to their property or home: "A hurricane in the Gulf would DEVASTATE us. Our properties and homes will be contaminated from the oil (Ruined)." As the following quotes illustrate, the predicted effects of oil on property also created anxiety surrounding evacuation and a timely return:

> I feel + worry if we have a storm and water/oil gets on our properties EPA will be involved and wonder how long before we could even come back home. . .

> If our property has damage will it be safe to come back in area to try to take care of it before more damage is done. If we have to evacuate this year I feel I may be saying goodbye to my home for a long time if not forever. Water can be cleaned up—not chemicals + oil on everything.

Two respondents specifically wondered about their insurance coverage in a na-tech disaster. One respondent showed the uncertainty surrounding impacts with the simple question, "Does our home owner insurance cover damage from oil in a hurricane?" Another respondent, who had experienced the complicated insurance claim process following Hurricane Katrina, was certain that na-tech disaster impacts would not be covered by insurance, as noted in the following quote.

> If we were to have another Katrina, the damage from the oil spill would *NOT* be covered under any type of insurance that we pay. We have already experienced

insurance companies that refused to pay for damages caused by hurricanes, wind or flooding.

CONCLUSION

As the 2010 hurricane season draws to a close, U.S. coastal residents seem to be breathing a collective sigh of relief. The season was active, as predicted, with nineteen named tropical storms (twelve of which became hurricanes) as of early November 2010. Although this season will be listed in the top ten for overall activity since 1851, remarkably, a major hurricane did not make landfall in the United States in 2010. The average annual probability of a hurricane of any category striking the U.S. is 83.1 percent and based on the high level of storm activity for 2010, this probability increased to 96.9 percent. These statistical predictions were generated by Brian McNoldy, a Colorado State University atmospheric scientist and hurricane expert.

In this chapter, we have illustrated how coastal residents in the midst of the BP oil spill, a catastrophic and ongoing technological disaster, framed this event differently than the hurricanes that pose an annual threat to them. We also discussed how coastal residents reacted to the potential na-tech disaster that would have occurred, had a hurricane occurred during the spill.

NATURAL, TECHNOLOGICAL, AND NA-TECH REFLECTIONS

Our results indicate that coastal residents experienced the fear, anger, and sadness that typically accompany technological disasters. These emotions emerged from the intense uncertainty and dread that the oil spill contributed to their everyday lives, which was further compounded by the regularly scheduled hurricane season. Their emotional responses were remarkably similar when speaking just of the oil spill as compared to the potential interaction with a hurricane. Moreover, the blame and contestation of responsibility was apparent throughout the written responses even though these survey respondents represented a general coastal population rather than a clearly demarcated geographic community. This blame extended beyond simply assigning responsibility for the initial oil spill; respondents also placed blame on the government, British Petroleum, corporations, and environmentalists for the delayed clean-up and for the predicted consequences if a hurricane amplified the spill's effects. Finally, our respondents clearly believed that spill impacts and potential na-tech impacts created an "un-ending" disaster that would hinder their resilience to a hurricane landfall. This study confirmed prior research through illustrating the ways that threatened populations distinguish, sometimes quite sharply, between natural and technological disasters.

Failures of the levees during Hurricane Katrina, combined with the numerous chemical and oil releases caused by the storm, have been framed as a na-tech disaster by researchers (see Picou and Marshall, 2007). Yet, the hurricane was clearly the catalyst for that disaster, which helps explain why so many coastal residents perceived Hurricane Katrina as an "act of God" first and foremost. In contrast, the BP oil spill was clearly a technological disaster *first*, and therefore the amplifying effects of a hurricane in 2010 were most frequently depicted by our respondents as also due to human failings. Respondents often mentioned that they were "praying to God" that a hurricane would not occur and make matters worse. Thus, the predicted but uncertain impacts from a na-tech disaster were much more clearly framed as the result of human error and subsequently generated heightened concern, anxiety, and blame for the na-tech disaster that, fortunately, did not occur in 2010. Because respondents framed the potential na-tech disaster as primarily due to human failings, their focus on the amplified destruction and long-term environmental, health, and economic impacts echo concerns surrounding "classic" technological disasters. These results indicate that if a hurricane had increased the impacts of the oil spill, then we would expect contestation over responsibility, blame, the corrosive community, and potential litigation similar to technological disaster effects.

Another important point from our study is that respondents expressed concern about the potential na-tech disaster and the oil spill, regardless of whether or not they were in the direct path or personally affected by the spill. This result shows how the spill, like other technological hazards, generated "generalized fear" that increases the worry and feelings of risk for large portions of a population and extends far beyond those that are personally affected.

We acknowledge that much of what the respondents said in response to our survey coverage was a reflection of what they had likely learned through the news media and scientific reports. This is no surprise, given the level of coverage of this disaster, the attention dedicated to both the spill and the hurricane season, *and* the fact that respondents received our survey at the height of the disaster. Our results then show that residents were generally aware of the information available and had interpreted (and internalized) this information to develop opinions, emotional responses, and conceptualize predictions for their own areas.

OPEN-ENDED SURVEY RESPONSES: LIMITATIONS AND RECOMMENDATIONS

Our results are limited based on the open-ended survey response method. First, because of the nature of the question wording, we have no way of assessing the level of representativeness of the sample beyond comparing demographics (which show no dramatic differences between oil spill respondents and the rest of the survey respondents). The question, compared to the close-ended questions in the survey, generated a much lower response rate, and was potentially seen as optional.

Second, because of the voluntary nature of the question we are unable to generalize from these findings. It is unclear what the opinions of the other 57 percent of those surveyed are and if they differ dramatically from those who responded to this question. It is possible that individuals did not respond because they were not concerned about the spill, had no opinion, or felt that it was not personally affecting them. Conversely, it is also possible that they did not respond due to time constraints, when they did actually have concerns about the spill.

Third, we relied solely on one open-ended qualitative question in the survey. No rating scale data, structured diagnostic interviews, or other quantitative data was collected on the oil spill topic. This was in part due to the decision to include an "oil spill question" in a survey otherwise dedicated to hurricane risk perception and evacuation behavior. Some respondents did not specifically address the issue as stated and used the space as a political platform and/or sounding board for general concerns. While this method provided then unexpected responses, many of these insights corroborated theoretical insights about technological and natural disasters that were not expected or targeted with the framing of the original question.

Fourth, as compared to in-depth qualitative research, the open-ended responses lacked the rich, narrative insight that is garnered from in-person interview data. Also, the open-ended method eliminated the ability to inquire for further information or clarification from respondents, which would have been useful in further developing and supporting our results. This method provided interesting, but brief insights into our respondents' feelings about the oil spill and the 2010 hurricane season.

Despite the aforementioned limitations, ours is one of the few available studies that draws upon the voices and perspectives of residents in the midst of a technological disaster and facing an impending na-tech disaster. Our findings suggest a general pattern of anxiety, fear, and uncertainty surrounding technological impacts spreading far beyond the impacted areas. Residents along the U.S. Gulf and Atlantic coastlines were aware and listening to media, government, and scientific reports about the oil spill and the predicted effects of a hurricane. They expressed extreme uncertainty about these impacts, indicating the need for continued information flows and specific discussion of any predictions. While it is accepted that the novel situation made it unclear what impacts may occur, clear, consistent messages could help calm fears by reducing uncertainty. For example, while our respondents were aware of the ability of a storm surge to push oil inland, they also foresaw oil swirling in hurricane winds—a prediction clearly falsified by NOAA (2010). This said, we recognize from our responses that the government was clearly in a "no win" situation, and generally not trusted, thus even clear and consistent reassurances would still leave a large portion of the population uncertain and fearful. Thus, the understanding that any hurricane interaction with the oil spill would be framed as human failings and generate increased blame should be considered during response and recovery preparation. This preparation includes assurances that resources are prepared to respond to na-tech disaster impacts.

REFERENCES

Barton, A.H. (1969). *Communities in Disaster: A Sociological Analysis of Collective Stress Situations*. Garden City, NY: Doubleday.

Erikson, K. (1994). *A New Species of Trouble: The Human Experience of Modern Disasters* New York: W. W. Norton and Company.

Fischer, H.W. (2008). *Response to Disaster: Fact Versus Fiction and its Perpetuation* (third edition). Lanham, Maryland: University Press of America.

Foster, R.P., Branovan, D.I. and Ukrainsky, G. (2003). *Surviving Chernobyl in America: Medical and Mental Health Consequences of the Chernobyl Nuclear Accident*. New York: Media Luna Ltd.

Freudenburg, W.R., Gramling, R., Laska, S. and Erikson, K.T. (2009). *Catastrophe in the Making: The Engineering of Katrina and the Disasters of Tomorrow*. Washington, DC: Island Press/Shearwater Books.

Fritz, C.E. (1961). Disaster. In R. K. Merton and R. A. Nisbet (Eds.), *Contemporary Social Problems*. New York: Harcourt, Brace, and World, pp. 651-94).

Gill, D.A. and Picou, J.S. (1998). Technological disaster and chronic community stress. *Society & Natural Resources, 11*(8), 795.

Kroll-Smith, J.S. and Couch, S.R. (1990). *The Real Disaster is Above Ground: A Mine Fire and Social Conflict*. Lexington: The University Press of Kentucky.

NOAA (2010). *Hurricanes and the Oil Spill*: National Oceanic and Atmospheric Administration.

Norris, F.H., Friedman, M.J., Watson, P.J., Byrne, C.M., Diaz, E. and Kaniasty, K. (2002). 60,000 Disaster Victims Speak: Part I. An Empirical Review of the Empirical Literature, 1981-2001. *Psychiatry, 65*(3), 207-60.

Picou, J.S., Gill, D.A. and Cohen, M.J. (Eds.). (1997). *The Exxon Valdez Disaster: Readings on a Modern Social Problem*. Dubuque, Iowa: Kendall/Hunt Publishing Company.

Picou, J.S. and Marshall, B.K. (2007). Introduction. In D.L. Brunsma, D., Overfelt and J.S. Picou (Eds.), *The Sociology of Katrina: Perspectives on a Modern Catastrophe* (pp. 1-22). Lanham, Maryland: Rowman & Littlefield Publishers, Inc.

Saldaña, J. (2009). *The Coding Manual for Qualitative Researchers*. Los Angeles: Sage.

Showalter, P.S. and Myers, M.F. (1994). Natural Disasters in the United States as Release Agents of Oil, Chemicals, or Radiological Materials between 1980-1989: Analysis and Recommendations. *Risk Analysis, 14*, 169-182.

CHAPTER TEN
THE NINETY-DAY STORM: MISSISSIPPI COMMUNITY RESPONSE TO THE DEEPWATER HORIZON OIL SPILL

JASON S. GORDON AND A. E. LULOFF

ABSTRACT

This research explored community response to the Deepwater Horizon Oil Spill (DWHOS) using a community-based approach. Key informants identified technological and social drivers that made the DWHOS a disaster. Discussions of vulnerabilities, including unemployment, urban sprawl, and social disruption, implicated their role in complicating the impacts of the oil spill. Many residents felt government agencies and BP ignored their communities' concerns; as a result, they felt helpless in preventing future accidents. At the same time, some communities were able to draw on local capacities and build on experiences from Hurricane Katrina to address the oil spill and redevelopment issues. In this chapter, two Mississippi gulf coast communities are compared. This effort indicated that resilience, arising long before this disaster, had enabled a foundation of empowerment to emerge which had been tested by previous community-wide challenges. Implications for community development in addition to theories of disaster response are advanced.

INTRODUCTION

Residents living along the Mississippi Gulf Coast are familiar with natural disasters. Over the last one hundred years, hurricanes have hit the area every five to ten years (Cumberpatch 2006). Despite this, they were not prepared for what they experienced during the British Petroleum Deep Water Horizon Oil Spill (DWHOS). While the majority of the 560 miles of oiled coastline occurred mostly in Plaquemines Parish, Louisiana, residents in Mississippi (and the other Gulf Coast states) were also forced to confront what Environmental Protection Administrator Lisa Jackson called "The worst environmental disaster in our nation's history" (EPA, 2010a).

This disaster presents a unique opportunity to explore the complex issues involved in assessing community response to these events. Even in the presence of significant effort expended on disaster research, questions about community level responses continue to unfold, especially when couched in trying to understand and differentiate between technological and natural hazards, assess vulnerability versus agency perspectives, and address therapeutic versus corrosive responses (Flint and Luloff, 2005). In the case of the DWHOS, community responses were influenced by multiple biophysical and social vulnerabilities of place, including recent Hurricane Katrina experiences and the long-term impacts of the current economic recession. In spite of these challenges, some communities engaged in actions that mitigated the oil spill's effects. It is precisely these communities, their interpretations of the disaster, and the emergence of such actions in response to disaster which need to be identified and understood so effective disaster response for similar communities in the future can be secured.

This paper explores the socioeconomic, sociocultural, and biophysical factors which complicated the DWHOS impacts on Mississippi gulf coast communities. We begin with a brief overview of disaster recovery research. The next section discusses the role of community capacity-building and development in disasters. There, a middle range theory is proposed that incorporates community and disaster theories. Then we apply this framework using a comparative case study of two coastal communities in Mississippi. These sites were chosen by proportion of racial/ethnic residents, income, and rural or urban geography; data were collected through semi-structured key informant interviews. In the fourth, and concluding section of the chapter, we discuss the implications of the study for managing community response and recovery while challenging some common notions of community disaster response.

BACKGROUND

A CONCEPTUAL OVERVIEW OF COMMUNITY-BASED DISASTER RESEARCH

Over the last half century, disaster research has largely been framed by two competing theoretical camps: those which view disasters as originating from environmental processes and those which see them as originating in social processes. Among the former, communities are typically depicted as victims of extreme natural or man-made events that disrupt the status quo (Blaikie et al., 1994; Burton, Kates, and White, 1978). Residents are depicted as helpless to respond without the assistance of outside expertise to control nature or technology. The implicit objective of the experts is to return local society to its former conditions, generally of dependency, without addressing extant issues of marginalization and vulnerability (Flint and Luloff, 2005; Hewitt 1998). On the other hand, the notion that disasters originate in social processes suggests preexisting societal conditions makes communities vulnerable to disaster, regardless of whether the event is natural or technological (Cannon, 1994; Hewitt, 1998; Kroll-Smith and Couch, 1991). This paradigm ignores both the capacity of communities to respond constructively to a disaster and the environmental characteristics of vulnerability.

Within these approaches, disaster research has portrayed experiences as being distinguishable on the basis of the hazards' natural and/or technological characteristics. This research suggests people tend to associate natural disasters with either ecological processes or acts of God (Erickson, 1976). These disasters are perceived to be swift, unexpected, and beyond human control (Fritz, 1961). Community members are said to galvanize because human error is not perceived to be a root cause; simply said, no one is to blame (Kroll-Smith and Couch, 1991; Picou, Marshall, and Gill, 2004). Studies have indicated victims find fewer explanations for technological disasters (Fritz, 1961). Not surprisingly, many of these disasters are met with anxiety, social disruption, and long-lasting feelings of vulnerability and uncertainty (Beck, 1999; Erickson, 1976; Fowlkes and Miller, 1987; Picou et al., 2009). The result is a disintegration of community.

A number of studies have addressed catastrophic marine oil spills as technological disasters. Much of this research examined the 1989 *Exxon Valdez* (EVOS) disaster. Among them, Dyer and colleagues (1992) and Palinkas and colleagues (1993) and were two studies that documented the oil spill's long-term effects. The authors focused on chronic stress and social disruption linked to dependency on natural resources for subsistence or commercial uses. Research also described the disintegration of social support networks, including collective management of resources, which in turn intensified environmental effects (Dyer, 1993; Gill and Picou, 1999). EVOS research tended to focus on Alaska Native communities and emphasized social vulnerability through resource dependency and the social construction of technological disasters (Gramling and Freudenburg, 1992; Picou et al., 2009).

More recent studies have noted that little empirical evidence exists to support a clearly dichotomous view of hazard types (Flint and Luloff, 2005; Gordon et al., 2010; Kumagai, Carroll and Cohn, 2004; Picou and Martin, 2006; Tootle, 2007). This research views disasters as a combination of environmental and social processes with technological and natural drivers (Kumagai, et al., 2004). Environmental factors include the community's proximity to the disaster, degree of damage, and ecological vulnerabilities (Mileti, 1999). Social factors include political processes, collective experience with previous disasters, socioeconomic inequalities, and proportion of the local population in need of special assistance (e.g., elderly, children, and disabled). Leadership, income, and education have played key roles in developing effective disaster response (Arnoff and Gunter, 1992; Cutter, Mitchell, and Scott, 2000).

Despite the evolution of disaster-related research, studies typically focus on individual or group level responses to immediate disaster impacts (Picou and Martin, 2006; Quarantelli and Dynes, 1977). Additional research is needed that addresses the relationship between community agency, response, and resiliency (Cutter et al., 2000; Flint and Luloff, 2005; Tootle, 2007). As Flint and Luloff (2005:408) suggest, research is needed that:

> incorporate[s] the perceptual and interactional processes that shape community response to problematic situations. . . .When disaster strikes, regardless of origin, local capacities for action and the different ways they manifest need to be incorporated in our studies. Further, we need to facilitate community interaction to help local residents define objectives and make and implement decisions.

BUILDING COMMUNITY

The term community is applied in many different ways. Community is usually referred to as being either place-based or place free. Two of the most widely used approaches to explain community embody this divide. The Social capital framework is framed around a place-based theory while the social network perspective allows for interpretations of community without linkage to a geographic territory (Flora, 1998; Wellman, 1988). Regardless, both the social capital and social network approach tend to discount feelings and emotion, a common life, and the intricate social processes necessary for residents to work together for a collective goal.

By contrast, Wilkinson (1991) describes community as an emergent process through which individuals and groups work together in pursuit of commonly held goals. As a process, community "draws its psychological strength from levels of motivation deeper than those of mere volition or interest" (Nisbet, 1953:77). These goals are framed by a shared territory. From this interactional perspective, the physical qualities of the local environment are seen as playing a critical role in social interactions.[1]

Capacities are characteristics that affect the community's ability to identify, mobilize, and address local problems (Frank and Smith, 1999). Capacities are built and strengthened through community development (Luloff and Swanson, 1995). Building and strengthening capacities means democratically cultivating skills and distributing knowledge and resources across the local landscape. This occurs through activities such as protecting and restoring the natural environment, strengthening participatory government, creating and supporting the local economy, and developing public educational opportunities (Walter, Arkin, and Crenshaw, 1992). And while economic development is essential to community development and capacity building, it could well be unsustainable if other aspects of community are ignored (Wilkinson, 1991; Luloff and Swanson, 1995).

Perhaps most importantly, capacities are also developed when residents are involved in the planning and implementation of local activities. Highly participatory projects promote networking, skills development, organization, and structure (Luloff and Swanson, 1995). In healthy communities, leaders and residents identify and focus local assets and talents towards a common goal. When effort is made to involve all residents, regardless of their individual background characteristics, both individuals and the community are empowered.

RESILIENT COMMUNITIES AND DISASTER RESPONSE

Some communities cope with challenges and exploit opportunities better than others (Kaufman and Wilkinson, 1967). This quality is especially evident during disaster response and has the potential to emerge as community resilience. Resilience involves the ability of communities to mitigate risks and resist, cope with, absorb, or cushion themselves against damage or loss from hazards (Mileti, 1999; Perrings, 1991). It includes predisaster and postdisaster activities and emerges from the effectiveness of individual response, public and private policies, and accessible physical resources (Mileti, 1999). Community resiliency invokes coordinating collective efforts towards enhancing and maintaining organizational, structural, socioeconomic, and ecological resilience (Holling, 1986). The construct is consistent with community development because resilient communities have diverse capacities and high levels of agency (Perrings, 1991; Mileti, 1999; Wilkinson, 1991). Capacity building is a continuous process in resilient communities.

To better understand the community interactional processes leading to disaster resiliency, we build on a framework proposed by Flint and Luloff (2005) integrating biophysical, socioeconomic, and sociocultural vulnerabilities with community capacity, local disaster response, and ultimately, community resiliency. The framework has its roots in a place-based conceptualization of community. As a result, it accounts for the local social interactions that grow into capacities and ultimately determine community-wide response. These same capacities can be cultivated and strengthened through development of community. Communities with high levels of agency are more likely to display resiliency than those with low levels.

As well, individual and collective experiences play important roles in capacities, response, and resiliency. Although disasters result in unique effects, communities that have responded to repeated disasters react differently than communities exposed to a single catastrophic event (Kumagai et al., 2004). Responses often range from apathy to panic to altruism, depending on local characteristics and disaster traits (Fritz, 1961; Quarantelli, 1960). Hazards and risk research have explained this based on individual probability assessment and heuristics, social norms, and communication about the disaster (Douglas, 1966; Slovic, 1987).

Interactional field theory also accounts for institutional influences on capacities, response, and resiliency. Local and extra-local institutions constantly moderate disaster response and processes leading to resiliency. Prominent local institutions include local government, media, churches, schools, and civic groups. External institutions include state and government agencies, large corporations, and national media. Particularly since Hurricane Katrina, each of these institutions has been shown to have important effects on community response (Fischer, 1998; Tootle, 2007). Institutional processes include recreancy, framing the disaster, and investigating its causes and consequences (Freudenburg, 1993; Tierney, Bevc and Kuligowski, 2006).

Using this framework, we explored two broad research questions. First, how did neighboring communities respond differently to the DWHOS? Second, what were the interactional factors affecting resilience? The following sections describe the study area, present findings, and offer implications based on the research.

METHODS

SITE SELECTION

Sites were chosen based on rural and urban populations, race, and income. Running Oak[2] is located in the most rural county of the Mississippi Gulf Coast and had a population of just over 8,000 residents in 2000. Over 80 percent of residents were white and 17 percent were African American. With a medium family income of $41,957, ten percent of families fell below the poverty level, slightly higher than the rest of the nation, but less than the state of Mississippi. Since the nineteenth and early twentieth Centuries, Running Oak has been a popular destination for New Orleans residents with vacation homes or for those who commuted to jobs in the city. Tourism continues to be an important part of the local economy. The community has been repeatedly devastated by hurricanes, including, most recently, Camille and Katrina.

Palmetto is a neighborhood in an urban center. In 2000, the city had a diverse population of slightly in excess of 50,000. Palmetto had just over 13,000 residents of which 40 percent were white, 40 percent African American, 15 percent Vietnamese-American, and 5 percent Hispanic. Nineteen percent of families fell below the poverty level with an average medium income of $30,852. The majority of residents were employed in services, reflecting the importance of casinos, restaurants, and tourism more generally to the local economy. Chemical manufacturing and the seafood industry were also important employers.

KEY INFORMANTS

Key informant interviews were conducted to identify attitudes, issues, and concerns of different community resident groups.[3] Key informants are community leaders and other residents knowledgeable about marine resources and/or community issues (Luloff, 1999). Initial informants were identified after reviewing directories of local leaders and newspaper articles. They were interviewed on July 8 and 9, 2010. Using a modified snowball procedure, the initial informants were asked to identify other potential participants (Luloff, 1999). The second set of interviews occurred during the first week of August, 104 days after the explosion that started the spill and seventeen days following the initial capping of the spill.

Participants represented perspectives from environmentalists, elected officials, media, nonprofit resident assistance organizations, commercial fishermen, the tourism industry, the petroleum/chemical industry, nonlocal government agencies, and active residents unaffiliated with a particular group. Eighteen informants were interviewed in each community. As well, an equal number of informants within each category were interviewed in each community. Table 10.1 describes informants' demographic characteristics. The key informants were nearly evenly divided by gender (with slightly more males interviewed), most were well educated (with at least an associates or four-year degree), and the vast majority had spent their whole lives in their communities.

A key informant instrument was used to guide the interviews through a series of open-ended questions. Reflecting the literature on hazards and community interactions, questions addressed: (1) description of the community, including important and divisive issues; (2) environmental concerns prior to the oil spill; (3) communication about the disaster between residents, government, and media; (4) definition of disaster; (5) who was to blame for the disaster and who had responsibility for mitigating future oil spills. Interviews were transcribed, coded, and analyzed line-by-line for emergent themes (Creswell, 1998).

Table 10.1: Key Informant Characteristics

Sex		Education		Residence	
Male	20	Less than 4-year degree	7	Lifelong	25
Female	16	Associate or 4-year degree	20	Intermittant or <5 years	4
		Graduate degree	9	10-20 years	5
				Non resident	2

FINDINGS

VULNERABILITIES

Residents acknowledged and described socioeconomic and biophysical vulnerabilities that amplified the impacts of the disaster. The informants discussed coastal erosion, exposure to hurricanes, population loss, unemployment, and infrastructure issues that affect many coastal places. A charter boat operator from Palmetto described vulnerabilities related to resource dependency, corporate and political influences, over-development, and a non-English speaking population:

> The fishing industry down here has been mismanaged for many years. The DMR [Department of Marine Resources] has allowed the menhaden industry to fish unchecked. Lots of rubber stamp permitting down here, offsetting of all the tidal wetlands, building of these seawalls. We have a lot of poorer residents who are being pushed out of their homes because of urban sprawl. This is mixed up with some terrible wastewater treatment problems that should be addressed. So the disaster really started way before the oil spill or even Katrina, and these just made things worse.

Highlighting the social construction of disasters, his comments depicted coastal Mississippi as a vulnerable region—not only because of hurricanes, but also as a result of socially-driven vulnerabilities that interacted and evolved over time. Thus, he hinted that the oil spill was a disaster of social origins, in addition to technological-environmental origins.

Several Running Oak residents emphasized a change in aesthetics and disruption of sense of place. Residents drew attention to the important role of natural and cultural heritage in vulnerability:

> Katrina destroyed so many of our precious live oaks. Everything was just so ugly, mildewy, smelled bad. Katrina assaulted our senses. Before the oil spill came, we used to joke about how you could ride down the road and look to the right out to the water to forget about Katrina for a little bit. After the oil spill, you found yourself still looking over the water but you're looking for dead birds. The oil spill

> was like a sucker punch. . .. It's impossible to think about the community in the same way as before these disasters (nonprofit organization employee).

Sociocultural vulnerability was associated with the collective identity of the community and attachments to the landscape as well as economic dependency on marine resources. Yet, sociocultural vulnerability was often overlooked by policymakers and emergency managers during the interviews. With few exceptions, decision-makers and emergency responders addressed vulnerabilities starkly in terms of exposure to natural disasters and economic conditions.

Palmetto residents similarly commented on the cultural importance of nearby marshlands and waterways. Residents regularly fished the waterways prior to contamination by a nearby chemical plant and faulty sewage system not long before the oil spill:

> Between the plant and the oil spill, I don't know if I'll ever be able to fish again. And the city has been making it worse for years. . .. The city would love to get its hands on this area [for development]. Everyone talks about losses to the commercial fishermen. . . . Now we don't fish for a living, but we fish because we've always fished and we eat what we catch (community activist).

Although not a subsistence activity, fishing in the same creeks and marshes had tied multiple generations of residents together for over a hundred years. When the activity changed, so did the type of connections between and among residents and between them and the land.

DISASTER EXPERIENCE

Informants from both communities used their past disaster experiences to understand and cope with the oil spill. In particular, two prominent themes related to experiences emerged during the interviews: (1) short-term disaster relief; and (2) uncertainty about the oil spill hazard. Community leaders in Palmetto often interpreted the disaster based on past relief and recovery efforts rather than with the physical characteristics of the disaster. Specifically, they discussed frustration with short-term disaster response by local and extra-local government. One resident had this to say:

> Every time there's a disaster, there's media coverage, they throw money at it, pictorial ops, and then all that goes away. And then the resources aren't there to respond to the level of creativity and resiliency of those who remain. And so they're left holding the bag until the next disaster (nonprofit organization employee).

Like this informant, residents' response to the oil spill incorporated anger and frustration with long-term impacts of disasters more generally. Others expressed apathy because they saw the oil spill as just another disaster which they felt powerless to address. Several Palmetto informants described how enthusiasm quickly diminished following the initial period of post-Katrina clean-up due to a lack of resources. The impacts of the oil spill were seen as further reinforcing marginalization associated with poverty, race, and social status.

The second theme was expressed strongly in Running Oak. Informants described Running Oak residents as experienced with hurricanes. They closely observed meteorological predictions from May through September. They knew about hurricane destruction, even to the extent caused by Hurricane Katrina. Previous experiences with hurricanes were used as a heuristic to interpret the impacts of the oil spill. The quote below illustrates how closely residents' immediate responses to the oil spill were connected to their responses to Katrina and other hurricanes:

> In a way, we thought of it like a hurricane, that imminent danger. But this was a 90 day hurricane. And it turned out to be a lot worse than a hurricane in many ways. You found yourself looking over the bay thinking, is it coming, is it coming, and it was this monster that was looming and they'd show you the pictures, and the wind would blow to the west and it would morph over here, and then it would come back and just be hovering right off shore. We knew something would happen, but we didn't know what (artist, community activist).

Residents' initial response to the oil spill as "just another disaster" was contradicted by a chronic hazard characterized by uncertainty, confusion, and heightened anxiety. Because the oil-spill resulted from human error and, as a result of uncertainty about the magnitude of the event, experience with hurricanes was negligible in lessening residents' anxiety. Residents worried when the oil spill would collide with the Mississippi coast, the exact location of impact, and the degree of damage. One of the biggest challenges was coping with the likelihood that, contrary to expectations, 2010 would not be an ideal year for fishing, shrimping, and/or tourism. These responses were intensified by the ways authorities communicated information about the spill.

INSTITUTIONS

Informants addressed the ways British Petroleum, extra-local agencies, state and federal officials, the national media, and local officials influenced community responses. Most informants said the local media and state government played minor roles in their interpretations of the disaster. The following is a brief overview of prominent themes revolving around trust in local government, crafting the story, and perceptions about responsibility for the disaster.

TRUST IN LOCAL GOVERNMENT

According to informants, Running Oak residents considered local government (including emergency responders) as capable of managing the disaster's effects. This perception was encouraged by regular updates through the Emergency Operations Center (EOC). Some informants expressed confidence in local government because "I think we assume that they [government] must be taking care of the problem because we haven't heard much complaining (business leader)." Conversely, informants said BP forced local government to exclude residents from local response activities. Informational updates were seen as window dressing for resident participation.

Distrust was strongly evident in Palmetto, which was often at odds with city government due to its development policies and perceptions that it ignores the needs of Palmetto residents. Informants said the relief activities and redevelopment following Katrina focused on assisting large industry, particularly casinos, rather than small businesses and lower income residents:

> Since the storm, disagreement stems not from within the community. . . Disagreement has stemmed from the people affecting policy outside the community. The conflict comes in government regulations and corporate double talkers (elected official).

As well, informants reported the lack of public EOC meetings such as those in Running Oak. Emergency officials explained the public was informed using alternative methods, such as question and answer sessions with the media. What is important, however, was the perception among residents that local government was operating without public input. Several informants suggested that local government was "working too closely with BP" (nonprofit organization employee). Following the initial capping of the well on July 15, a tour operator said, "The only euphoria over capping the well is BP's euphoria. The rest of us are going 'whatever.' There is so much distrust—who's telling the truth?"

CRAFTING THE STORY: BP AND THE NATIONAL MEDIA

Unsurprisingly, informants tended to describe residents as having higher levels of trust in local government than in BP, the federal government, and the national media. Although informants had observed oil on the Mississippi beaches, residents from both communities complained that media stories were unbalanced. A commercial shrimper and Vessel of Opportunity[4] contractor from Running Oak commented:

> It's not so much that the media is doing a bad job. But maybe they don't cover it the way they should. It's an environmental catastrophe—you just can't say it

> didn't happen. Now the main impact for us has been economic. So it would be good if they could feature some of the other attractions here other than the beaches.

A slightly different theme in the interview data concerned perceptions that media story lines pitted BP against the natural environment. One informant, a volunteer with a Vietnamese assistance organization in Palmetto, remarked that the national news organizations:

> Constantly talk about BP as this powerful, corporate monster with no regard for the environment—which is true. But then they [the media] depict coastal people as weaklings. I know a lot of families who endured a lot to get here [United States] and I can tell you that they're very resilient people.

This resident perceived the media focusing attention on BP, the environment, and economic consequences over other social impacts, such as mental health and social disruption. Moreover, she said, the media failed to sufficiently highlight the community's assets. This perception was strongest among Palmetto residents which, with the help of a number of nonprofit organizations, were involved in various local initiatives.

Informants were specifically asked for their opinions regarding how BP communicated with residents about the disaster. Some business leaders, informants involved with the chemical industry, and one politician described BP as "doing all they possibly can. . . . There has never been an oil spill like this before." Most informants from both communities, however, described BP's reaction to the spill as intended to sway public opinion as much as possible to sympathize with continued off-shore oil production. Informants perceived this framing occurring two ways.

First, they complained about BP's control of clean-up efforts and communication of information.[5] In particular, they were concerned that BP was too closely involved with disseminating information about the causes of the spill, its immediate response, and measures of impact. Residents complained that contractors hired by BP to staff beach cleanup crews were bussing in workers from other areas instead of hiring locals. Informants from Palmetto and Running Oak suggested that the Vessels of Opportunity program was a simple way to manipulate public opinion and mitigate anger by those who directly lost their livelihoods: "It's interesting, the day the well got capped, all these people got laid off" (political leader in Palmetto).

As well, several informants alleged that those who applied for cleanup work were forced to sign documents prohibiting discussion of working conditions and forfeiting legal redress for lost livelihoods.[6] Many of the crew members who signed these documents reportedly could not speak English and had low literacy levels. Finally, informants were frustrated that BP prohibited volunteers from contributing to the clean-up on their beaches. BP claimed the process was too hazardous; however, several informants wondered if the policy was another way to strengthen control over the disaster response.

> People were willing to come from all over to help, just like after Katrina. We could have had voluntourism here. But BP wanted to limit involvement from the people affected by the spill in order to maintain as much control as they could (local journalist in Palmetto).

RESPONSIBILITY OF EXTRA-LOCAL INSTITUTIONS

After asking if residents considered the DWHOS a disaster, the interview instrument addressed residents' perceptions of responsibility for causing the disaster. According to informants, residents primarily blamed BP and the federal government, with BP receiving the majority of blame. Respondents' initial response to the question tended to reflect this statement by a Running Oak politician: "Obviously, BP caused the spill. One hundred percent of the people here think it's BP's fault." However, with few exceptions, informants criticized other institutional actors as well. In Running Oak, a politically conservative area, several business leaders blamed federal environmental policies. Some informants suggested the disaster may have been avoided if drilling was allowed in more geographic areas.

Alternatively, informants from the liberal-oriented, underserved community of Palmetto made comments such as this:

> Fifteen to twenty years ago the laws changed to allow a few people to own all of media so whatever we get right now is their spin on whatever's going on. They own TV, radio, and newspaper. Guess who owns stock in BP. Guess who's in Washington. . . The government, the news media, and BP, they're all part of this problem (community activist).

Although the tone is conspiratorial, it illustrates a wide-spread perception that extra-local institutions were negligent in their responsibility to protect residents. Nearly every informant in the two communities blamed BP for creating the spill, the federal government for its failure to provide oversight, and the media for its complicity in not communicating the problem fairly. In all cases, informants felt major social institutions had failed in their responsibility to provide the public with clear and truthful information:

> People here would have responded better if the agencies would have started off by saying they didn't know [about the oil spill's impacts]. . .rather than trying to pretend and try and make people feel okay in ways that end up making people feel worse and then makes us distrustful (business owner in Running Oak).

This informant also noted that residents were concerned about the ecological effects from dispersants. According to her, information about the dispersants had not been properly communicated to residents. Informants with this perspective further noted the oil spill was not the first or the last time these institutions would be at fault for creating a disaster.

As a follow-up question, informants were asked if their communities could have done anything to mitigate the risk of an oil spill disaster. The question was intended to explore a sense of local responsibility for the environment. In response, informants said there was nothing that could have been done at the local level. They suggested the problem was too big and too complex for local action. A typical response went something like: "This is something the federal government needs to take care of—they have the knowledge and the resources" (commercial fisherman from Running Oak). Although informants placed much of the blame on the federal government, they continued to ascribe full responsibility for risk mitigation to federal government.

Exceptions to this trend were more likely to emerge from Palmetto than Running Oak. One resident from Palmetto responded that the community needs to "be more vigilant of our government and demand that it protects us from companies that are only interested in the bottom line" (elected official). Similarly, a local activist said, "What people can do at a local level is protest, scream as loud as they can. At the end of the day, power concedes nothing without the demand." These quotes reflect the shared vulnerabilities, experiences with disaster, and relationships with local and external institutions that characterize Palmetto as a community. These traits interact with local capacities to explain the community's response to the oil spill.

COMMUNITY CAPACITIES

Palmetto and Running Oak provide opportunities to explore how capacities affected community-wide response to the oil spill. All communities have capacities, but not all are aware of them or can exploit them in a positive way for the equal benefit of all residents (Kaufman and Wilkinson, 1967). Capacities materialized in different ways and to different degrees in both places. As well, barriers to effectively utilizing capacities affected response in both communities.

Despite the challenges facing Palmetto residents, informants described many positive qualities suggestive of a resilient community, even when long-term mitigation was difficult. Among these, most of the eighteen informants discussed local knowledge in addition to attachments to place and community through local culture and the natural landscape. Although not necessarily resource dependent, there was a clear and palpable legacy of resource use and pride associated with coastal life. This translated to a strong local knowledge of the natural resources surrounding the community. A charter boat operator suggested:

> It would have really been interesting if some of the federal agencies had gathered up some of the local folk. These fishermen have information that's so internalized that the scientists don't have. And the scientists could have made good use of that in terms of guidance about where to do their research, about where the baseline could have been. And I think that's where an awful lot of the problem lies. That the people with the titles and the positions assume that they know what is needed.

Other capacities emerged from the community's experience in disaster response and addressing environmental and social problems. Informants described local initiatives against city policies that encouraged unwanted land uses adjacent to the neighborhood. Residents were also galvanized over the threat of toxic dumping and septic waste problems. Some racial problems existed prior to Katrina; however, a unified stance against perceived extra-local hostility has replaced internal racial conflict since the storm. Describing this, a local elected official said:

> The biggest problem with poor communities is the community itself. I get into trouble for saying this, but black, white, Asian, Hispanic, it don't matter—we're all poor here. We can't let them [outsiders] kick us anymore. We have to learn to not be the path of least resistance. Like Miss Rosa said, 'I ain't havin it no more!' We don't have the lawyers, we don't have the money, but we got the people.

The leader continued by describing the loosely formed coalition of races and groups in the neighborhood as "the seedling, the new growth."

Although residents frequently described a negative relationship with city government, capacity emerged from a number of nonprofit organizations operating in Palmetto. The organizations included local and extra-local groups seeking to accomplish various objectives by providing organizational and physical resources to residents. One resident-operated organization helped to plan and coordinate the hurricane recovery and community development activities of other organizations. Other organizations built and rehabilitated nearly 700 homes of the 80 percent of homes destroyed by Hurricane Katrina (nonprofit organization employee). Addressing the benefits provided by the various groups, a business leader said, "If you take away people like the Mississippi Center for Justice and other groups that are here, then we'd really be in big trouble."

Running Oak residents tended to be more racially homogeneous, better educated, and had higher incomes than residents in Palmetto. However, interviews indicated a very different local activism than its counterpart. Informants noted a limited number of activist or resident assistance organizations. A local government agency employee explained the small, dispersed population accounted for a lack of resident groups: "There are not very many of those here. . . . It's a rural community." A commercial fisherman explained that resident participation was negatively affected by outmigration following Hurricane Katrina.

Many occasions for local interactions existed, but they failed to center on purposive community-wide initiatives as in Palmetto. A small number of organizations focused on providing adequate housing for Katrina victims. One state-wide group located in Running Oak advocated for conservation of marine resources. Informants listed a number of strong organizations focused on hobbies, such as history, gardening, and sailing. Civic groups and religious organizations provided assistance and organized community-wide recreational activities. Downtown, destroyed during the hurricane, tended to have few visitors; but, the public library was one of the busiest places in the community where countless interactions occurred.

One of the greatest assets of Running Oak was its strong city and county governments. Informants were generally satisfied with their elected officials and local agencies, even if they differed on political views. One informant said, "There's never a shortage of disasters here, so the local government has learned how to deal with it" (active resident). The government had aggressively undertaken redevelopment initiatives, including a "fresh look for tourists" by means of a beautification project.

DISCUSSION

This case study responds to Flint and Luloff's (2005) call for research that accounts for the interactional processes of community prior to and following a disaster event. To this end, we sought to compare responses between neighboring localities and explore interactional factors affecting resiliency. As well, our study challenges approaches highlighting dichotomies in hazards and risks. Informants' responses indicated the complexity of the oil spill event with the disaster rooted in technological, biophysical, and social origins, all of which were exacerbated by a number of vulnerabilities (Flint and Luloff, 2005; Tootle, 2007). More importantly, residents demonstrated capacities to assess their vulnerabilities and assets, their knowledge of the coastal landscape, and a strong desire to be involved with mitigation activities.

Community disaster response was strongly influenced by local and extra local institutions, which were the chief communicators of hazard information and risk. The control of information and framing characteristics of the event were used to manipulate public response. Like Katrina, competing narratives by national media and BP increased confusion, anxiety, and mistrust (Barnes et al., 2008; Elder et al., 2007; Tierney et al., 2006). Residents acknowledged benefits of the national media's spotlight on the problem, but criticized messaging that focused on vulnerability while overlooking cultural impacts and resilience. BP's control of response activities led to increased anger and distrust, especially in the underserved area of Palmetto.

Interviews underscored the notion of recreancy, reflecting Freudenburg and Gramling's (1994) study of coastal residents' views about the U.S. Mineral Management Service. Recreancy emerged from residents' perceptions that the federal government failed to impartially and objectively regulate BP and protect the public. Although recreancy typically refers to government agencies, our study illustrates how the concept applies to other institutions as well. Residents attributed varying degrees of blame to BP and the national media for their failure to fulfill societal responsibilities such as providing a product or service free of harm (Davis, 2001).

Our study challenges a dichotomous view of social versus environmental vulnerabilities. Informants acknowledged biophysical and socioeconomic vulnerabilities discussed in the literature on coastal hazards (e.g., Picou and Martin, 2006). Repeated hurricanes, coastal erosion, and toxic dumping intersected with social vulnerabilities such as poverty, unemployment, development, and environmental racism, each of which intensified the impacts of the oil spill. Informants also highlighted cultural disruptions, many caused by Hurricane Katrina or man-made environmental threats, which made communities more vulnerable to the impacts of the oil spill. Informants from both communities discussed disruptions in sense of place that were worsened by the oil spill. Cultural vulnerabilities were not always associated with resource dependency as examined in previous research (Dyer, 1993; Gill and Picou, 1991). Nonetheless, they were significant barriers to the emergence of community resiliency.

Previous disaster experience was interwoven throughout the interviews and strongly influenced community response to the oil spill. Running Oak informants used hurricanes as a baseline to understand the process of the oil spill moving from deep water to shoreline. However, this heuristic was not especially useful, and led to anxiety, confusion, and anger. In Palmetto, a lack of resources prevented full recovery from past disasters, making even a perspective about long-term recovery increasingly difficult to foster. The process has made residents pessimistic towards extra-local assistance, while simultaneously galvanizing them towards community-wide response activities. Local leaders carried out these activities from the perspective of shielding the community from external forces.

Much of the disaster literature has portrayed communities as helpless victims of disaster (Flint and Luloff, 2005). In many ways, this was played out in the study communities as extra-local institutions frequently choreographed relief and recovery. However, community responses illustrated capacities inherent to both localities. Although much of our conversations with informants focused on organizational assets such as nonprofit organizations or resident groups, many other capacities existed in the localities. For instance, local knowledge about coastal life and natural resources was mentioned as an important local asset, and something that was ignored during the disaster.

Vulnerabilities, past disaster experiences, and social institutions influenced the ability of informants to acknowledge local capacities. They also impacted the ability of the community to exploit its assets for community-wide benefit. Running Oak exemplified communities with capacities perhaps not fully realized due to an active, but not fully participatory government response. By contrast, Palmetto was able to utilize its capacities, ultimately leading to a basis for resiliency. Driving the differences were variations in how residents in each community shared understanding of vulnerabilities, their shared past disaster experiences, and the communities' institutional relationships. Unlike previous studies (Cutter et al., 2000; Paton and Johnston, 2006), income was less important than social interactions. In fact, income had an opposite effect in Running Oak where wealthy mobile residents were able to leave the community, thereby increasing vulnerability. Leadership and participatory action via nonprofit organizations, rather than local government, were critical in developing capacities for Palmetto.

Disaster research has examined therapeutic and corrosive community responses, depending on the origin of the disaster (Kroll-Smith and Couch, 1991).

In this study, neither community evidenced a corrosive reaction to the oil spill. Although Running Oak government was aggressively mitigating disaster effects, residents were disengaged and tended towards indifference or blinded trust towards local government. Such a response is not representative of community agency from an interactional perspective (Luloff and Swanson, 1995).

Comments suggested a unified response among Palmetto residents, despite perceptions that city government provided little long-term assistance. The response was subtle and largely underscored shared perceptions of risk, blame, and perspectives about social vulnerabilities. However, the characteristics were far from therapeutic, thereby serving to ameliorate disruption. Some apathy existed; however, by and large, residents and local leaders portrayed community-wide perceptions strengthening resentment towards extra-local institutions and processes that marginalized residents.

IMPLICATIONS FOR COMMUNITY DEVELOPMENT

Too often, disaster response is focused on relief and recovery because it is built upon the traditional hazard paradigm. Recovery without harnessing capacities for redevelopment implies a return to the status quo. Instead, planners, local decision-makers, and recovery organizations should emphasize redevelopment throughout the response process. Needs and capacities must be incorporated into a community redevelopment model suited to the locality. Such a model necessarily integrates an "interactional community" (Wilkinson 1991) assessment of the social and biophysical environments (including interactions between the two), institutional relationships, and experiences and local knowledge of residents.

For example, local knowledge must be incorporated in disaster impact assessment, redevelopment, and risk mitigation each through democratic processes that directly incorporate residents. Through participation, communities can identify their capacities and needs for the times in-between disasters that need to be addressed so that residents are in a better position to respond to disaster events. Creating opportunities for resident input is not enough. Residents must be empowered and actively engaged in decision-making for disaster response and redevelopment projects. They must have some control over the structure of the process.

Incorporating community development in discussions about disaster-prone places is nothing new (Burton et al., 1978; Mileti, 1999; Paton and Johnston, 2006; Perrings, 1991). Yet there continues to be a disconnect between research advocating holistic response to disasters and actual responses. Our study suggests discourses surrounding disaster response continue to take place within a hazards paradigm relying on technical solutions, bypassing the role of social interaction and local capacity for action. Similarly, Deborah Tootle (2007:23) commented in her reflective account of Hurricane Katrina response in southern Louisiana:

> The organizations with the most responsibility for disaster recovery are working under the assumption that communities and the people within those communities are incapable of addressing issues related to community redevelopment. They are, in essence, ignoring the complexities of the community development process.

It is therefore critical to educate local decision makers and community leaders about community development processes in the context of disaster response.

On a final note, while a redevelopment model for disaster response may have short-term goals, it is embedded in a long-term process. As a result, there is a need to study the long-term community impacts of the DWHOS spill. The literature on the Exxon Valdez Oil Spill provides an example of such research (Gill and Picou, 1991; Palinkas et al. 1993). Study of the long-term impacts in Mississippi should account for the many residents not economically dependent on marine resources, although culturally dependent on these resources. As well, long-term impact assessments should address other vulnerabilities unique to the area, such as marine resources that were stressed prior to the oil spill. Combined with an interactional community assessment, long-term processes of redevelopment will improve the social and ecological dimensions of the communities.

NOTES

1. Other fundamental elements of community encompass family, government and religious institutions, civic and business organizations, the environment, and solidarity (Wilkinson, 1991).
2. The names presented here are fictional to help protect the anonymity of research participants.
3. Facilitated group discussions were conducted to verify key informant data and provide additional insights. This paper focuses only on the key informant data.
4. A BP response program that employed displaced fisherman and charter boat operators in a variety of clean-up activities, including transporting supplies, performing wildlife rescue, and towing and deploying oil containment booms.
5. BP was legally obligated to control clean-up activities under the *Oil Pollution Act* of 1990 (EPA 2010b).
6. The lead author of this paper was denied repeated requests for interviews with BP employees and contractors, citing company policy against speaking publically about the spill. However, these informants *were* interviewed once their commitments with BP or BP contractors were completed.

REFERENCES

Aronoff, M. and Gunter, V. (1992) Defining disaster: local constructions for recovery in the aftermath of chemical contamination, *Social Problems*, 39, 4, 345-65.

Barnes, M.D., Hanson, C.L., Novilla, L.M., Meacham, A.T., McIntyre, E., and Erickson, B.C. (2008). Analysis of media agenda setting during and after Hurricane Katrina: Implications for emergency preparedness, disaster response, and disaster policy. *American Journal of Public Health*, 98, 604-10.

Beck, U. (1999). *World Risk Society*. Cambridge: Polity Press.

Blaikie, P., Cannon, T., Davis, I., and Wisner, B. (1994). *At risk: Natural hazards, people's vulnerability, and disasters*. London: Routledge.

Burton, I., Kates, R.W., and White, G. F. (1978). The environment as hazard. New York: Guilford.

Cannon, T. (1994). Vulnerability analysis and the explanation of 'natural' disasters. In A. Varley (Ed.), *Disasters, development, and the environment* (pp.13-30). New York: Wiley.

Creswell, J.W. (1998). *Qualitative inquiry and research design: Choosing among five traditions*. Thousand Oaks, CA: Sage.

Cumberpatch, M.L. (Ed.) (2006). *Gulf coast hurricanes: Selected resources in the NOAA libraries and information network*. NOAA Central Library.

Cutter, S.L., Mitchell, J.T., Scott, M.S. (2000). Revealing the vulnerability of people and places: A case study of Georgetown County, South Carolina. *Annals of the Association of American Geographers*, 90, 713-737.

Davis, K. (2001). The case for and against business assumption of social responsibilities. *Academy of Management Journal*, 16, 2, 312-322.

Douglas, M. (1966). *Purity and danger: Concepts of pollution and taboo*. London: Routledge and Kegan Paul.

Dyer, C.L., Gill, D.A., and Picou, J.S. (1992). Social disruption and the Valdez oil spill: Alaskan Natives in a natural resource community. *Sociological Spectrum*, 12(2), 105-126.

Dyer, C.L. (1993). Tradition loss as secondary disaster: The long-term cultural impacts of the Exxon Valdez Oil Spill. *Sociological Spectrum*, 13, 65-88.

Elder, K., Xirasagar, S., Miller, N., Bowen, S., Glover, A., and Piper, C. (2007). African Americans' decisions not to evacuate New Orleans before Hurricane Katrina: A qualitative study. *American Journal of Public Health*, 97(S1).

EPA. (2010a). Administrator Lisa P. Jackson, Remarks at the 2010 Small Business Environmental Conference, As Prepared. Accessed 10/9/2010: http://www.yosemite.epa.gov/ opa/ admpress.nsf.

EPA. (2010b). Oil Pollution Act. Accessed 9/2/2010: http://www.epa.gov/oem/ content/lawsregs/ opaover.htm.

Erikson, K. (1976). *Everything in its path: Destruction of community in the Buffalo Creek Flood*. New York: Simon and Schuster.

Fischer, H.W., III. (1998). *Response to disaster: Fact versus fiction and its perpetuation: The sociology of disaster* (second edition). New York: University Press of America.

Flint, C.G. and Luloff, A. E. (2005). Natural resource based communities, risk and disaster: An intersection of theories. *Society and Natural Resources*, 18, 399-412.

Flora, J.L. 1998. Social capital and communities of place. *Rural Sociology*, 63, 481-506.

Fowlkes, M. and P. Miller. (1987). Chemicals and community at Love Canal. In B. Johnson and V. Covello (Eds.), *The social and cultural construction of risk*. Boston: D. Reidel Publishing Company, pp. 55-78.

Frank, F. and Smith, A. (1999). *The community development handbook: A tool to build community capacity*. Ottawa: Minister of Public Works and Government Services Canada.

Freudenburg, W.R. (1993). Risk and recreancy: Weber, the division of labor, and the rationality of risk perceptions. *Social Forces*, 71,909-32.

Freudenburg, W.R. and Gramling, R. (1994). *Oil in troubled waters: Perceptions, politics, and the battle over offshore drilling*. New York: State University of New York Press.

Fritz, C.E. (1961). Disasters. In R.K. Merton and R.A. Nisbet (Eds.), *An introduction to the sociology of deviant behavior and social disorganization* (pp. 651-691). New York: Harcourt, Brace, and World.

Gill, D. and Picou, S. (1991). The social psychological impacts of a technological disaster: Collective stress and perceived health risks. *Journal of Hazardous Materials*, 27, 77-89.

Gill, D. and Picou, S. (1999). The day the water died: Cultural impacts of the Exxon Valdez oil spill. In J.S. Picou, D. Gill, and M. Cohen (Eds.), *The Exxon Valdez disaster: Readings on a modern social problem* (pp. 167-191). Dubuque, Iowa: Kendall/Hunt.

Gordon, J.S., Stedman, R.C., Matarrita-Cascante, D., and Luloff, A.E. (2010). Fire perception in rapid growth communities. *Rural Sociology*, 75, 3, 455-477.

Gramling, R., and Freudenburg, W.R. (1992). The Exxon Valdez oil spill in the context of US petroleum politics. *Industrial Crisis Quarterly*, 6, 175-195.

Hewitt, K. (1998). Excluded perspectives in the social construction of disaster. In E. L. Quarantelli (Ed.), What is a disaster? Perspectives on the question (pp. 75-91). New York: Routledge.

Holling, C. (1986). Resilience and stability of ecological systems. *Annual Review of Ecology and Systematics*, 4, 1-23.

Kaufman, H., and Wilkinson, K. (1967). Community structure and leadership: An interactional perspective in the study of community. *Research Bulletin 13*. Mississippi State, MS: Mississippi State University Social Science.

Kroll-Smith, S. and Couch, S. (1991). "What is a Disaster?" An ecological symbolic approach to resolving the definitional debate. *International Journal of Mass Emergencies and Disasters*, 9, 355-66.

Kumagai, Y., Carroll, M.S., and Cohn, P. (2004). Coping with interface wildfire as a human event: Lessons from the disaster/hazards literature. *Journal of Forestry*, 102, 28-32.

Luloff, A.E. and Swanson, L.E. (1995). Community agency and disaffection: Enhancing collective resources. In L.J. Beaulieu and D. Mulkey (Eds.),

Investing in people: The Human capital needs of rural America Boulder, CO: Westview Press, pp. 357-72.

Luloff, A.E. (1999). The doing of rural community development research. *Rural Society*, 9, 1, 313-27.

Mileti, D. (1999). *Disasters by design: A reassessment of natural hazards in the United States*. Washington, D.C.: Joseph Henry Press.

Nisbet, R. (1953). *The quest for community: A study in the ethics of order and freedom*. Oxford University Press.

Palinkas, L.A., Downs, M.A., Petterson, J.S., and Russell, J. (1993). Social, cultural, and psychological impacts of the *Exxon Valdez* oil spill. *Human Organization*, 52, 1, 1-13.

Paton, D. and Johnston, D. (2006). Identifying the characteristics of a disaster resilient society. In D. Paton and D. Johnston (Eds.), *Disaster resilience: An integrated approach* (pp. 11-18). Springfield, IL: Charles C. Thomas.

Perrings, C. (1991). Resilience and sustainability. In H. Folmer, H.L. Gabel, and A. Rose (Eds.), *Frontiers of Environmental Economics*. Cheltenham, UK: Edward Elgar, pp. 319-41.

Picou, J.S., Formichella, C., Marshall, B.K., and Arata, C. (2009). Community impacts of the Exxon Valdez oil spill: A synthesis and elaboration of social science research. In S.R. Braund and J. Kruse (Eds.), *Synthesis: Three decades of research on socioeconomic effects related to offshore petroleum development in coastal Alaska*. MMS OCS Study 2009-006, pp. 278-310.

Picou, J.S. and Gill, D. (2000). The Exxon Valdez disaster as localized environmental catastrophe: Dissimilarities to risk society theory. In M.J. Cohen (Ed.), *Risk in the modern age: Social theory, science and environmental decision-making*. New York: St. Martin's Press, pp. 143-70.

Picou, J.S. and Martin, C.G. (2006). Community impacts of Hurricane Ivan: A case study of Orange Beach, Alabama. *Quick Response Report 190*. Natural Hazards Center.

Picou, J.S., Marshall, B.K., and Gill, D.A. (2004). Disaster, litigation, and the corrosive community. *Social Forces*, 82, 1493-522.

Quarentelli, E.L. and Dynes, R. (1987). Community Conflict: Its absence and its presence in natural disasters. *Mass Emergencies*, 1, 139-52.

Quarantelli, E.L. (1960). Images of withdrawal behavior in disasters: Some basic misconceptions. *Social Problems*, 9, 68-79.

Slovic, P. (1987). Perceptions of risk. *Science*, 236, 280-85.

Tierney K., Bevc C., and Kuligowski E. (2006). Metaphors matter: disaster myths, media frames, and their consequences in Hurricane Katrina. *Annals of the American Academy of Political Social Science*, 604, 57–81.

Tootle, D.M. (2007). Disaster recovery in rural communities: A case study of southwest Louisiana. *Southern Rural Sociology*, 22, 6-27.

Walter, B., Arkin, L., and Crenshaw, R. (Eds.) (1992). *Sustainable cities: Concepts and strategies for eco-city development*. Los Angeles: Eco-Home Media.

Wellman, B. (1988). Structural analysis: From method and metaphor to theory and substance. In B. Wellman and S.D. Berkowitz (Eds.), *Social structures: A network approach* (pp. 19-61). Cambridge: Cambridge University Press.

Wilkinson, K.P. (1991). *The community in rural America*. Middleton, WI: Social Ecology Press.

CHAPTER ELEVEN
DISASTER VULNERABILITY: THE DIFFERENTIAL IMPACT OF THE DEEPWATER HORIZON DISASTER AMONG ALABAMA'S GULF COAST RESIDENTS

JAMES HAWDON AND JOHN RYAN

ABSTRACT

On April 20, 2010, the Deepwater Horizon oilrig exploded, killing eleven crewmembers and eventually becoming the largest offshore-oil spill in U.S. history. While the environmental and economic effects of the Deepwater Horizon disaster will likely be severe and widespread, it is widely recognized in disaster research and vulnerability science that disasters do not affect populations equally. Instead, class, ethnicity, gender, health, age, and the nature of the social networks in the affected community influence vulnerability. In this paper, we analyze survey data collected from 296 residents of Bayou La Batra, Gulf Shores, and Mobile Alabama in August 2010 that address the affect of the BP oil spill on the residents' economic, social, and community wellbeing using a series of binary logistic regression analyses. Our results do not support the frequent finding that disasters disproportionately affect the disenfranchised. We offer a possible explanation for these findings, derive hypotheses based on this explanation, and outline possible implications for disaster research.

INTRODUCTION

On April 20, 2010, the Deepwater Horizon oilrig exploded and killed eleven crewmembers. When the rig sank two days later, the well continued to spew oil. With 35,000 to 60,000 barrels of oil leaking from the well each day for over three months (Gillis 2010), Deepwater Horizon is the largest offshore-oil spill in U.S. history (USA Today 2010). As the oil continued to leak from the ocean floor, Deepwater Horizon became catastrophic. By early June, oil slicks threatened the GulfCoast's fragile ecology and already frail economy (see MSNBC 2010; Morgan 2010). The disaster's economic effects were widespread as the federal government suspended deepwater oil drilling, states prohibited fishing in waters off the coastline, and vacationers canceled their GulfCoast trips (Morgan 2010). Although BP (British Petroleum) stopped the spill, it had already affected the regional economy, which is heavily dependent on tourism and the oil and fishing industries, and will likely result in as much as $1.6 billion of annual economic activity and services in the region (McKinney 2010).

While the environmental and economic effects of the Deepwater Horizon disaster will likely be severe and widespread (see McKinney 2010), it is widely recognized in disaster research, vulnerability science, and critical incident analysis that disasters do not affect populations equally. Instead, class, ethnicity, gender, health, age, and the nature of the social networks in the affected community influence vulnerability (Bullard 2000; Fradkin 2005; Peacock et al. 1997; Wisner et al. 2003). Based on numerous studies, scholars generally recognize that disenfranchised individuals, groups, and communities disproportionately suffer the brunt of disasters, both in terms of their likelihood of being a disaster victim and in access to resources necessary to recover from their victimization (e.g., Barry 1997; Bolin and Stanford 1998; Brinkley 2006; Dash, Peacock, and Morrow 1997; Fradkin 2005; Girard and Peacock 1997; Gotham and Greenberg 2008; Horne 2006; Jenkins and Phillips 2008; Klinenberg, 2002; Morrow 1997; Norris et al, 2001; Ryang 2003). However, is this always the case? Do the disenfranchised always suffer more, or does the likelihood of being a disaster victim depend on the nature of the disaster? We address these questions in this paper.

We begin by reviewing the literature on the impact of disasters on different populations. We review evidence concerning both natural and technological disasters and derive hypotheses concerning how the likelihood that a disaster will adversely affect the disenfranchised or the relatively affluent. We then analyze survey data collected from 296 residents of Bayou La Batra, Gulf Shores, and Mobile Alabama in August 2010 that address the affect of the BP oil spill on the residents' economic, social, and community wellbeing using a series of binary logistic regression analyses. Finally, we conclude by outlining the implications of our findings for disaster research.

LITERATURE REVIEW

Disasters can happen anywhere and to anyone. Regardless of their race, class, gender, religious affiliation, age, ethnicity, sexual orientation, or political allegiance, no one in the path of a Category 5 hurricane, at the epicenter of a 7.1 earthquake, or in the shaft of an exploding coalmine is immune from danger. Yet, while disasters kill, wound, traumatize, and disrupt the lives of all those in its destructive wake, it would be mistaken to argue that everyone has an equal probability of being in the wake of a disaster or that everyone who survives a disaster has an equal probability of recovering from it. Disaster research and vulnerability science has clearly established that disenfranchised individuals, groups, and communities disproportionately suffer the brunt of disasters, both in terms of their likelihood of being a disaster victim and in access to resources necessary to recover from their victimization.

For example, relative to the more socially affluent in the afflicted communities, the disadvantaged were more adversely affected by the 1906 San Francisco earthquake (Corbett 2010; Fradkin 2005), the 1923 Great Kanto earthquake (Ryang 2003), the Northridge earthquake (Bolin and Stanford 1998), hurricanes Andrew (Morrow 1997; Girard and Peacock 1997; Dash, Peacock, and Morrow 1997), Katrina and Rita (see Brinkley 2006; Gotham and Greenberg 2008; Horne 2006; Jenkins and Phillips 2008;), the 1927 Mississippi River flood (Barry 1997), the Pakistani floods of 2010 (Ahmed 2010), and the 1999 Venezuelan mudslides (Quarantelli 2003). The clustering of low-income populations in flood-prone areas in Buenos Aires and numerous other cities in Argentina (Clichevsky 2003) as well as in Dhaka Bangladesh (Huq and Alam 2003) increase the poor's vulnerability to disastrous floods there just as poverty increased vulnerability during Hurricane Katrina[1].

Other statuses also pattern vulnerability, and these too typically result in the relatively disenfranchised facing greater vulnerability. For example, the elderly, especially those with limited economic means and those with limited social-support networks, were the most likely to perish in the 1995 Chicago heat wave (Klinenberg, 2002) and the 2003 European heat wave (Larsen 2003). Likewise, street children comprised a disproportionate number of the victims when Hurricane Mitch killed nearly 19,000 people throughout Central America in 1998 (General Secretariat Central American Integration System 1999). Throughout American history, African-Americans have frequently suffered disproportionately in disasters (see, for example, Brinkley 2006; Rivera and Miller 2007). Women are also more vulnerable to disasters than are men (Fordham 1999). For example, the 2004 tsunami cost more women than men their lives, largely due to cultural constructions of gender (Oxfam International 2005). Similar patterns are found for numerous statuses and in numerous other disasters. In short, "class, ethnicity, caste, gender, disability status, health, age" and the "nature and extent of social networks" influence vulnerability (Wisner, et al 2003, p. 11), and those in the higher status positions or the socially dominant group face less risk than their less advantaged neighbors.

In addition, just as the more disenfranchised within a community have a greater probability of being a victim of a disaster than do the more socially affluent, as compared to developed nations, developing nations experience a disproportionate amount of disasters (see OAS 1991; CRED 2010; Vos et al. 2010; Noris et al. 2001; Albala-Bertrand 1993; ECLAC 1991). Although total property damage tends to be more extensive when a disaster strikes in the developed world, the economic costs of disasters in poor countries often exceeds three percent of the nation's gross national product and can exceed 20 percent of the nation's GDP (Berke 1995). In contrast, even massively expensive disasters such as hurricanes Andrew and Katrina represent an almost undetectable proportion of the country's economy (see Quarantelli 2003). More dramatically, as seen in the recent Haiti earthquake, China earthquake, or the Pakistan floods, the loss of life can reach dramatic proportions when disasters affect developing countries (in general, see Bowonder and Kasperson 1988; Quarantelli 2003).

Thus, natural disasters appear to disproportionately affect the disenfranchised; yet, is this also the case for technological disasters?[2] A *technological disaster* is the breakdown of modern systems, equipment, or engineering which results in harm to people, their surroundings, and the environment. Examples of technological disasters include chemical or nuclear plant explosions, mining accidents, and major train derailments involving hazardous materials. Based on the evidence from several disasters, it also appears that the disenfranchised are more vulnerable to technological disasters.

For example, the 1984 Bhopal chemical disaster, which occurred when a chemical reaction in a tank at a Union Carbide pesticide factory created a poisonous gas, killed an estimated 3,800 people immediately (Dhara and Dhara 2002; Fortun 2001; Broughton 2005; Koplan, Falk and Green 1990). Government of Madhya Pradesh confirmed a death toll of 3,598, but by 1994 the toll was estimated at over 6,000. It is now believed that 15,000 and 20,000 have died because of the disaster (Dhara and Dhara 2002). The victims of the disaster were mostly poor, with 80 percent having monthly incomes of less than $6.00 per month. The majority of the victims lived in a slum colony adjacent to the Union Carbide plant, and the majority of those exposed to the toxic chemicals "were close to the government-defined poverty level" (Dhara and Dhara 2002, p. 392). Similarly, the Buffalo Creek disaster, where a Pittston Coal Company's mine-waste impediment damn broke on February 26 1972, killed 125, injured more than 1,100, and left over 4,000 residents homeless (Erikson 1976). The areas along Buffalo Creek were coal towns that once provided employment for thousands; however, beginning in the 1950s and 1960s, mechanized mining forced many out of work. As a result, the county was relatively poor with over 23 percent of the county living below poverty. In addition, over one-third of the victims were children, and nearly 15 percent of the victims were over the age of 65 (West Virginia Division of Culture and History 2010). Again, the relatively disenfranchised disproportionately shouldered the brunt of the disaster's force.

In general, the disenfranchised are vulnerable to technological disasters because they live in close proximity to dangerous technologies while the more advantaged are able to avoid living in these hazardous environments. Toxic-waste facilities are often located in communities that have high percentages of poor, elderly, young, and minority residents, and uncontrolled toxic-waste sites are disproportionately found in black and Hispanic urban communities (Bullard 2000; Greenberg and Anderson 1984). For example, in Los Angeles, Hispanic's are nearly twice as likely to live in waste-site areas as are whites, and large commercial hazardous-waste landfills and disposal facilities are disproportionately found in poor, rural African-American communities throughout the south (Bullard 2000; Greenberg and Anderson 1984). Other polluting industries, such as dangerous chemicals and paper mills, are also disproportionately located in minority communities (Bullard 2000). Recently, the Tennessee Valley Authority shipped toxic coal ash from the massive spill that occurred in December 2008 at the Kingston power plant in east Tennessee to landfills in Georgia and Alabama, and the communities to which the ash was sent have large black populations and high poverty rates (Sturgis 2009). Given that toxic sites are more likely to be found near poor communities, poor citizens face greater risks of suffer from an environmental disaster than does the general population. As Benson and Clay (2003, p. 12) state, "poverty is the single most important factor determining vulnerability, in part reflecting location of housing (e.g., on floodplains, riverbanks, steep slopes, or contaminated land previously occupied by industrial facilities)."

Based on this review, we would predict that the BP disaster adversely affected the relatively disenfranchised more than the relatively affluent. Therefore, at the community level, we predict that the two resource-dependent communities (Bayou La Batra and Gulf Shores) will be more adversely affected than was Mobile. At the individual level, we predict that those with lower levels of civic participation and trust, those who rent their homes, women, African Americans, those with less education, those who are older, and those who are not employed would be more likely than others to experience adverse effects from the Deepwater Horizon disaster. Was this indeed the case? We now turn to some results of a survey conducted while the disaster was still unfolding.

METHODS

We conduct a series of analyses to determine upon whom the Gulf disaster had the greatest effect. The Center for Survey Research (CSR) at Virginia Tech fielded a telephone survey in Gulf Shores, Bayou La Batra, and Mobile Alabama between August 2 and August 11, 2010. The survey primarily focused on community cohesion and social capital; however, several items concerning the Gulf disaster were also included.

To obtain our sample, we first select three Alabama communities to study. Although we had the generous support of Virginia Tech's Institute for Society, Culture, and the Environment, our resources were limited; therefore, we selected three contrasting areas that would likely experience the Gulf disaster very differently because they vary in size, economic diversification, and, most importantly, their relationship with the Gulf of Mexico. The three communities were Bayou La Batra, Gulf Shores and Mobile Alabama.

Alabama's coastal area has a substantial seafood and tourism industry, and the Gulf contributes to both the economic vitality and overall quality of life for coastal residents. Bayou La Batre, the "Seafood Capitol of Alabama" (http://gulfinfo.com/bayoulabatre/), is a small fishing town of less than 3,000 people. It is a seafood-processing harbor and is ranked nineteenth in the nation in value of landings and the fifth most important port in the Gulf of Mexico (Burrage 2010). In 2000, over eight percent of Bayou La Batra's workforce worked in the fishing industry (U.S. Census 2010a). Oil extraction and shipbuilding are also major industries in the area. Gulf Shores, with a population of over 5,000 people, is one of Alabama's three seaside resorts. With numerous beachfront hotels, condominiums, and vacation houses, tourism and real estate dominate the local economy. Over 20 percent of the workforce works in the "Arts, entertainment, recreation, accommodation, and food services" sector (U.S. Census 2010b). As a "fishing village" and "resort town," these two small towns are both "resource-dependent" communities that rely heavily on the Gulf's ecological health; however, they provide a contrast in terms of which aspects of the Gulf on which they depend. Bayou La Batra is dependent on the Gulf's ecological vitality while Gulf Shores depends on the Gulf's aesthetic qualities. The disaster damaged both the Gulf's ecological vitality and aesthetic qualities and likely negatively affected both areas. Yet, because they depend on different aspects of the Gulf's resources, our study will allow us to determine if the disaster's community-level impacts vary depending on the community's relationship with its local region and the basis of its economy.

In contrast to the two resource-dependent communities, Mobile is a major metropolitan area with a diversified economy. With a 2006 population of approximately 193,000, Mobile is the third largest city in Alabama (US Census 2010c). Located on the Gulf at the junction of the Mobile River and Mobile Bay, the city is the only seaport in Alabama and the ninth largest by tonnage in the United States (US Corp of Engineers 2008). Mobile also has aerospace, retail, services, construction, medicine, and manufacturing as major industries. Although the Gulf disaster has likely adversely affected Mobile's economy, there are important differences between Mobile and the coastal communities. First, Mobile's economy is more diversified and therefore less dependent on the Gulf. Second, Mobile's economic relationship with the Gulf of Mexico is based on shipping more than tourism or fishing, and the disaster did not seriously affect the shipping industry. Thus, the three communities offer a contrast with respect to their size, economic diversification, and relationship with the Gulf of Mexico.

Once these communities were selected, we then sampled households within each community. CSR used systematic randomization from the telephone exchanges available for the Federal Information Processing Standards (FIPS) code identifiers associated with each of the localities. CSR randomly selected households from each community using random-digit dialing (RDD), and both listed and unlisted telephone numbers will be included in the samples. Because some telephone exchanges were outside the targeted geographic areas, a screener question confirmed residence in one of the targeted communities.

In total, 907 residents of the targeted areas were contacted, and 303 completed surveys were obtained for a total response rate of 33.4 percent. Missing data on key variables eliminated seven respondents; thus, the analyses are conducted on 297 residents. Although the response rate is not ideal, it is reasonably high for community telephone surveys (see AAPOR 2008 for a discussion of total response rates). More importantly, the data accurate reflect the communities from which the samples were selected on a number of important demographic characteristics. For example, the sample was within five percent of the population figures derived from census data for percent minority, percent with at least a high school degree, and percent home ownership in all three communities. As is often the case with telephone surveys that sample households, females were over-represented in the sample. The analyses are based on weighted data that corrects for the over-sampling of women.

MEASURES

We investigate several measures of the disaster's effect on residents of the three communities. First, respondents were asked if "the oil spill negatively affected your work situation in any way" (responses were 0 "no" or 1 "yes"). Next, respondents were asked if they knew of anyone who moved away from the area because of the oil spill (responses were 0 "no" or 1 "yes"). Finally, respondents were asked "Since the Gulf oil spill, would you say people in your neighborhood seem closer together, more divided, or about the same."

Possible predictors of the disaster-impact variables included age, length of residency in the community, education level, gender, ethnicity, homeownership, and work status. Age and length of residency were coded as continuous variables. Education included the responses "did not complete high school," "high school degree or equivalent," "trade school," "some college," college degree," and "graduate or professional degree." The remaining variables were coded as indicator variables. Females, African-Americans, and homeowners are compared to males, non-African-Americans, and renters, respectively. Work status included the possible responses of "working," "unemployed," "temporarily unemployed," "retired," "disabled," "homemaker," or "student." We use several different categories as reference categories in the analyses that follow, and these will be identified in each analysis. Finally, since social capital can mitigate the effect of a disaster on individuals (see Wisner, et al 2003; Bolin and Stanford 1998; Hawdon and Ryan forthcoming), we include measures that reflect the two major dimensions of social capital: trust and civic participation (see Putnam 2000). As is common in research on social capital, we measure the "trust" dimension with the item, "In general, would you say that most people can be trusted or that you can't be too careful?" Responses "you can't be too careful" and "it depends" were coded as 0, while the response "most people can be trusted" was coded as 1. To measure the civic participation dimension of social capital, respondents were asked, "In the past year, approximately how many times have you volunteered for a religious, political, social, neighborhood, or other organization." Responses on this item ranged from 0 (never) to 8 (more than once a week). Descriptive statistics for all variables used in the analyses are reported in the Appendix.

ANALYSES

We begin by comparing our measures of the disaster's impact across the communities. We compare respondents from Bayou La Batra and Gulf Shores with those from Mobile on the issues of the disaster affecting the respondents' work situation and if they knew someone who relocated because our initial investigation indicated that the two coastal communities did not significantly differ on these issues. Thus, the two "resource dependent communities" are compared to Mobile, which is economically less dependent on the Gulf. On our third measure of the disaster's impact (if the disaster affected the community), we compare the three communities because Bayou La Batra and Gulf Shores significantly differ.

WORK SITUATION

We first investigate if the oil spill negatively affected the respondents' work situation. For this analysis, we eliminated respondents who identified their work situation as retired, a student, a homemaker, or disabled and selected only those respondents who were working, unemployed, or temporarily unemployed. We assume here that those who identified themselves as retired did not retire because of the oil spill. Here, it is clear that the oil spill affected those in the communities that are most dependent on the Gulf for their economic vitality. Two-thirds of the respondents in Bayou La Batra and Gulf Shores responded affirmatively to the question. By comparison, only 30 percent of respondents from Mobile said the spill negatively affected their work situation. The relationship between community and the disaster's impact on respondents' work is statistically significant ($\chi^{2(df=1)} = 18.1$; $p < .001$). Table 11.1 reports this crosstabular analysis.

Table 11.1: Crosstabulation of Spill Negatively Affected Work Situation by Community

		Community		
		Mobile	**Bayou La Batra and Gulf Shores**	**Total**
Did the oil spill negatively affect your work situation in any way?	No		23	71
		69.6%	33.3%	51.4%
	Yes	21	46	48
		30.4%	66.7%	48.6%
Total		69	69	138
		100%	100%	100%

Turning to a multivariate analysis, we now focus on the disaster's effect on individuals. Since the data are clustered (i.e. respondents are nested within communities), we conduct binary logistic regressions using clustered robust standard errors to control for the clustering of the data by community. The model accounted for approximately 13 percent of the variance in the dependent variable (pseudo $R^2 = .134$). Based on the logistic regression, community was the only significant predictor. As seen in the community analysis, relative to those in Mobile, residents of the two resource-dependent communities were nearly five times more likely to have their work situations adversely affected by the Gulf disaster (odds ratio = 4.53, $p < .001$). None of the other variables in the model approached statistical significance except age. Age was inversely related to the dependent variable (odds ratio = 0.97) at the $p = .059$ level. Table 11.2 on page 214 reports the results of the analysis.

RELOCATION DUE TO SPILL

Next, we investigate if the oil spill resulted in someone the respondent knew moving from the community. This analysis was conducted on all respondents regardless of work status. Again, the disaster's impact was not equally experienced in the three communities. As seen in Table 11.3 on page 214, 33.8 percent of respondents in the resource-dependent communities said someone they knew moved from the area because of the oil spill. In contrast, only 4.7 percent of respondents from Mobile knew someone who had moved because of the disaster. These differences are statistically significant ($\chi^{2(df-1)} = 40.7$; $p < .001$).

Table 11.2: Logistic Regression of if Spill Negatively Affected Work Situation

	Coef-ficient	Clustered Robust Standard Error (Clustered on Community)	Odds Ratio
Resource-Dependent Community	1.51 ***	.217	4.52
Civic Participation	-.006	.043	0.99
Social Capital / General Trust	-.127	.106	0.88
African-American	-.170	.276	0.84
Female	.335	.477	1.39
Education	.029	.091	1.03
Homeowner	-.304	.339	0.74
Age	-.031	.016	0.97
Length of Residency in Community	-.015	.025	0.98
Constant	1.33 ***	.338	

N = 127 * p < .05 ** p < .01 ** p < .001
Log pseudolikelihood = -76.23 Pseudo $R^2 = .133$

Table 11.3: Crosstabulation of Knowing Someone Who Moved Because of Spill by Community

		Community		
		Mobile	Bayou La Batra and Gulf Shores	Total
Do you know of anyone who moved from the area because of the oil spill?	No	142	100	242
		95.3%	66.2%	80.7%
	Yes	7	51	58
		4.7%	33.8%	19.3%
Total		149	151	300
		100%	100%	100%

In the multivariate analysis, we investigate if the oil spill caused anyone the respondents knew to move from the community. Again, we use logistic regression with clustered robust standard errors. In addition to the predictors used in the disaster's effect on work situation analysis, we included work status (those who were employed contrasted with all others) in the model. The model was significant and accounted for 25.7 percent of the variance in the dependent variable (pseudo $R^2 = .257$). Respondents from one of the resource-dependent communities were over nine times more likely to know someone who had moved because of the spill than were respondents from Mobile (odds ratio = 9.18; $p < .001$). Those who were working were over twice as likely to know someone who had moved because of the spill (odds ration = 2.32; p = .037). Age (odds ration = 1.04; p = .001) and education (odds ration = 1.39; p =.017) were also positively associated with knowing someone who moved because of the oil spill. Civic participation (odds ration = 1.09; p =.065) was positively related to knowing someone who moved because of the disaster at liberal levels of significance. Homeownership, length of residency in the community, race, gender, and the trust dimension of social capital were unrelated to knowing someone who moved from the community due to the disaster. Table 11.4 reports these results.

Table 11.4: Logistic Regression of Knowing Someone Who Moved from Community due to Oil Spill

	Coefficient	**Clustered Robust Standard Error (Clustered on Community)**	**Odds Ratio**
Resource-Dependent Community	2.22 ***	.062	9.18
Working	.843 *	.405	2.32
Civic Participation	.086	.046	1.09
Social Capital / General Trust	-.292	.248	0.75
African-American	-.813	.643	0.44
Female	-.140	.281	0.87
Education	.262 *	.119	1.29
Homeowner	.403	.309	1.49
Age	.027 *	.014	1.03
Length of Residency in Community	-.029	.018	0.97
Constant	-5.83	.433	

N = 289 * p < .05 ** p < .01 ** p < .001

Log pseudolikelihood = -105.57 Pseudo $R^2 = .257$

AFFECT ON COMMUNITY SOLIDARITY

Finally, we investigate resident's perceptions of how the oil spill affected their communities. Again, the relationship between community and the disaster's impact is statistically significant ($\chi^{2(df = 4)} = 51.3$; $p < .001$). While one-fourth of the respondents in Bayou La Batra believed the disaster divided the community, only 2.4 percent of Gulf Shore respondents and 1.4 percent of Mobile's respondents thought the disaster was divisive at the community level. At the same time, the disaster was also more likely to be seen as bringing residents closer together in Bayou La Batra than in the other communities. Slightly more than one-third of the respondents in Bayou La Batra (37.5 percent) said the disaster brought residents closer together in their community compared to 23.4 percent and 10.2 percent in Gulf Shores and Mobile, respectively. Table 11.5 reports the results of this analysis.

Table 11.5: Crosstabulation of Effect of Spill on Community Solidarity by Community

		Community			
		Bayou La Batre	**Gulf Shores**	**Mobile**	**Total**
Since the Gulf oil spill, would you say people in your neighborhood seem closer together, more divided, or about the same?	Closing Together	9	29	15	53
		37.5%	23.4%	10.2%	18.0%
	About the Same as Before	9	92	130	231
		37.5%	74.2%	88.4%	78.3%
	More Divided	6	3	2	11
		25.0%	2.4%	1.4%	3.7%
Total		24	124	147	295
		100.0%	100.0%	100.0%	100.0%

For the multivariate analysis, we investigate what predicts believing the disaster created a "corrosive community" (see Gil (2007) for a discussion). That is, the model predicts which respondents believed that the disaster had caused members of their community to become divided. As seen in the bivariate analysis, respondents from Gulf Shores (odds ratio = .080; p < .001) and Mobile (odds ratio = .057; p < .001) were significantly less likely than those from Bayou La Batra to believe that the oil spill had caused the members of their community to become divided. Civic participation (odds ratio = 1.35; p < .001) and length of residency in the community (odds ratio = 1.06; $p < .001$) were also positively related to believing the disaster had created a "corrosive community." African Americans were less likely than were others to hold this belief (odds ratio = 0.37; $p = .048$). The model accounted for approximately 29 percent of the variance (pseudo R^2 = .290). Table 11.6 presents the results of this analysis.

Table 11.6: Logistic Regression of Oil Spill Led to Corrosive Community

	Coefficient	Clustered Robust Standard Error (Clustered on Community)	Odds Ratio
Gulf Shores	-2.52 ***	.464	0.08
Mobile	-2.86 ***	.429	0.06
Working	-.443	1.12	0.64
Civic Participation	.302 ***	.068	1.35
Social Capital / General Trust	.550	.718	1.73
African-American	-.987 *	.499	0.37
Female	-.473	.323	0.62
Education	-.169	.318	0.84
Homeowner	-1.97	1.14	0.14
Age	-.007	.017	0.99
Length of Residency in Community	.061 ***	.008	1.06
Constant	-.527	1.41	

N = 284 * $p < .05$ ** $p < .01$ ** $p < .001$
Log pseudolikelihood = -33.57 Pseudo R^2 = .290

DISCUSSION

Based on our analyses, our general hypothesis that the disaster would disproportionately affect the relatively disadvantaged was supported at the community level; however, it was not supported at the individual level. At the community level, respondents in Bayou La Batra and Gulf Shores, two communities highly dependent on the Gulf for their economic vitality, were far more likely than were Mobile respondents to say that the disaster adversely affected their work situation. In our model, two-thirds of the respondents from resource-dependent communities who were working or were unemployed said the disaster had negative effects on their work situation, and they were four times more likely than Mobile respondents to have their work affected. Is striking as this effect is, it may under-represent the disaster's total impact on these communities. Several respondents from these communities noted that while the disaster did not affect them personally, it had affected a relative. For example, a male from Bayou La

Batra said that the spill "did not affect (my) work, but I know of many people who did experience work problems because of the oil spill." Similarly, a women from Gulf Shores said, "I'm a Real Estate broker and it has almost shut me down because no one wants anything on the beach anymore." Other respondents noted that while they personally had not been affected, many of their family members, particularly those in the fishing industry, had been adversely affected.

Similarly, respondents from the two resource-dependent communities were significantly more likely than were respondents from Mobile to say that the disaster had caused someone they knew to move from the community. Those from Bayou La Batra were also far more likely than others to say that the disaster had corrosive effects on their community. Thus, it is clear that the disaster was devastating to the communities that rely on the Gulf of Mexico for their economic vitality. Not only did it disrupt residents' work, it led to higher-than-normal rates of outmigration. Moreover, at least in one of the communities, the disaster appears to have damaged levels of community solidarity. This confirms the argument that technological disasters are more likely to create conflicts in the community than are natural disasters (see Gil 2007). Given the importance of social solidarity for community recovery to disasters (see Hawdon and Ryan 2010), it may take the residents of Bayou La Batra may years or even decades to recover from the adverse effects of the disaster.

Although the Gulf disaster appears to have adversely affected those communities that are relatively disenfranchised in the sense that they are more economically dependent on a limited number of industries than are a more diversified city such as Mobile, there is little evidence that the disaster adversely affected disenfranchised individuals more than their more affluent and privileged co-residents. First, and most importantly, we should recognize that the primary victims of the disaster were the eleven rig workers who were killed by the explosion. These workers, while not members of the upper class, were relatively skilled labors who earned relatively good wages. According to the Bureau of Labor Statistics and Oil Career.Com, as of September 2009, oilrig workers make between $40,000 and $69,000 per year, with an industry average of $54,640 per year (US Bureau of Labor Statistics 2010; Oil Career.Com 2010). While this average wage does not make one "rich," it is above the U.S.'s median family income of $52,029 (US Census Bureau 2010d). It is difficult to argue that anyone who makes more than the median income is "disenfranchised."

Next, once the disproportionate affects associated with living in a resource-dependent community were accounted for, none of the individual-level factors were related to having the disaster disproportionately affect the respondents' work situation. Therefore, in terms of employment and wages, the disaster equally affected those of different ages, homeownership status, races, gender, and levels of social capital. With respect to knowing someone who moved from their community because of the disaster, the relatively affluent were more likely than were the relatively disenfranchised to suffer such a disruption to their social networks. Those who were active in the community, those who were employed, and those with higher levels of education were more likely to know someone who had moved because of the disaster. Those who were active in the community were also more likely to believe that the disaster had corrosive effects on their community. Thus, our data do not support the general hypothesis found in the literature that disasters disproportionately affect the community's disenfranchised members.

IMPLICATIONS

While we do not want to over-generalize these findings, these findings make us question the generalizability of the hypothesis and may have implications for future disaster and vulnerability research. Although the evidence from natural disasters appear to clearly establish that the disenfranchised are more vulnerable than are the relatively affluent, this may not be as true for technological disasters. In the Deepwater Horizon disaster, it appears that the "relatively socially affluent" disproportionately experienced more adverse effects. Could this be the case for technological disasters more generally? We believe there may be an important distinction between natural and technological disasters that influences the general relationship between vulnerability and social status.

We should recognize that, like natural disasters, the impact of technological disasters varies tremendously. The geographic scope of devastation in floods and hurricanes are usually greater than are the scope of devastation of tornados or earthquakes. Similarly, while a mine explosion typically kills only those in the mine, the Bhopal disaster affected residents throughout the city and neighboring communities (Dhara and Dhara 2002), and the Romanian gold-extracting plant's dike break created cyanide-polluted water that flowed hundreds of miles into Hungary and Yugoslavia (Quarantelli 2003). As with all disasters, we should consider vulnerability as a function of the geographic scope of the disaster. If we consider the scope of the disaster, another important distinction between natural and technological disasters emerges with respect to the relative vulnerability for different social groups.

In natural disasters, the affluent can purchase relative safety by buying better constructed homes, less vulnerable properties, less densely populated living areas, and greater resources to assist in evacuating the afflicted area; thus, all else being, the disenfranchised tend to be more vulnerable to natural disasters. Thus, in a tornado, the affluent person's well constructed home is less likely to be destroyed than is the less affluent person's poorly constructed home. Similarly, when flood-waters begin to rise, they typically first destroy the homes of those who live in high-risk areas (i.e. the disenfranchised); however, as they continue to rise, the probability that the more affluent will also be affected increases. Thus, as the disaster's scope of destruction increases, the lower the probability that the affluent person's resources can protect him or her from harm. It is therefore likely that the "vulnerability gap" between the relatively disenfranchised and relatively affluent narrows as the destructive scope of a natural disaster increase. This relationship is depicted in Figure 11.1.

Figure 11.1: Hypothesized Relative Vulnerability of Affluent and Disenfranchised for Natural Disasters

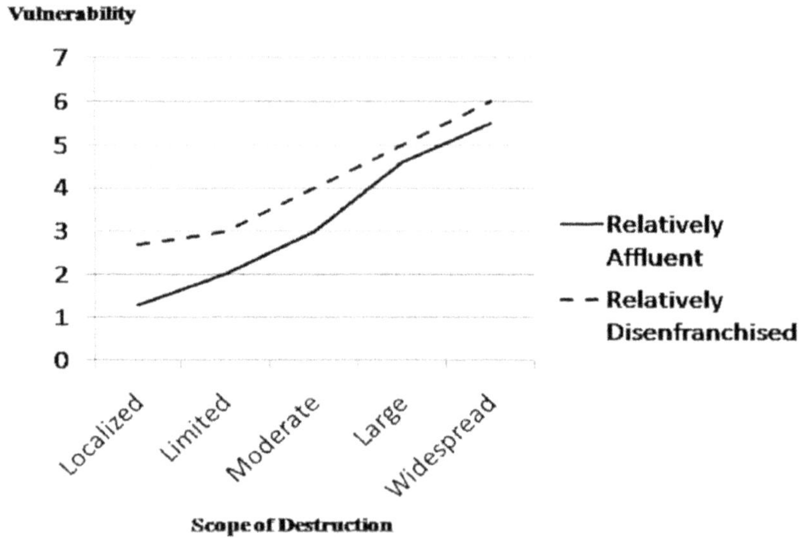

Technological disasters, however, produce a different vulnerability curve as a function of social affluence. Technological disasters usually affect those who work most closely with the failing technology. The mineworkers, the airplane pilots, the nuclear energy plant workers, and the oil well drillers are the ones who have the highest probability of dying from a mine explosion, plane crash, reactor meltdown, and well explosion, respectively. For example, when biological toxins were accidentally released at a Soviet research center, approximately 1,000 workers died (see Quarantelli 2003). The Chernobyl nuclear disaster provides an additional example. The disaster killed most of the reactor crew within three weeks of the disaster, and 237 fire and rescue workers suffered from acute radiation sickness. Of these workers, twenty-eight died within the first four months following the disaster (see Rosenthal 2005; Medvedev 1991; UNDP 2002). Similarly, the deaths directly attributable to the Deepwater Horizon disaster thus far are the eleven workers killed in the initial rig explosion.

Thus, in technological disasters with a relatively limited geographic scope of destruction, those who work with the technology are the most vulnerable, and these workers are not typically the most socially disadvantaged. While they are unlikely to be "elites" in any sense, they are typically highly skilled labors and often "middle class." As the scope of the disaster spreads, however, those most geographically proximate to the disaster, who tend to be the disenfranchised, become most vulnerable, and the vulnerability of the more affluent would likely decrease. Thus, the "vulnerability gap" between the relatively disenfranchised and relatively affluent is inverted with limited-scope technological disasters. That is, the relatively affluent are more vulnerable than are the truly disenfranchised. The very characteristics such as being skilled laborers, males, or members of the ethnic majority that provide them advantages in obtaining the technological jobs increases their vulnerability to a technological disaster. Then the vulnerability gap narrows as the destructive scope of the disaster increases and the vulnerability of the disenfranchised increase while the vulnerability of the affluent decrease. Eventually, the relative vulnerability of the affluent and disenfranchised reverses as the disaster's scope widens further. This relationship is depicted in Figure 11.2.

Figure 11.2: Hypothesized Relative Vulnerability of Affluent and Disenfranchised for Technological Disasters

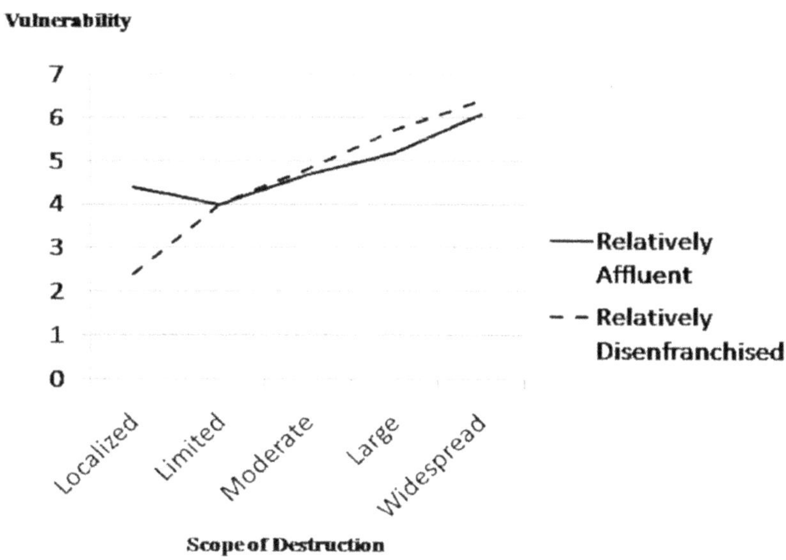

In addition to those who are killed or injured in a disaster (the primary victims of the disaster), we also need to consider the spillover effects of disasters; most notably, we should consider the economic effects of disasters. Disasters, regardless of their trigger, can be economically devastating, and we certainly do not want to dismiss the massive destruction that natural disasters can cause to economy of the afflicted area. However, the economic effects of technological can be more widespread and disrupt industries beyond the immediately afflicted community. For example, the Three Mile Island (TMI) nuclear disaster virtually halted the expansion of nuclear power in the United States. Beginning in the early 1960s, the number of nuclear reactors under construction globally increased nearly every year until the TMI disaster. Following TMI, the number of reactors under construction in the US declined every year for the next two decades. In addition, the total capacity of nuclear power plants decreased substantially after TMI as numerous American utilities continued to cancel orders they had placed prior to the incident (see Fisher 1997). The effects of the Chernobyl disaster on the nuclear power industry were even more dramatic (Fisher 1997). Although less dramatic than the TMI disaster, the Deepwater Horizon disaster had repercussions throughout the oil industry. For example, federal government stopped all deepwater drilling, not just those operations in the Gulf. In addition, the tourism industry far beyond the area most directly affected was adversely affected (see Proctor 2010).

Consequently, technological disasters can, at times, call into question the technologies at the center of the disaster. Whether these technologies are nuclear reactors, airplane engines, mining or oil-drilling technologies, or chemical industry protective technologies, when our technology fails, we often question the wisdom of its continued use. When this occurs, the economic repercussions can be felt globally. Even if we do not question, limit, or halt the use of the technology (we will not, for example, start to question the wisdom of flying), when a technological disaster strikes, we often require the technology involved to be inspected, repaired, or replaced. These actions, well often well warranted, can have widespread economic consequences as the industries involved scramble to meet new policy requirements.

Thus, the economic consequences of technological disasters can be far more widespread than those associated with natural disasters. When this occurs, it is once again those who rely on the technology in question for their economic livelihood that disproportionately carry the disaster's burden. In the Deepwater Horizon disaster, for example, it was the oil-rig workers, the fishermen and fisherwomen, the realtors, hotel owners and workers, restaurateurs, and small business owners who rely on tourists whose economic situation was most adversely affected. While these people are unlikely to be among our country's most privileged, they are also not the most disenfranchised.

Finally, the difference between Gulf Shores and Bayou La Batra with respect to the emergence of a corrosive community is worthy of mention. While technological disasters are more likely than natural disasters to lead to conflict and a corrosive community (see Gil 2007), it is interesting that the BP disaster's effect was not similarly experienced in the two communities that rely on the Gulf. Unfortunately, our data cannot address this issue. Given the importance of community solidarity for long-term community and individual well-being (see Hawdon and Ryan forthcoming), further investigation into this issue could provide valuable insights.

CONCLUSION

The effects of the Deepwater Horizon disaster are only now being fully understood, and the long-term effects may be felt for decades. Like many disasters, this technological disaster caused a wake of destruction that included lost lives, disrupted social networks, and economic hardship. These effects were particularly felt in the resource-dependent communities along the Gulf coast; however, they were also felt, albeit to a lesser degree, in the more economically diversified cities throughout the region. In this way, the Gulf disaster was similar to many other disasters. Yet, it appears to have differed in other respects. At least based on our data, the disaster did not disproportionately affect the most disenfranchised members of the region. Unlike Hurricane Andrew, Katrina, and Hugo; unlike the Great San Francisco, Northridge, or Haiti earthquake; unlike the massive Indian Ocean tsunami; the Deepwater Horizon disaster had the most profound effects on the "middle class" and "well integrated."

Although these findings may be limited to this one disaster, we believe they are not. We already know that technological and natural disasters have important differences (see Gil 2007), and we argue that another important distinction is the relative relationship between the disaster's scope of destruction and vulnerability. In natural disasters, the relationship is relatively straight forward: as the scope of the disaster increases, the gap in vulnerability between the disenfranchised and the relatively affluent narrows. However, with technological disasters, the relationship is more complex. In narrow-scope technological disasters, the vulnerability gap actually favors the relatively disempowered. This gap quickly disappears and becomes inverted as the geographic scope of devastation increases. It then slowly narrows in a similar manner as the vulnerability gap in natural disasters does. Moreover, the economic scope of the devastation may frequently be more widespread and long lasting than the economic scope of natural disasters. We believe these hypotheses deserve closer inspection.

NOTES

1. Prior to Katrina, 36.4 percent of the population of the Lower Ninth Ward lived in poverty and 63.6 percent lived below or slightly above the poverty line (US Census 2010; see Landphair (2007) for a discussion of the complex history of the Lower Ninth Ward).
2. Beginning in the late 1970s, debates about the value and validity of distinguishing natural from technological disasters emerged (see, for example, Couch and Kroll-Smith 1985; Quarantelli 1985). We recognize that it is difficult to disentangle natural from technological disasters because all disasters are largely the result of human agency (see, for example, Pelling 2003; Steinberg 2003; Stallings, 1995), and that the same event, such as Hurricane Katrina, can be both a natural disaster and a technological disaster

(Gil 2007). However, we argue that there are important distinctions among disastrous events, and these distinctions are significant enough to maintain and elaborate the differences.

APPENDIX

Table 11.7: Descriptive Statistics

	Mean	Standard Deviation
Did the oil spill negatively affect your work situation in any way?	.28	.45
Do you know of anyone who moved from the area because of the oil spill?	.19	.40
Did the oil spill lead to a corrosive community?	.04	.19
	.42	.49
Mobile	.50	.50
Bayou La Batra	.08	.27
Civic Participation	3.36	2.80
Social Capital/Trust	.40	.49
Currently Working	.38	.49
Temporarily Unemployed	.05	.22
Unemployed	.03	.17
Age	49.46	15.67
Own Home	.84	.36
Education	4.02	1.53
Female	.50	.50
Length of Residency in Community	13.98	8.78
African American	.17	.38

REFERENCES

Ahmed, Issam. (2010-08-12) "Pakistan floods strand the poor while rich go to higher ground." Christian Science Monitor. http://www.csmonitor.com/World/Asia-South-Central/2010/0812/Pakistan-floods-strand-the-poor-while-rich-go-to-higher-ground. Retrieved 2010-09-04.

Albala-Bertrand, JM. (1993). *The political economy of large natural disasters: With special reference to developing countries.* New York: Oxford University Press.

American Association for Public Opinion Research (AAPOR). 2008. "Response Rates—An Overview." Retrieved September 29, 2008 (http://www.aapor.org/responseratesanoverview).

Barry J. M. (1997). *Rising tide: The great Mississippi Flood of 1927 and how it changed America.* New York: Simon and Schuster.

Benson. C. and Clay, E. J. (2003). Disasters, vulnerability, and the global economy. In A. Kreimer, M. Arnold, and A. Carlin (Eds.). *Building safer cities: the future of disaster risk* (3-31). Washington, D.C.: The World Bank Disaster Management Facility.

Berke, Philip R. 1995. "Natural-hazard reduction and sustainable development: A global assessment." *Journal of Planning Literature* 9: 370–75.

Bolin, R., and Stanford, L. (1998). *The Northridge earthquake: Vulnerability and disaster* London: Routledge.

Bowonder, B. and Jeanne Kasperson. 1988. "Hazards in developing countries: Cause for global concern." *Risk Abstracts* 5: 103–09.

Brinkley D. 2006. *The Great Deluge: Hurricane Katrina, New Orleans, and the Mississippi Gulf Coast.* New York: HarperCollins.

Broughton, E. (2005). The Bhopal disaster and its aftermath: A review. *Environ Health.* 4: 6. Published online 2005 May 10. doi: 10.1186/1476-069X-4-6.

Bullard, R. (2000) *Dumping in Dixie: Race, class and environmental quality.* third edition. Boulder, CO.: Westview Press.

Burrage, D. (2010). Marine resources in Mississippi and Alabama. Mississippi State University Extension Services. Retrieved 6/15/2010 from http://msucares.com/crec/marine.html.

Clichevsky, Nora (2003). Urban land markets and disasters: Floods in Argentina's cities. In A. Kreimer, M. Arnold and A. Carlin (Eds.), *Building safer cities: the future of disaster risk* .Washington, D.C.: The World Bank Disaster Management Facility, pp. 165-76..

Centre for Research on the Epidemiology of Disasters (CRED, 2010) *The International Disaster Database.* Retrieved 9/27/2010 from http://www.cred.be/.

Corbett, G. (2010, July). Disaster and displacement: A historical perspective viewed through the lens of four conflagrations. Paper presented at the meeting of the Academy of Critical Incident Analysis, New Orleans, Louisiana.

Couch, Stephen R. and J. Stephen Kroll-Smith. 1985. "The Chronic Technical Disaster: Toward a Social Scientific Perspective." *Social Science Quarterly* 66(3):564–75.

Dash, N., Peacock, W. G., and Morrow, B. H. (1997). And the poor get poorer: A neglected Black community. In Peacock, W. G., Morrow, B. H., and Gladwin, H. (Eds), *Hurricane Andrew: Ethnicity, gender, and the sociology of disasters*. New York: Routledge, pp. 206-25.

Dhara, V. R. and Dhara, R. 2002. "The Union Carbide Disaster in Bhopal: A Review of Health Effects." *Archives of Environmental Health*. September/October 57(5):391-404.

Economic Commission for Latin America and the Caribbean Programme Planning and Operations Division (ECLAC). (1991). *Manual for estimating the socio-economic effects of natural disasters*. United Nations Economic Commission for Latin America and the Caribbean. Santiago, Chile.

Erikson, K. (1976). *Everything in its path: Destruction of community in the Buffalo Creek flood*. New York: Simon and Schuster.

Fischer. D. (1997). *History of the International Atomic Energy Agency: the first forty years*, IAEA, Vienna: Austria.

Fordham, M. (1999). The intersection of gender and social class in disaster: Balancing resilience and vulnerability *International Journal of Mass Emergencies and Disasters* 17, 15-36.

Fortun K (2001). *Advocacy after Bhopal*. Chicago: University of Chicago Press.

Fradkin P. L. (2005). *The great earthquake and firestorms of 1906: How San Francisco nearly destroyed itself*. Berkeley: University of California Press.

General Secretariat Central American Integration System. (1999). *Reconstruction and transformation of Central America after Hurricane Mitch*. Central American Integration System.

Gil, Duane A. (2007). Secondary trauma or secondary disaster? Insights from hurricane Katrina *Sociological Spectrum*, 27: 613-32.

Gillis, J. (6-15-2010). Estimates of Oil Flow Jump Higher. *New York Times*. Retrieved, 6-15-2010 from http://www.nytimes.com/2010/06/16/us/16spill.html.

Girard, C. and Peacock, W. G. (1997) Ethnicity and segregation: post-hurricane relocation. In Peacock, W. G., Morrow, B. H., and Gladwin, H. (Eds), *Hurricane Andrew: Ethnicity, gender, and the sociology of disasters* (191-205). New York: Routledge.

Gotham, K. F., and Greenberg, M. (2008). From 9/11 to 8/29: Post-disaster recovery and rebuilding in New York and New Orleans. *Social Forces*, 87: 1039-62.

Greenberg, M. R., and Anderson, R. F. (1984). *Hazardous waste sites: The credibility gap*. New Brunswick, N.J.: Rutgers University Center for Urban Policy Research.

Hawdon, J. and Ryan. J. (Forthcoming). "Space, sentiment, and social structure after critical incidents: Toward a theory of recovery," Schwester, R. (Editor), *Handbook of Critical Incident Analysis*.

Horne, Jed. (2006). *Breach of Faith: Hurricane Katrina and the Near Death of a Great American City*. New York: Random House.

Huq, S., and Alam, M. (2003). Flood management and vulnerability of Dhaka City. In A. Kreimer, M. Arnold and A. Carlin (Eds.), *Building safer cities: the future of disaster risk*. Washington, D.C.: The World Bank Disaster Management Facility, pp. 121-35.

Jenkins, P. and Phillips, B. (2008). Battered women, catastrophe, and the context of safety after Hurricane Katrina." *NWSA Journal*, 20(3), 49-68.

Klinenberg, Eric. (2002). *Heat wave: A social autopsy of disaster in Chicago*. Chicago: Chicago University Press.

Koplan, J. P., Falk, H., Green, G. (1990). Public health lessons from the Bhopal chemical disaster. *JAMA*. 264:2795-96.

Landphair, J. (2007). The forgotten people of New Orleans: Community, vulnerability, and the Lower Ninth Ward. *Journal of American History*, *94*, 837-45.

Larsen, J. (2003). "Record Heat Wave in Europe Takes 35,000 Lives," *Eco-Economy Update*, 9 October 2003.

McKinney, L. (2010). *The Deepwater Horizon oil spill – Putting a price on the priceless* Harte Research Institute for Gulf of Mexico Studies. Retrieved 7/24/2010 from http://www.harteresearchinstitute.org/images/press_releases/spill_valuations.pdf.

Medvedev, G. (1991). *The truth about Chernobyl*. New York: Basic Books.

Morgan, D. S. (2010, 6-15-2010) Oil hit-or-miss on Gulf beaches, as are tourists *CBS NEWS*. Retrieved 6-15-2010 from http://www.cbsnews.com/stories/2010/06/15/national/main6584524.shtml.

Morrow, B. H. (1997a) Disaster in the first person. In Peacock, W. G., Morrow, B. H., and Gladwin, H. (Eds), *Hurricane Andrew: Ethnicity, gender, and the sociology of disasters*. New York: Routledge, pp. 1-19.

MSNBC (2010, 4-30-2010) Oil from massive Gulf spill reaching La. coast. Retrieved 6/15/2010 from http://www.msnbc.msn.com/id/36800673/ns/us_news-environment/.

Norris, F. H., Byrne, C. M., Diaz, E. and Kaniasty, K. (2001). 50,000 Disaster victims speak: An empirical review of the empirical literature, 1981 – 2001. The National Center for PTSD and The Center for Mental Health Services (SAMHSA). Retrieved 10/18/2010 from http://www.dhss.mo.gov/SpecialNeedsToolkit/General/disaster-impact.pdf.

Oil Carreers.Com (2010). *Offshore oilrig jobs*. Retrieved 11/20/2010 from http://www.oilcareer.net/jobs-other.html).

Organization of American States (OAS) (1991). *Primer on natural hazard management in integrated regional development planning*. Washington, D.C.: Department of Regional Development and Environment Executive Secretariat for Economic and Social Affairs Organization of American States.

Oxfam International. (2005). *Briefing Report: The tsunami's impact on women* Retrieved 10/12/2010 from http://www.oxfam.org.uk/what_we_do/issues/conflict_disasters/downloads/bn_tsunami_women.pdf.

Peacock, W. G., Morrow, B. H., and Gladwin, H. (Eds). (1997) *Hurricane Andrew: Ethnicity, gender, and the sociology of disasters.* New York: Routledge.

Pelling, M. (2003). *The vulnerability of cities: Natural disasters and social resilience.* Sterling, VA.: Earthscan Publications.

Proctor, C. (2010-08-01). "Big price tag for recovery of Gulf Coast." *Pensacola News Journal.* http://www.pnj.com/article/20100801/BUSINESS/8010313/Carlton-Procter-Big-price-tag-for-recovery-of-Gulf-Coast. Retrieved 2010-11-01.

Putnam, R. D. (2000). *Bowling alone: The collapse and revival of American community.* New York, N.Y.: Simon and Schuster.

Quarantelli, E. L. 1985. "What is a Disaster? The Need for Clarification in Definition and Conceptualization in Research." in Disasters and Mental Health: Selected Contemporary Perspectives, edited by B. Sowder. Washington, DC: U.S. Government Printing Office, pp. 41-73.

Quarantelli, E. L. (2003). Urban vulnerability to disasters in developing countries: managing risks. In Alcira Kreimer, Margaret Arnold and Anne Carlin (Eds.), *Building safer cities: the future of disaster risk.* Washington, D.C.: The World Bank Disaster Management Facility, pp. 211-31.

Rivera, J. D., and Miller, D. S. (2007). Continually neglected: Situating natural disasters in the African American experience. *Journal of Black Studies*, 37, 502-22.

Rosenthal. E. (6 September 2005). "Experts find reduced effects of Chernobyl." *New York Times.* http://www.nytimes.com/2005/09/06/international/europe/06chernobyl.html. Retrieved 11 September 2010.

Ryang S. (2003). The great Kanto earthquake and the massacre of Koreans in 1923: notes on Japan's modern national sovereignty. *Anthropological Quarterly*, 76, 731-48.

Stallings, R. A. (1995). *Promoting risk: Constructing the earthquake threat.* Chicago: Aldine de Gruyter.

Steinberg, T. (2003). *Acts of God: The unnatural history of natural disaster in America.* Oxford: Oxford University Press.

Sturgis, S. (2009, May 12) Dumping in Dixie: TVA sends toxic coal ash to poor black communities in Georgia and Alabama" *Facing South.* Retrieved 11/2/2010 from http://www.southernstudies.org/2009/05/tva-sends-spilled-coal-ash-to-impoverished-black-communities-in-georgia-and-alabama.html.

UNDP (2002). *The human consequences of the Chernobyl nuclear accident: A strategy for recovery.* Retrieved 11/10/2010 from http://www.unicef.org/newsline/chernobylreport.pdf.

United States Army Corps of Engineers (2008). Waterborne commerce of the United States: Waterborne commerce statistics. Part five: National summaries. Retrieved 10/8/2010 from http://www.ndc.iwr.usace.army.mil//wcsc/pdf/wcusnatl08.pdf.

United States Bureau of Labor Statistics. (2010). http://www.bls.gov/oes/current/oes_nat.htm#00-0000.

United States Census. 2010a Bayou La Batre Profile http://censtats.census.gov/data/AL/1600104684.pdf.

United States Census. 2010b Gulf Shores Profile. http://censtats.census.gov/data/AL/1600132272.pdf.

United States Census. 2010c Mobile Profile. http://quickfacts.census.gov/qfd/states/01/0150000.html.

United States Census. 2010d. United States Profile. http://quickfacts.census.gov/qfd/states/00000.html.

USA Today (Thursday, 27 May 2010). "Gulf oil spill now largest offshore spill in U.S. history as BP continues plug effort." http://www.usatoday.com/news/nation/2010-05-27-oil-spill-news_N.htm?csp=34news. Retrieved 9 June 2010.

Vos F, Rodriguez J, Below R, Guha-Sapir D. *Annual Disaster Statistical Review 2009: The Numbers and Trends*. Brussels: CRED; 2010."

West Virginia Division of Culture and History. The Buffalo Creek Disaster. http://www.wvculture.org/history/buffcreek/mem.html.

Wisner, B., Blaikie, P., Cannon, T., and Davis, I. (2003). *At risk: Natural hazards, people's vulnerability and disasters.* New York: Routledge.

CHAPTER TWELVE
DISASTER PHASES, STRUCTURAL VULNERABILITY AND CRIME

KELLY FRAILING AND DEE WOOD HARPER

INTRODUCTION

Disasters are by no means a new phenomenon. Researchers have long recognized the different types of disasters and the different phases through which they progress, but only recently has their effect on crime received scholarly attention. In fact, most of what is known about the effects of disasters on crime comes from the study of natural disasters, such as earthquakes and floods and na-tech disasters, such as Hurricane Katrina. The effects of technological disasters on crime have gone largely uninvestigated. The 2010 Deepwater Horizon oil spill is one such technological disaster. It disrupted normal social and economic activities across the Gulf of Mexico, including commercial fishing and tourism, and this disruption has very likely had an effect on various crime rates.

After summarizing what is known about the effects of natural and na-tech disasters on crime, this chapter explores the anecdotal evidence of the Deepwater Horizon oil spill's effect on crime, criminal opportunity and mental health along the Gulf Coast and sets out a plan to further investigate changes in various crime rates in the parishes and counties most directly affected by the oil spill in an effort to determine which crimes are associated with which phases of this disaster, as well as what social structural characteristics make people more or less vulnerable to crime. Based on previous research on disasters and crime, we are particularly interested in changes in drug and alcohol abuse rates, domestic violence, sexual assault, theft and fraud.

The broader impact of this chapter is rooted in its focus on a heretofore uninvestigated aspect of crime and disaster, namely what types of crime characterize the stages of a technological disaster and who is affected by these crimes.

Determining what these characterizing crimes are not only expands knowledge in the fledging area of crime and disaster, it also provides crucial information for emergency response planners, law enforcement and service providers, who can anticipate these crimes in times of disaster and use their resources more effectively to combat their occurrence. Knowing what types of crime characterize the disaster at hand and being able to plan for them is useful in stopping the revictimization of those already victimized by the disaster.

DISASTERS AND CRIME

For decades, disaster research has focused on the effects of natural and manmade disasters on the environment and those who live in the affected areas. The aftermath of major catastrophes of recent decades, including Hurricane Katrina, has led researchers to call for a reorientation and redirection for future disaster research (Picou, 2007). Part of this new direction has been the effects of disaster on crime. As noted by Decker, Varano and Greene, "despite their rich theoretical and practical importance, criminologists have paid scant attention to the patterns of crime and the responses to crime during exceptional times" (2007: 89).

The criminological consequences of disaster have received limited attention in the literature. In April of 1906, an 8.3 magnitude earthquake struck San Francisco. About 1,500 people were killed and over 300,000 were left homeless. Anecdotal reports of crime after the quake include a number of people attempting to steal jewelry off of corpses, plundering for artifacts in Chinatown and price gouging (Morris, 2002). In September of 1923, a 7.9 magnitude earthquake struck Kanto, Japan. Over 91,000 people were killed. Rumors of an uprising by Koreans living in the affected area spread and in the week after the earthquake, about 6,000 Koreans were hunted down and killed by the Japanese (Ishiguro, 1998). In February of 1972, the dam holding back 132 million gallons of water used in coal mining broke in Logan County, West Virginia. The ensuing flood washed through the 16 towns in Buffalo Creek. One hundred twenty five people were killed and over 4,000 were left homeless. The loss of community caused by the flood was associated with an increase in crime, including drug use and theft (Erikson, 1976). In June of 1972, Hurricane Agnes brought near constant rain to the Wilkes Barre, Pennsylvania area. The Susquehanna River flooded, plunging 30 percent of the city underwater. Property crime increased significantly following the flood, as did drinking-related offenses. Price gouging was also observed, as was fraud perpetrated both by and against the Small Business Administration (Siman, 1977).

In July of 1976, the deadliest earthquake in the last four hundred years struck Tangshan, China. Over 240,000 people were killed and an additional 800,000 were injured by the 7.6 magnitude quake. Property crimes, including mass looting, occurred in the aftermath. The crime rate in Tangshan was higher in 1976 than in either 1975 or 1977. Tangshan's crime rate was also higher in 1976 than that of two nearby but unaffected cities (Zhou, 1997). In September of 1989, Hurricane Hugo, a Category 4 storm, struck the island of St. Croix. Looting followed in the wake of the storm, perpetrated first by delinquent youth who took consumer goods, then by otherwise law abiding citizens who took items from hardware and other stores and finally by a third group who took survival items from grocery stores (Quarantelli, 2007). In December of 2004, an earthquake in the Indian Ocean created a wall of water that washed over Indonesia, Sri Lanka, India and Thailand. The bodies of many of the hundreds of thousands killed were never recovered. Various types of crime were observed after the Boxing Day tsunami, including looting of corpses, hotels and relief convoys, rape of survivors, fraudulent claims against the tsunami fund, theft from the homes and bank accounts of the dead and pilfering of relief money and goods by military and government officials (Teh, 2008).

As a single event, the Hurricane Katrina Disaster has probably received the most attention from criminologists. In August of 2005, Hurricane Katrina struck New Orleans. This third strongest hurricane of all time breached three of the city's levees, which plunged 80 percent of it underwater for two weeks. Crime followed in short order, including sexual assault (Thornton and Voigt, 2010a), looting (Frailing and Harper, 2007) and fraud (Frailing, 2010). Despite these accounts, little is known about crime that occurs during disasters and the complexities that surround social structural vulnerability. Complicating this further is the fact that there are several types of disasters, several phases through which each disaster progresses and the variable impact of disaster on different segments of communities. In short, we lack an overarching theoretical framework to explain the problem of crime in disaster.

Types of disasters include purely natural disasters, such as earthquakes and tsunamis, na-tech disasters such as Hurricane Katrina, where the origins are both natural and technological, technological disasters such as the *Exxon Valdez* oil spill and terroristic disasters such as the September 11, 2001 attacks, where the origins are politically motivated groups (Gill, Picou and Ritchie, 2010). While not without its critics (e.g. Neal, 1997), the disaster phase model is useful in understanding how disasters progress. Killian's (2002) widely used phase model includes the warning, impact, emergency and recovery phases. The warning phase is the period in which information about the impending disaster is known, prior to its impact. The impact phase is the actual disastrous event. The emergency phase is the immediate post-impact phase in which urgent rescue, removal and care are provided. The recovery phase has short and long term components; the latter is sometimes called the reconstruction phase and includes reestablishing crucial infrastructures.

Research on crime during and after the Hurricane Katrina Disaster has revealed important insights about what crimes occurred during the different phases of this na-tech disaster and how the social structural inequalities of class, race, ethnicity, gender and poverty (Fothergill and Peek, 2004) predicted differential patterns of victimization. Sexual violence against women and children occurred during the warning phase, while looting, drug and alcohol use and sexual assaults occurred during the impact and emergency phases. The looting that was observed during the impact and emergency phases of Katrina was associated with the difficult socioeconomic conditions that had characterized New Orleans for decades prior to the disaster (Frailing and Harper, 2007). A wide variety of crimes, including violence against recovery workers, violence against those displaced to temporary housing, architectural looting, fraud, domestic violence and prostitution, occurred during the recovery and reconstruction phases (Thornton and Voigt, 2010). In fact, sexual assault and domestic violence were observed during all phases of the Katrina Disaster (Thornton and Voigt, 2010a) and following hurricanes in Florida (Zahran, Shelley, Peek and Brody, 2009). The differential vulnerability of women during disasters has been observed for some time (e.g. Fothergill, 1996; 1998).

THE DEEPWATER HORIZON OIL SPILL

Based primarily on studies of Hurricane Katrina, we have information on what types of crime characterize the different phases of a na-tech disaster and why they do, namely poor socioeconomic conditions preceding the disaster and differential vulnerability of some populations. However, most of the research on crime that occurs in conjunction with disasters has been done on natural or na-tech disasters, as seen above. What we do not have and the void this study aims to fill is knowledge of the crimes that characterize the phases of a technological disaster, in this case the Deepwater Horizon oil spill.

On April 20, 2010, the Deepwater Horizon oil rig in the Gulf of Mexico exploded, killing eleven people. Two days later, the rig sank and two days after that, the oil leak in a well owned by BP was detected a mile below the water's surface and was estimated to be spewing 1,000 barrels of oil per day into the Gulf. Louisiana governor Bobby Jindal declared a state of emergency on April 29, the same day on which the Coast Guard increased the leak estimate to 5,000 barrels a day (Elliott and Horsley, 2010). By the end of May, the oil spill had become the worst offshore spill in U.S. history (NPR, 2010d). Initially, BP attempted several strategies to contain the oil leak, but all of them failed. Diversion of oil from the leak onto containment ships was the first partially successful containment endeavor and in mid July, a cap on the well fully contained the leak until the permanent relief well was completed five months and 206 million gallons of oil after the rig exploded. (NPR, 2010; NPR, 2010a; NPR, 2010c; Associated Press, 2010). The oil that washed ashore from Louisiana to Florida negatively impacted those people whose livelihoods are dependent on the Gulf, from fishermen to hoteliers (NPR, 2010b; Allen, 2010; Elliott, 2010). Initial losses to the fishing industries in Louisiana, Mississippi and Alabama were estimated at $2.5 billion and initial losses to the tourism industry in Florida alone were estimated at $3 billion (Walsh, 2010). Shrimpers and fishermen are since back to work, but are suffering from the perception that Gulf seafood is tainted by oil (Fulton, 2010; Kaufman, 2010; Associated Press, 2010).

THE OIL SPILL'S EFFECTS ON CRIME, CRIMINAL OPPORTUNITIES AND MENTAL HEALTH

While systematically collected and analyzed data on the change in crime rates in Gulf parishes and counties before and after the oil spill is still lacking, there is anecdotal evidence that the oil spill is responsible for changes in both crime and opportunities for crime and has had a negative effect on Gulf residents' mental health.

CRIME

Media reports reveal that a variety of crime, including assaults, rape, robbery and domestic violence have been observed in conjunction with the oil spill:

> The first stabbing in recent memory in Grand Isle, Louisiana occurred after the oil spill. Both the victim and perpetrator were believed to be cleanup workers. Grand Isle locals also suspect the cleanup workers when property goes missing. In fact, BP fired some workers for unspecified misconduct and began requiring that cleanup workers pass drug tests (Smith, 2010). Drug tests may have had an inadvertent effect on domestic violence. Some fishermen who smoke marijuana but who wanted to work for BP cleaning up oil stopped smoking in order to pass the drug test. Instead of using marijuana, some of these fishermen drank alcohol and committed domestic violence while intoxicated (Whitney, 2010).

In Pascagoula, Mississippi, a cleanup worker reportedly raped another. The alleged offender is on the national sex offender registry and has a criminal record dating back to 1991, but was permitted to work on the oil spill, thanks to BP neglecting to conduct background checks on cleanup workers, opting instead to only conduct drug tests (Boudreau, Zamost and Joseph, 2010).

An oil spill cleanup worker in Panama City Beach, Florida confessed to robbing two juveniles at gunpoint as well as to burglarizing a nearby condominium (Press-Register, 2010).

A Slidell, Louisiana National Guardsman assigned to work on the oil spill attempted to rob a fast food restaurant at knifepoint and killed himself while subsequently fleeing from police (Adelson, 2010).

Calls to the Bayou La Batre, Alabama police department nearly doubled two months after the spill as compared to two months before the leak began (Andrews, 2010). The mayor of Bayou La Batre claims domestic violence has increased 320 percent in his town since the oil spill (BBC, 2010). The Mississippi Department of Health also reported an increase in domestic violence and substance abuse, while homelessness spiked in Pensacola, Florida (McGill and Breen, 2010).

CRIMINAL OPPORTUNITIES

The oil spill may also have created opportunities for crime, especially fraud. Cognizant of this, the Federal Trade Commission issued a statement to Gulf Coast residents, warning them about potential scams, including people offering to expedite both cleanup of oil and claims for monetary compensation for a fee, people who offer to assist those interested in cleanup work with obtaining a job for a fee and people soliciting for nonexistent environmental charities (FTC, 2010).

In May of 2010, BP promised to pay legitimate claims from those whose livelihoods had been adversely affected by the oil spill in the Gulf but declined to explain what would constitute a legitimate claim. This lack of explanation is worrisome to Walter Parker, who was the chair of the Alaska Oil Spill Commission over twenty years ago when the *Exxon Valdez* ran aground and its oil devastated commercial fishing and peripheral industries. Parker claims Exxon refused to pay many claims and cases stayed in court for years (Noguchi, 2010; for more on the litigation that followed the *Exxon Valdez* oil spill, see Gill, Picou and Ritchie, 2010). Later in May, some fishermen and hoteliers were claiming that BP refused to pay them more than $5,000, which was a fraction of what they would have made had the spill not occurred (Mann, 2010). Workers in industries peripheral to fishing and tourism in particular were uncertain whether BP would pay any part of their claims (Smith, 2010a). In June, BP agreed to expedite its claims process, in response to criticisms that it refused to pay larger claims without substantive documentation (NPR, 2010b). Later that month, government officials called on BP to establish a $20 billion compensation fund (NPR, 2010e) and by mid August, BP had paid about $368 million in claims, with Gulf Coast Claims Facility director Kenneth Feinberg poised to start repaying claimants from the $20 billion compensation fund. Feinberg, who administered payouts to victims of the September 11, 2001 terrorist attack, assured individual claimants they would be paid within 48 hours of filing a claim and businesses would be paid within a week (Fleming, 2010).

Despite Feinberg's promises, these accounts indicate that obtaining compensation, especially direct compensation from BP, is a difficult and uncertain process. The fact that it took government pressure for BP to establish a compensation fund in the first place indicates the company may not be committed to reimbursing losses caused by the oil spill. BP's lack of commitment, combined with the difficulty and uncertainty of the process may cause frustration among claimants, some of whom might engage in fraud against the compensation fund in order to fully recover their losses. Conversely, Feinberg's promise to pay claims so quickly may encourage people with no entitlement to the money to make fraudulent claims on the compensation fund, as some did after Hurricane Katrina (Frailing, 2010). In fact, those who committed fraud against the Federal Emergency Management Agency (FEMA) and the Red Cross in the wake of Katrina are still being prosecuted more than five years after the hurricane struck New Orleans (Associated Press, 2010a).

To date, about 98,000 claims have been filed with the Gulf Coast Claims Facility. A total of about $1 billion has been paid to some 50,000 claimants, though complaints continue that payouts are much less than what is requested by those individuals and businesses who have lost their livelihoods due to the oil spill. There are currently about 35,000 claims that require more documentation before payouts can be made. Complicating and slowing the compensation process for many are claims that the Claims Facility suspects of being fraudulent (Skoloff, 2010).

MENTAL HEALTH

Reports indicate that the oil spill has had a deleterious effect on the mental health of Gulf Coast residents. A Centers for Disease Control survey of 128 households in two counties in Alabama affected by the oil spill revealed increased symptoms of depression and anxiety among residents (Andrews, 2010a). A larger survey of 2,600 Gulf Coast residents by Gallup revealed that diagnoses of depression have increased by 25 percent across the region since the oil spill and that residents are experiencing more stress, worry and sadness than those who live further inland, confirming findings by smaller surveys (Associated Press, 2010b). If the aftermath of the *Exxon Valdez* oil spill is any guide, mental health professionals can expect to see an increase in psychiatric disorders, an increase in symptoms of anxiety, post traumatic stress disorder and depression and an increase in drug and alcohol abuse and domestic violence among Gulf Coast residents as a result of the oil spill (Farris, 2010). In perhaps the most serious example of the oil spill's negative effect on mental health, a charter boat captain committed suicide two weeks after he was hired by BP to assist in cleaning up the oil spill. William Kruse's suicide took friends and colleagues by surprise, as he gave no indication he was any more upset by the oil spill than others in the community. They postulated the oil spill had been "too much" for Kruse to bear (AOL, 2010).

PROPOSED METHODOLOGY FOR FURTHER STUDY

The main thrust of the proposed research is to further determine the utility of disaster phase analysis as a framework to better understand how different phases of different types of disasters lend themselves to vulnerabilities or opportunities for certain types of crime to occur. In more detail, we will investigate changes in various crime rates in the parishes and counties most directly affected by the oil spill in an effort to determine which crimes are associated with which phases of this disaster and who is most vulnerable to crime. In so doing, we will glean information that can be useful for emergency responders and criminal justice practitioners in preparing for and preventing these crimes during future technological disasters.

We are interested in examining various indicators of crime and how they have changed since the Deepwater Horizon's explosion on April 20, 2010 and since oil reached the shores of Louisiana, Mississippi, Alabama and Florida on May 20 as compared to their base rates, one year prior to the rig's explosion. Using what is known about crimes that occurred during the different phases of the Katrina Disaster (Thornton and Voigt, 2010) as a guide, we are establishing base rates for the following:

- Sexual assaults calls for service and arrests
- Domestic violence calls for service and arrests
- Thefts calls for service and arrests
- Drug possession and alcohol intoxication arrests
- Census of domestic violence shelters
- Sexual assault and domestic violence hotlines calls received and
- Emergency room intakes for drug and alcohol issues

With the base rates established, we will continue to gather these data through the remainder of the disaster's phases to determine whether they indicate an increase in these crimes as a result of the disaster. Again, if the Katrina Disaster is any guide, and we do not know without the proposed research if it is, we would also be looking for other issues such as fraud and the victimization of emergency workers as the disaster's phases unfold. Of course, this necessitates defining the phases of the disaster. We maintain that the Deepwater Horizon oil spill disaster is currently in the early part of the recovery phase, following an almost nonexistent warning phase a prolonged impact phase, in which the oil was leaking but had not yet washed ashore and an even longer emergency phase, from the time the oil first washed ashore to the completion of the relief well.

We are gathering these data from the coastal parishes and counties most directly affected by the oil spill disaster in terms of disruption of normal social and economic activities, including commercial fishing and tourism. In Louisiana, this includes Vermilion, Iberia, St. Mary, Terrebonne, Lafourche, Jefferson, Plaquemines, St. Bernard, Orleans and St. Tammany parishes, in Mississippi, it includes Hancock, Harrison and Jackson counties, in Alabama, it includes Mobile and Baldwin counties and in Florida, it includes Escambia, Santa Rosa, Okaloosa, Walton and Bay counties. It is no coincidence that in Louisiana, these include the parishes on which many workers rely for their employment (Plyer and Campanella, 2010) and that across the region, these are among the parishes and counties with more employees in the leisure and hospitality industries (BLS, 2010). We are also gathering other data from the parishes and counties of interest, namely that on sociodemographics and employment, using sources such as the U.S. Census and the Bureau of Labor Statistics. These data will be particularly useful in assessing the impact of the disaster on the lives and livelihoods of people across the region.

Cognizant of the challenges of gathering a variety of criminological data from a variety of sources in a variety of areas, we have not embarked upon this data collection unaided. We have longstanding connections in the New Orleans Police Department (Orleans Parish) and the Jefferson Parish, St. Tammany Parish and Lafourche Parish Sheriff's Offices who are providing a great deal of assistance with gathering data for base rates and who will continue to provide assistance with data gathering as the phases of the disaster unfold. Moreover, our professional contacts in the field of disaster and crime research can provide guidance on disaster phases, data gathering, statistical techniques and the interpretation of results.

The media reports of crime detailed above highlight many of the issues we are dealing with in the proposed research in a more systematic and quantitative manner, including substance abuse, sexual assault and domestic violence. We are collecting baseline data on arrests for DUI and drug possession as well as on arrests and calls for service for sexual assault, domestic violence and thefts from April, 2009 through April, 2010. We will collect the same data for the time periods after the rig's explosion and after oil began washing ashore through the present. From that point forward, we will collect the same data from the parishes and counties of interest for the year after the oil first came ashore, through May, 2011. Our law enforcement contacts will also assist us in gaining access to sexual assault hotlines, domestic violence shelters and hotlines and hospital emergency rooms so that we can measure changes in calls for assistance and service use in the parishes and counties in the study.

Timely data collection is critical to ensure consistent availability of the same data over time and to ensure that the project is well-known and viewed as an integral part of law enforcement's and agencies' record keeping in the parishes and counties of interest. Moreover, ingratiating our project and ourselves to law enforcement and other specified agencies across the Gulf Coast puts us in a good position to observe other changes in crime that may present themselves, particularly fraud and crime against emergency workers, as the disaster phases continue to unfold.

Our analysis of quantitative data will be primarily a time series design with baseline observations serving to establish "normal" crime rates and normal distributions of victimization prior to the oil rig explosion compared with changes in these rates and distribution of victimization after the oil rig explosion and after the oil began washing ashore. An event count model that analyzes the number of events or rates of the event since some starting time, such as Poisson regression, is appropriate for our data; we may also use logit, probit and logistic models. We also anticipate using time series analysis to reveal how non-crime variables, such as population and job market characteristics, change over time across the region.

Newly developed models to analyze zero-inflated clustered count data using random effects (see Lee, Joo, Song and Harper, 2011) are particularly appropriate for count data with excess zeroes. These data are often encountered in criminological research when the event of interest is rare. An advantage of these models is their ability to use conditional models for serial association of responses while still structuring the marginal mean as a function of covariates directly. Moreover, a high percentage of zeroes does not disqualify Poisson or negative binomial models.

BROADER IMPACTS OF THE RESEARCH

As alluded to above, disasters affect different people differently. For example, research has shown that women are at greater risk for victimization in the wake of disasters (Fothergill, 1996; 1998). In addition to determining changes in crime rates associated with the oil spill, we are interested in how this disaster has affected different groups of people. The Gulf coast region provides us with enormous cultural diversity. The bounty of the sea has attracted a diverse ethnic gumbo beginning with Native Americans and later Isleños (natives of the Canary Islands), French, Croatians, African Americans and most recently, Vietnamese. In the course of our investigation into crime, we are paying close attention to how each of these groups individually and collectively deals with the disruptions in their lives and livelihoods brought on by the Deepwater Horizon oil spill and in so doing, are able to provide a richer and more comprehensive picture of the effects of this technological disaster.

The original Native Americans of the lower Mississippi River region were the Chitimacha that in 1492, may have numbered as many as 20,000. While they had little direct contact with Europeans for the next 200 years, their population was indirectly decimated by European diseases spread from the Spanish missions in Florida. Their numbers had been halved by 1699, when the French founded Louisiana as its colony. Currently, Chitimacha tribal enrollment is 900. The most dominant Native American group of the region today is the United Houma Nation, which numbers about 11,000. Beginning in the 1800s, the Houma intermarried with French Creoles and blacks, as well as absorbed a number of sub-tribes. The Houma have retained a tribal identity, but the Houma language was replaced by Cajun French and most Houma became Catholic. In the twentieth century, the Houma continued to support themselves by hunting and fishing and during the 1930s, began to work in the oil fields (Sultzman, n.d.).

Three hundred years ago, the Spanish colonial government brought 3,000 Canary Islanders to colonize the region of south Louisiana that is presently St. Bernard Parish. They were to serve as militia against the British. Because of their isolation, they retained much of their Isleños culture. While the flood waters of Hurricane Katrina dispersed much of the population of St. Bernard Parish, the Isleños of the Delacroix Island community got back to their business of fishing, crabbing, shrimping and oyster fishing fairly quickly. Their resilience is thought to be a result of the closeness of their community (Monteleone, 2007). The oil spill poses a greater threat to their economic way of life than even Hurricane Katrina.

The French arrived in south Louisiana shortly before the Isleños. The area was claimed by France in 1682 and in 1699, Pierre Le Moyne d'Iberville and a small group of settlers founded Louisiana as a French colony. Even during the period of

Spanish control (1762-1799), Acadians, descendants of French colonists in Canada, were welcomed to the wilderness that was Louisiana, with over 1,000 arriving between 1764 and 1769. In 1785, seven more ships with approximately 1,600 Acadians arrived in Louisiana. These Acadians would become the ancestors of the French speaking Cajuns along Bayou Lafourche and other parts of the state (Brasseaux, 1987). Presently, over 400,000 people in Louisiana claim Acadian-Cajun ancestry. These people are economically heavily involved in both the shrimping and the oil and gas industries of the region. The oil spill is likely to doubly impact their way of life.

Croatians have been present in the New World since the eighteenth century. The Croat people of Dalmatia were seafarers who visited the Gulf of Mexico coast. A wave of Croatian immigration to the United States occurred between 1900 and 1914 (Shapiro, 1989) and Croatians who came to Louisiana settled primarily in Plaquemines Parish, just south of New Orleans. Commercial oyster fishing is a traditional occupation among Croatians and in Louisiana, they largely developed this industry. Some third and fourth generation Croatians still work as oystermen. With the oyster season opening on November 15, 2010, there is great trepidation among the fishermen about the condition of the oysters and the reefs on which they live (Kirkham, 2010). Some of the reefs in public waters in the Gulf have already been reported to be dead, primarily as a result of fresh water that was pumped into the area in order to keep the oil out.

African Americans have been in south Louisiana since the eighteenth century. As a result of the Atlantic slave trade, 3.5 million slaves lived in the southern United Sates by the middle of the nineteenth century and another half million free persons of color lived across the United States. Following the Haitian revolution in 1805, a wave of immigrants from there came to Louisiana including French creoles, French creoles of color and African slaves. This immigration greatly expanded the population of New Orleans, making the city a majority black city for a time (Brasseaux, 2007). Presently, the urban parishes of the state and the rural parishes along the banks of the Mississippi River in north Louisiana have the heaviest concentration of African Americans and the coastal parishes have the least. The coastal parish of St. Bernard, for example, has an African American population of 12.5 percent, while the urban parish of Orleans has an African American population of 60.5 percent (U.S. Census, 2008). Even though African Americans make up a small percentage of the population in the coastal parishes and counties along the Gulf, they are likely to be severely impacted by the oil spill because they are overrepresented among the impoverished of these areas.

Vietnamese people began arriving in Louisiana in 1975 after the fall of Saigon. The Catholic Dioceses of Louisiana helped with the resettlement of Vietnamese refugees and presently, Vietnamese number nearly 25,000 in Louisiana. The Vietnamese population is concentrated in the southern part of the state, with access to the Gulf of Mexico. They brought with them a tradition of fishing and shrimping; about 10 percent of the Vietnamese men in Louisiana work as fishermen or shrimpers and Vietnamese account for five percent of all workers in the Louisiana fishing industry (Bankston, 2007). Bayou La Batre, Alabama is home to a large Vietnamese community. Unable to fish commercially and without the skills in English to get hired by BP for the oil spill cleanup, this community has been hit hard. The fact that the Bayou La Batre food bank has passed out over 65,000 pounds of food mostly to Vietnamese is indicative of the devastation the oil spill has caused them. The Bayou La Batre Vietnamese have also been targeted by scam artists, who offer to arrange to have BP rent their boats for the oil cleanup for a fee and reports tell of an "army of lawyers" that has descended on the town and is trying to get the Vietnamese to sign on to law suits though they are unable to read the attendant English language documents (Di Paola, 2010).

The social/structural characteristics of each of these groups make them more or less vulnerable to the consequences of the oil spill. Across the board, they are all sharing in the negative economic impact brought on by the oil spill as well as the uncertainty of their future in the region. We then can expect them to behave similarly with regard to the problems we will be assessing. However, we can also anticipate differences in the degree to which resources in their respective communities, especially their cohesiveness and resilience, can mitigate the harmful consequences of the oil spill.

* * *

The proposed research can shed light on the types of crime that characterize the different phases of a technological disaster and whether those crimes are the same as those that characterize the phases of na-tech disasters, such as Hurricane Katrina. In short, this research will reveal how some individuals criminally respond to technological disasters, when they do so and in what ways. It will also reveal who is more and less vulnerable to criminal victimization. Knowing more about what crimes occur during the different phases of different types of disasters is crucial information for emergency response planners as well as law enforcement and service providers, who can anticipate these crimes in times of disaster and use their resources more effectively to combat their occurrence. We took a similar position after Katrina with regard to looting and observed that as a result of New Orleans preparing itself for the probability of post-storm looting, much less took place in the wake of Hurricane Gustav in 2008 (Frailing and Harper, 2010). The knowledge generated by this study can also help to prevent revictimization of those already victimized by the disaster.

Moreover, if the data reveal that crimes occurring during the phases of the oil spill disaster mirror those that occurred during the phases of Katrina, we can reasonably anticipate that other crimes, fraud in particular, will characterize the later part of the recovery phase of the oil spill, as it did during Katrina. Anticipating fraud is also useful in devising strategies for prevention, such as target hardening, and in allocating preventive resources (Frailing, 2010).

REFERENCES

Adelson, Jeff. (2010). Slidell robbery suspect who committed suicide was National Guardsman. *The Times-Picayune*, August 4 Available online at: http://www.nola.com/crime/index.ssf/2010/08/slidell_armed_robber_who_commi.html.

Allen, Greg. (2010). Florida fishing town on edge over Gulf spill. *National Public Radio*, May 31. Available online at: http://www.npr.org/templates/story/story.php?storyId=127246469&ps=rs.

Andrews, Casandra. (2010). Gulf oil spill spews anxiety and despair. *The Press-Register*, June 27. Available online at: http://blog.al.com/live/2010/06/gulf_oil_spill_spews_anxiety_a.html?utm_source=twitterfeed&utm_medium=twitter.

Andrews, Casandra. (2010a). After oil spill, depression and stress levels rise in coastal Alabama. *The Press-Register*, September 27. Available online at: http://blog.al.com/live/2010/09/after_oil_spill_depression_and.html.

AOL. (2010). Oil spill 'too much' for skipper who killed self. *AOL News*, June 24. Available online at: http://www.aolnews.com/gulf-oil-spill/article/oil-spill-too-much-for-skipper-who-committed-suicide/19529447

Associated Press. (2010). Gulf oil well is dead, but the pain will remain. *National Public Radio*, September 20. Available online at: http://www.npr.org/templates/story/story.php?storyId=127901771.

Associated Press. (2010a). Hurricane Katrina, Rita and Wilma fraudsters indicted. *The Times-Picayune*, September 14. Available online at: http://www.nola.com/newsflash/index.ssf/story/feds-have-charged-1360-with-hurricane-fraud/253bff456d484a78858dec7adff5fbea.

Associated Press. (2010b). Gallup: Depression up 25 percent on Gulf after oil. *The Associated Press*, September 28. Available online at: http://www.al.com/newsflash/index.ssf/story/gallup-25-percent-hike-in-depression-along/30359242e0e549df9b6d882169affb7a.

BBC. (2010). Bayou mayor: Domestic violence 'up 320%' since oil spill. *British Broadcasting Corporation*, June 15. Available online at: http://news.bbc.co.uk/1/hi/programmes/world_news_america/8740073.stm.

BLS. (2010). Gulf Coast leisure and hospitality employment and wages fact sheet. Bureau of Labor Statistics, June. Available online at: http://www.bls.gov/cew/gulf_coast_leisure_hospitality.htm.

Bankston, Carl L. III. (2007). The Vietnamese population of Louisiana. Center for Cultural and Eco-Tourism. University of Louisiana at Lafayette. Available online at: http://ccet.louisiana.edu/tourism/cultural/The_People/vietnamese.people.html.

Boudreau, Abbie, Zamost, Scott and Joseph, Jessi. (2010). Investigation: Could background check have prevented alleged rape? *CNN*, August 12. Available online at: http://www.cnn.com/2010/CRIME/08/12/gulf.cleanup.worker charged/index.html?hpt=T2.

Brasseaux, Carl A. (1987). *The Founding of New Acadia: The Beginnings of Acadian Life in Louisiana, 1765-1803.* Baton Rouge, LA: Louisiana State University Press.

Brasseaux, Carl A. (2007). Haitians. Center for Culture and Eco-Tourism. University of Louisiana at Lafayette. Available online at: http://ccet.louisiana.edu/tourism/cultural/The_People/haitian.html.

Decker, Scott H., Varano, Sean P. and Greene, Jack R. (2007). Routine crime in exceptional times: The impact of the 2002 Winter Olympics on citizen demand for police services. *Journal of Criminal Justice, 35* (1): 89-101.

Di Paola, Mike. (2010). Gulf's Vietnamese community hit by unemployment, boat scams after oil spill. *The Press-Register*, August 19. Available online at: http://blog.al.com/live/2010/08/gulfs_vietnamese_community_hit.html.

Elliott, Debbie. (2010). Oil spill means dismal summer for Gulf beach town. *National Public Radio*, July 1. Available online at: http://www.npr.org/templates/story/story.php?storyId=128248167.

Elliott, Debbie and Horsley, Scott. (2010). How an oil spill spread into a national crisis. *National Public Radio*, May 5. Available online at: http://www.npr.org/templates/story/story.php?storyId=126508979.

Erikson, Kai T. (1976). *Everything in Its Path: Destruction of the Community in the Buffalo Creek Flood.* New York: Simon and Schuster.

Farris, Meg. (2010). Doctors say stress over oil spill to linger for years. *Houma Today*, July 15. Available online at: http://www.houmatoday.com/article/20100715/FEATURES12/100719543.

Fleming, Eileen. (2010). Last day to file claim with BP for individuals and businesses. *WWNO.* Available online at: http://www.publicbroadcasting.net/wwno/news.newsmain/article/7054/0/1689436/Gulf.Oil.Spill/Last.Day.To.File.Claim.With.BP.For.Individuals.and.Businesses.

Fothergill, Alice. (1996). Gender, risk and disaster. *International Journal of Mass Emergencies and Disasters, 14*: 33-56.

Fothergill, Alice. (1998). The neglect of gender in disaster work: An overview of the literature. In E. Enarson and B. H. Morrow (eds.). *The Gendered Terrain of Disaster: Through Women's Eyes.* Westport, CT: Praeger: 11-26.

Fothergill, Alice and Peek, Lori A. (2004). Poverty and disasters in the United States: A review of recent sociological findings. *Natural Hazards, 32*(1): 89-110.

Frailing, Kelly. (2010). Fraud following the September 11, 2001 and Hurricane Katrina disasters. In D. W. Harper and K. Frailing (eds.). *Crime and Criminal Justice in Disaster.* Durham, NC: Carolina Academic Press: 139-160.

Frailing, Kelly and Harper, Dee Wood. (2007). Crime and hurricanes in New Orleans. In D. L. Brunsma, D. Overfelt, and J. S. Picou (eds.). *The Sociology of Katrina: Perspectives on a Modern Catastrophe.* Lanham, MD: Rowman & Littlefield: 51-68.

Frailing, Kelly and Harper Dee Wood. (2010). Fear, prosocial behavior and looting. In D. W. Harper and K. Frailing (eds.). *Crime and Criminal Justice in Disaster*. Durham, NC: Carolina Academic Press: 89-105.

FTC. (2010). FTC warns of oil spill scams. Federal Trade Commission, May. Available online at: http://www.ftc.gov/bcp/edu/pubs/consumer/alerts/alt058.shtm.

Fulton, April. (2010). Consumers still skittish about Gulf seafood. *National Public Radio*, August 5. Available online at: http://www.npr.org/blogs/health/2010/08/05/129000765/consumers-skittish-on-gulf-seafood-safety.

Gill, Duane A., Picou, J. S. and Ritchie, Liesel. (2010). When the disaster is a crime: Legal issues and the *Exxon Valdez* oil spill. In D. W. Harper and K. Frailing (eds.). *Crime and Criminal Justice in Disaster*. Durham, NC: Carolina Academic Press: 61-82.

Ishiguro, Y. (1998). A Japanese national crime: The Korean massacre after the Great Kanto Earthquake of 1923. *Korea Journal, 38*(4): 331-355.

Killian, L. M. (2002). An introduction to methodological problems of field studies in disasters. In R. A. Stallings (ed.). *Methods of Disaster Research*. Xlibris Corporation: 49-93.

Kirkham, Chris. (2010). Oyster harvester trying to find his footing 5 months after Gulf oil spill. *The Times Picayune*, October 6. Available online at: http://www.nola.com/news/gulf-oil-spill/index.ssf/2010/10/oyster_fisherman_trying_to_fin.html.

Kaufman, Wendy. (2010). Gulf seafood industry works to wipe oil off image. *National Public Radio*, August 20. Available online at: http://www.npr.org/templates/story/story.php?storyId=129322243.

Lee, Keunbaik, Yongsung, Joo, Joon, Jin Song and Harper, Dee Wood. (2011) Analysis of zero-inflated clustered count data: a marginalized model approach. *Computational Statistics and Data Analysis, 55*(1): 824-37.

Mann, Brian. (2010). Critics: 'BP' needs to stand for 'bills paid.' *National Public Radio*, June 8. Available online at: http://www.npr.org/templates/story/story.php?storyId=127551098.

McGill, Kevin and Breen, Tom. (2010). Gulf residents struggle in aftermath of oil spill. *Houma Today*, August 24. Available online at: http://www.houmatoday.com/article/20100824/FEATURES12/100829694?p=3&tc=pg&tc=ar.

Monteleone, Diana C. (2007). Isleños. Center for Cultural and Eco-Tourism. University of Louisiana at Lafayette. Available online at: http://ccet.louisiana.edu/tourism/cultural/The_People/isleno.html.

Morris, Charles. (2002). *The San Francisco Calamity by Earthquake and Fire*. Urbana, IL: University of Illinois Press.

Neal, D. M. (1997). Reconsidering the phases of disasters. *International Journal of Mass Emergencies and Disasters, 15*: 239-64.

Noguchi, Yuki. (2010). BP hard to pin down on oil spill claims. *National Public Radio*, May 24. Available online at: http://www.npr.org/templates/story/story.php?storyId=127038021&ps=rs.

NPR. (2010). BP says tube is diverting some oil from Gulf well. *National Public Radio*, May 16. Available online at: http://www.npr.org/templates/story/story.php?storyId=126852086.

NPR. (2010a). BP recasts spill size as oil fouls Louisiana marshes. *National Public Radio*, May 20. Available online at: http://www.npr.org/templates/story/story.php?storyId=127012041.

NPR. (2010b). New estimates put oil flow rate significantly higher. *National Public Radio*, June 10. Available online at: http://www.npr.org/templates/story/story.php?storyId=127740491.

NPR. (2010c). Well cap holds, but tests raise issues. *National Public Radio*, July 16. Available online at: http://www.npr.org/templates/story/story.php?storyId=128577564.

NPR. (2010d). Gulf disaster is now worst U.S. offshore oil spill. *National Public Radio*, May 27. Available online at: http://www.npr.org/templates/story/story.php?storyId=127207628.

NPR. (2010e). White House: BP poised to create spill victims' fund. *National Public Radio*, June 14. Available online at: http://www.npr.org/templates/story/story. php?storyId=127835826.

Picou, J. S. (2007). Introduction: Katrina as paradigm shift: Reflections on disaster research in the twenty-first century. In D. L. Brunsma, D. Overfelt, and J. S. Picou (eds.). *The Sociology of Katrina: Perspectives on a Modern Catastrophe*. Lanham, MD: Rowman & Littlefield: 1-20.

Plyer, Allison and Campanella, Richard. Coastal employment before the 2010 Deepwater Horizon oil disaster: Employment maps and data from 2008. Greater New Orleans Community Data Center. Available online at: https://gnocdc.s3.amazonaws.com/reports/GNOCDC_CoastalEmployment.pdf.

Press-Register. (2010). BP oil spill contract worker in Panhandle charged with armed robbery. *The Press-Register*, July 20. Available online at: http://blog.al.com/live/2010/07/bp_oil_spill_contract_worker_i.html.

Quarantelli, Enrico L. (2007). The myth and the realities: Keeping the "looting" myth in perspective. *Natural Hazards Observer, 31*(4): 2-3.

Shapiro, Ellen. (1989). *The Croatian Americans*. New York: Chelsea House.

Siman, B. A. (1977). *Crime During Disaster*. University of Pennsylvania PhD diss. Ann Arbor, MI: University Microfilms International.

Skoloff, Brian. (2010). Bogus claims slow oil-spill payments, chief says. *Houma Today*, October 5. Available online at: http://www.houmatoday.com/article/20101005/ARTICLES/101009702?p=4&tc=pg.

Smith, Vicki. (2010). Oil cleanup brings 'strangers,' tension to towns. *The Associated Press*. July 25. Available online at: http://www.msnbc.msn.com/id/38406839/ns/disaster_in_the_gulf.

Smith, Tovia. (2010a). Many in Gulf on road to uncertain compensation. National Public Radio, July 28. Available online at: http://www.npr.org/templates/story/story.php?storyId=128830924.

Sultzman, Lee. (n.d.) Chitimacha history page, available online at: http://www.dickshovel.com/chi.html. Houma history page available online at: http://www.dickshovel.com/hou.html.

Teh, Y. K. (2008). The abuses and offenses committed during the tsunami crisis. *Asian Criminology, 3*: 201-11.

Thornton, William E. and Voigt, Lydia. (2010). Disaster phase analysis and crime facilitation patterns. In D. W. Harper and K. Frailing (eds.). *Crime and Criminal Justice in Disaster*. Durham, NC: Carolina Academic Press: 27-60.

Thornton, William E. and Voigt, Lydia. (2010a). Disaster rape: Vulnerability of women to sexual assaults during Hurricane Katrina. In D. W. Harper and K. Frailing (eds.). *Crime and Criminal Justice in Disaster*. Durham, NC: Carolina Academic Press: 107-38.

U.S. Census. (2008). American Community Survey detailed tables on population by race for St. Bernard and Orleans Parishes. Available online from the homepage at: http://factfinder.census.gov/servlet/DatasetMainPageServlet?_lang=en&_ts=304686914686&_ds_name=ACS_2008_3YR_G00_&_program=.

Walsh, Bryan. (2010). Gulf of Mexico oil spill: No end in sight for eco-disaster. *Time*, May 1.

Whitney, Michael. (2010). Are BP's drug tests fueling domestic violence in the Gulf? *Work in Progress*, July 1. Available online at: http://workinprogress.firedoglake.com/2010/07/01/are-bps-drug-tests-fueling-domestic-violence-in-the-gulf/.

Zahran, Sammy, Shelley, Tara O., Peek, Lori and Brody, Samual D. (2009) Natural disasters and social order: Modeling crime outcomes in Florida. *International Journal of Mass Emergencies and Disasters, 26*(1): 26-52.

Zhou, D. (1997). *Disaster, Disorganization, and Crime*. University of Albany State University of New York PhD diss. Ann Arbor, MI: University Microfilms International.

CHAPTER THIRTEEN
HAZARD, OUTRAGE AND LOCALITY: AN ANALYSIS OF TWO OIL SPILLS

AMANDA K. GODDARD, KENNETH A. LACHLAN AND PATRIC R. SPENCE

Risk communication has been defined as "all those integrated processes and procedures: a) that involve and inform all interested parties within the risk analysis process; b) that assist the development of transparent and credible decision making processes; and c) that can instill confidence in risk management decisions" (FAO/WHO, 1988, p. 5). Covello (1992) defines risk communication as "the exchange of information among interested parties about the nature, magnitude, significance or control of a risk" (p. 359). The goal of risk communication is to provide a public (the receiver) with messages containing useful information about the type and magnitude of an outcome from a behavior or context. The information provided is to be meaningful, relevant, accurate, clear, and in comprehensible terms to the target audience. Usually risk communication involves messages about some event or the probability of the event manifesting. Risk communication attempts to disseminate information that will prepare people for a possible threat and then provide the steps individuals should take if that risk manifests.

Risk communication becomes more difficult in situations that are unfamiliar to the public, or unfair, or where the negative effects are delayed. This is characteristic of the BP Deep Water Horizon Spill. Because the risk associated with the event were unfamiliar, unfair and not immediate, communicating how the spill was a risk became a difficult task. At the same time, near Marshall Michigan a more localized oil spill occurred which had several of the characteristics. This chapter examines the risk communication used in both technical breakdowns (Coombs 2009). First examined is a model of risk communication that is uniquely designed to meet the needs of such a novel event. Then the characteristics of the two events are described. Two newspapers are used to determine if elements of hazard and outrage emerge in reporting of these stories. Finally recommendations and suggestions are offered concerning risk communication in such extreme events.

Communication in such a circumstance should be made up of two facets: "scaring people" and "calming people down." Stated differently, messages should alert people, reassure them and give them direction. Risk communication messages should contain two components, hazard and outrage. Thus messages should ideally attempt to create a level of outrage that is appropriate given the level of hazard (a formula Sandman labels "Risk = Hazard + Outrage"). A helpful feature of the model is how it can be useful in aiding experts in defining and communicating risk in a manner that is more congruent with the public's perception of risk (Lachlan and Spence, 2007; 2010, Sandman, Weinstein and Hallman, 1998). Experts often communicate risk as expected annual mortality; however, the public views risk as much more than this and in the case of an Oil Spill, annual mortality is not a good indicator. The public tends to focus on outrage, what to do and who to blame, whereas the experts focus on hazard. Outrage factors are not distortions in the public's perception of risk but are more intrinsic parts of what is meant by risk. Because experts fail to acknowledge outrage in their communication messages, these efforts become less useful to the public. To better understand the effectiveness of newspaper stories and public response manifestations of hazard and outrage will be analyzed in the media.

HAZARD

If the public is responding incorrectly to a risk, it may be beneficial to address the hazard with the goal of modifying their understanding of the risk. Hazard is the technical seriousness of a risk. This is conceptualized by Lachlan and Spence (2007, 2009, 2010) as the magnitude (the size of the potential threat and its potential damage) and likelihood of the event occurring and its associated undesirable outcomes. The construct of the model examines both cognitive and affective responses to risks through examining elaboration by the message receiver, in contrast to other approaches that depend on fear alone as a motivator. In addition to likelihood, magnitude, and fear, other components of hazard include dread (the idea that the risk requires a respect or reverence helping people to live with and think about the risk reasonably), mortality (the number of deaths expected), and morbidity (the prevalence of the hazard). Risk messages therefore need to provide people with information explaining the dangers and likelihood of the event.

This multifactor construct is important in commutating risk effectively because focusing on just one dimension of the possibility of a hazard could underestimate the true risk. For example focusing simply on the likelihood of an oil spill is not an adequate way to communicate the potential risk. Oil Spills, particularly ones caused by the explosion of an Oil Rig are uncommon. Thus focusing on the likelihood is not guaranteed to effectively communication the real hazard. Additionally to gain the attention of various publics issues or mortality and morbidity may become salient as the long term effects of the risk will contribute to the understanding of the hazard. Finally, dread, or having a respect or reverence for the risk is quite power and related to each of the many hazard factors. Hazard has been shown to motivate people to take action and increase their attention to a risk. Therefore, news stories that address that the risk is real, influenced publics should be fearful of the effects, that those effects include death and destruction and finally that the long term effects are something they should be respectful of, a news story containing these elements should garner public attention and help the public obtain a true estimate of the risk.

OUTRAGE

Outrage is the cultural seriousness of a risk, or how people react to the risk information received. This can include both specific actions taken and emotional factors such as need for control, trust, and responsiveness. Emotional factors include worry, concern, anxiety or alarm; these include the suspense caused by an awareness of danger and hostility. Such emotional factors have the potential to motivate or impede actions; therefore, clear directions must be given to facilitate outrage. Messages and news features need to address outrage to motivate the public to indentify and partake in the best response to the threat and therefore should ideally include specific steps the public should take to avoid dangers or reassurance that no action is necessary. Although the assessment of risk may be fairly consistent across subpopulations, affective responses to that risk may vary greatly across different groups of people. Messages and news reports should also include explicit conclusions when addressing outrage to promote adoption of the intended behaviors, as this information will lead to higher levels of comprehension and attitude change in receivers.

Four potential scenarios of risk susceptibility exist within the framework of the model. First, it is possible that risks exist that may potentially harm people but fail to upset many, described as high hazard/low outrage risk. One such example is avoiding taking measures to protect one from sexually transmitted diseases. The hazard is high because there is a clear risk of contracting a sexually transmitted disease through unprotected sex; however, many people perceive the risk as remote and avoid taking preventative steps or getting tested (Becker and Janz, 1987).

Second, there are potential risks that upset people despite the fact that the likelihood of harm is low, this risk perception may motivate the public to take unnecessary precautions; this is a low hazard/high outrage risk. The potential outcome of risk messages focusing on the Oil Spill could result in people stock-piling gasoline, or bottled water because of a belief is price spikes and contamination. Messages should induce enough fear that the audience will take the desirable, if not essential action but not induce too much fear that could lead to unnecessary panic or the avoidance of low risk actions. Moreover, it is important that enough factual information is given so people do not over estimate the risk. Even though the hazard is low, it must be addressed with the goal of reducing the unnecessarily high outrage. Low hazard/high outrage events can be extremely harmful in that they have the potential to take time and resources away from other more immediate risks.

Third, risks may manifest that are serious, high in potential harm, and create outrage among those at risk. This is characterized as high hazard/high outrage. While a strong outrage response to a highly threatening situation may seem ideal on surface, problems can exist with this type of reaction. In the best situations if the risk is high and people act, harmful consequences can be avoided. This is how the possibility of Oil Spills should be perceived by the public. In some extreme instances, however, outrage can extend beyond the ideal level to induce remedial responses and actually become socially negative. If this occurs a type negative outrage can emerge, creating inaction. This could happen when a person believes that environmental impact of the spill can be cleaned up or the economic impact of the spill is will cause continued harm on the region and therefore any of the recommended actions are useless. This can be a problem with messages or new stories that focus only on fear appeals. If the receiver becomes too fearful, he/she could believe that nothing can be done.

Finally, the fourth potential scenario is that there are low hazard/low outrage risks. In this situation a risk has little chance of causing harm and does not require special actions to be taken from the population. Thus if the oil spills were small and contained, there would be no impact on the public or actions for them to take. Thus, hazard and outrage remain low, given the fairly benign nature of the threat.

RESPONDING TO THE RISK

The identification of these types of crises naturally begs the question of how to subsequently respond through varying communication efforts. A substantive body of communication scholarship has examined message strategy, primarily through the assumption that the optimal outcome of these messages is the retention of public image. These studies have largely focused on the reduction of hostility among affected audiences and the processes of image restoration (Benoit, 1997). Benoit and others have contributed to a useful corpus of research concerning Image Restoration Theory (IRT), drawing on earlier research in crisis management (Scott and Lyman, 1968; Ware and Linkugel, 1973). In IRT, the general public can be expected to respond in certain ways when events take place in which blame can be attributed to an organization. Hostile responses can be neutralized through the use of particular message strategies (see Benoit, 1997; 2006). In the aftermath of a crisis or emergency, perceptions may be more important than realities in terms of managing negative affective responses and hostility stemming from those affected.

Image Restoration Theory offers numerous messages strategies for dealing with hostile audiences, which can be broadly categorized under one of five headings (Brinson and Benoit, 1996). The first category is denial; organizations may state that they are not responsible for the event that took place, or attempt to shift the blame and target another individual or organization. The second is evasion, in which organizations committing an act but refuse to acknowledge responsibility for negative ramification; one common tactic in this rubric is to attempt to minimize the offensiveness or degree of harm posed by the incident. Finally, organizations may hostile audience responses through corrective action or simply by issuing an apology (called *mortification* in IRT terms) (see Benoit, 2006 for a lengthy explanation of the specific message strategies under these headings). .

Lachlan and Spence (2010) argue that there is a logical connection between hazard, outrage, and the message strategies specified by IRT. In their multi-study evaluation of the constructs, they conclude that outrage may be more accurately conceptualized in two dimensions. They label the first "Event Outrage," otherwise known as negative affect or hostility directed at an event itself. They extend this conceptualization from Sandman's (2003) definition of the cultural seriousness of risk to include emotional considerations such as need for control, responsiveness, and trust. In this subtype of outrage, affected publics need information in order to assess the cultural seriousness of the risk and make behavioral decisions.

Lachlan and Spence (2010) also argue for a second, critical type of outrage: "Blame Outrage." As stated above, individuals often have the need to blame individuals or organizations for crises and other non-routine events. In instances in which there may be personal or organizational culpability, this manifests as Blame Outrage.

There appear to be clear connections between IRT and the consideration of outrage, and blame outrage in particular. Considering whether to use denial, shift blame, or engage in corrective action or mortification may stem from considering the degree and source of blame outrage and the likelihood of success of any one of these strategies. When blame outrage is strong, people are not looking for information related to recovery from a risk or crisis, nor are they seeking information about how to prepare for occurrence similar crisis in the future. Rather, they are articulating negative affect toward those likely responsible. While blame outrage can be prosocial if it motivates individuals to take action, it can also be anti-social if it leads to irrational behavior or inhibits remedial responses. If nothing else, it at least serves to seriously threaten the reputation and viability of the organization or individuals that appear to be responsible.

Thus we are presented with multiple considerations concerning the types of hazard and outrage that may follow a crisis, and the message strategies that should be used subsequently in order to best serve the public and maximize benefit for the organizations and/or individuals involved. Messages concerning an impending high risk event must clearly state that the hazard is real, followed by specific information concerning remedial steps that can be taken to reduce potential harm. Once this immediate need for information is satisfied, individuals will continue to scan for information, seeking repetition of the initial message, clarification, and new developments (Lachlan, Spence and Seeger, 2009). Therefore, it is critical for crisis communication practitioners to create messages that adequately address hazard, and to repeat them. These messages must also be crafted with consideration for the degree of outrage experienced by the affected public, and the relative prominence of event outrage and blame outrage. It is at this point that the typologies specified by IRT (see below) become critical. Thus, the crisis communication practitioner is faced with balancing factual and informative information concerning remedial responses with specific IRT driven message strategies that may be applicable to crises that beget different levels and types of outrage.

Of course, little is known to this point about the balance of hazard and outrage considerations in typical responses to organizational crises. While a few studies have attempted to explore these constructs in audience responses (Lachlan and Spence 2007; 2010), to date none have attempted to systematically quantify appeals to these concerns in a series of mediated responses. This is a critical shortcoming in the literature, as speculation concerning appeals and audience responses could be addressed through the exploration of the information actually received by the general public. An understanding of these messages may also help create a roadmap for crisis communication practitioners in developing future response strategies. These considerations require an examination of media content in order to evaluate what types of messages are typically provided, and how well they meet the expected goals and needs of the audience.

THE CURRENT STUDY

The goal of any message should be to reduce any error between the true nature of the risk, and the public's assessment of the risk. Therefore, if the true risk is something that will not harm the public and requires no actions by the public (low hazard/low outrage) but the public misunderstands the risk and is reacting in correctly, the model can help solve this problem. Using the Risk = Hazard + Outrage model, the risk of surrounding the BP Oil Spill and the Marshall Oil Spill should be high hazard and high outrage however, the question is to what extent did news papers promote this as a high hazard high outrage event? Because many people did not view it as a hazard that could specially affect them they did not take actions. Although a large part of the public may have not lived on the Gulf Coast or near the Kalamazoo River, effects such as economic and wildlife impact were real concerns which may not have been communicated well.

In addition to issues of Risk, also imbedded in many messages is an attempt to repair and organizations image. Image repair strategies can be placed into five broad categories: denial, evasion of responsibility, corrective action, reducing offensiveness, and mortification (Benoit, 1995). The strategies are each a general label and have variants.

Denial is an attempt by the accused individual to deny all wrongdoing. This can be accomplished either through denying that the act occurred or through denying that he/she did the act. A second option of denial is to shift the blame away from himself/herself. This is helpful because it answers the obvious question: If you didn't do it, then who did? In this situation shifting the blame may not answer the question but it could be used to set up another more successful strategy for the situation.

Evasion of responsibility occurs when the accused doesn't deny committing the offensive act or can't deny committing the offense but attempts to evade or reduce his/her responsibility for the act. Four variants of this strategy are identified: scapegoating/provocation, defeasibility, accident and good intentions.

Scapegoating or provocation occurs when the accused argues that the act was a reaction to another offensive act, therefore, the instigator of the original act should be held responsible. A second variant is defeasibility. In this strategy the accused claims that he/she had a lack of information or a lack of control over the situation. The accused may also attempt to evade responsibility through suggesting that the offensive act committed was an accident. Individual's hold others responsible only when the circumstance appears to be one that could have been controlled. Therefore if some type of accident occurred or something was done by accident, it is likely that the individual will not be held responsible for the act. Thus the accused could claim that he/she never tried to hide information about a prior relationship from the relational partner, he/she just forgot to disclose it. Thus the accused provides information that may aid in reducing his/her apparent responsibility for the offensive act. Lastly, the accused may claim that the offensive act was done with good intentions. The offensive act is not denied but the message given is that the accused should not be held completely responsible because the intentions behind the act were those of good and not evil. When a person does bad while trying to do good it is believed that they should not be blamed as much as one who had tried to do bad. The accused can make the argument that details of the past relationship were of no relation to the present relationship and bringing up the details would just hurt the relational partner. Therefore failure to disclose the information was done with the intentions of avoiding hurt to come.

Another way in which an individual can attempt to repair his/her image is through reduction of offensiveness. The individual in this situation will admit to the offensive act, but attempt to reduce the perceived offensiveness of the act, with the goal of leading to a repair of the image of the accused. This strategy has six variants (Benoit and Hanczor, 1994). The strategies are bolstering, minimization, differentiation, transcendence, compensation and/or attacking the accuser.

Bolstering is an attempt to strengthen the audience's positive affect toward the accused to counteract negative feeling about the individual. Another possibility is to minimize the negative perception of the offensive act. If the accused can be persuaded to believe that the act isn't as bad as it originally appeared, the amount of negative feeling associated with the act can be reduced. The accused may engage in differentiation by distinguishing the act from other similar but less beneficial acts. The goal of differentiation is to make a contrast effect. Through transcendence the accuser paints a picture of the offensive in a broader context in order to reduce the offensiveness of the act. Compensation is an offer to reimburse the victim of the act. It is believed if the compensation is accepted then the negative effect from the wrongful act will be eliminated, restoring the individual's image. It is difficult to put a monetary price on a person's emotions but the accused could attempt to purchase a gift to compensate for the offensive act, if the gift is accepted the wrongful act will be seen as eliminated. The last variant of reduction of offensiveness is attacking the accuser. The attempt is to discredit the source of the allegations in an attempt to restore one's own image.

The fourth category of image repair is corrective action. In this strategy the accused attempts to correct the problem. The willingness to correct the problem as well as help prevent any reoccurrence of the problem is believed to help the accused repair his/her image.

The final act is mortification. This strategy requires the accused to admit responsibility for the offensive act and ask for forgiveness.

THE EVENTS

The British Petroleum oil spill in the Gulf of Mexico has been labeled the worst oil spill in history, and the worst human created environmental disaster in known history. An oil rig explosion and pipeline rupture sent 90,000 gallons of oil spilling into the gulf during the initial stages of the crisis. BP shut down the pipeline when it became aware of the spill and notified environmental agencies. The company initially said that the spill was large but manageable, and they felt there was a fast response to the spill. Following the initial rupture, BP spokespersons gave more details to the crisis, stating that 210,000 gallons of oil a day were leaking into the gulf and it was reported later that eleven workers were missing and presumed dead. Following the first week of this disaster government officials stated, "I do have concerns that BP's current resources are not adequate to meet the three challenges we face...and they may need more help from the federal government and from others," (Wold, Davis and Lemoine, May 1, 2010). Only two months later, another oil spill would affect the area of Marshall, Michigan.

The second oil spill, although of smaller scale, occurred at a similar point in time. More than 800,000 gallons of oil had been released into a creek in Marshall Michigan. The creek which was contaminated feeds into the Kalamazoo River. Officials' state that oil began to flow into the creek after a leak had developed in the pipeline. After Enbridge, the oil company, learned of the spill, the pipeline was shutdown and isolation valves were closed which stopped the flow of oil. A few hours later the valves were turned back on because the problem was believed to be fixed. However, later that afternoon, after the spill had been detected, there was a thick layer of oil that was visible and many small fish had washed up onto the riverbanks. Officials later decided to evacuate homes located near the river due to the gases being distributed from the oil spill, and wildlife would be affected due to the crude oil in the water. Many residents stated that they were very worried, and they felt a sense of hopelessness after the spill. Businesses were also affected due to this spill, a canoeing company had to shut down business indefinitely and sales were dropping on fishing licenses.

Each of the oil spills were reported on in daily in newspapers not only in the United States, by throughout the world. From the inception of the BP oil spill, the company was being looked at through a microscope by the media, and considering the Kalamazoo River oil spill happened soon after the BP oil spill, Enbridge Oil Company was also being examined very closely.

To examine the use of the constructs of hazard and outrage in news reporting a content analysis was conducted of one paper in each of the affected regions. Two local papers were chosen in order to give the best possibility of coverage of hazard, outrage and IRT. It was believed that the local papers would provide more information about the scope and impact of the crisis. The two papers analyzed were *The Advocate* (a regional paper), which is based out of Baton Rouge, Louisiana. The newspaper consists of a Monday-Friday Advocate, as well as the weekend editions Saturday and the Sunday Advocate. As of 2004, it had a daily circulation of nearly 100,000 and a Sunday Advocate circulation of nearly 125,000. The paper maintains bureaus throughout south Louisiana as well as in bordering states. For the conditions and reporting of the Kalamazoo River oil spill was *the Kalamazoo Gazette (*a local paper) was selected. The Kalamazoo Gazette is the oldest continuously publishing newspaper in Michigan outside of Detroit. The Kalamazoo Gazette is a daily paper with a circulation of over 30,000.

Two weeks of papers were content analyzed for each area. For the Advocate, all articles addressing the oil spill for a two week period were included in the study. The articles selected began with the seventh day of the event. The sample began with day seven due to the time it took for the public to be warned about the true scope of the oil spill. For the Kalamazoo Gazette, the sample began the first day. The Advocate had seventeen articles that mentioned the spill over the selected fourteen-day period. The Kalamazoo Gazette produced eighteen articles on the topic (See Tables 13.1 and 13.2). One individual content analyzed all articles in the sample and another individual analyzed 20 percent of the articles. Cohens's kappa (1968) was calculated to assess intercoder reliability, "the extent to which independent coders evaluate a characteristic of a message or artifact and reach the same conclusion" (Lombard et al., 2002, p. 589).

To measure the occurrence of hazard and outrage the following themes were examined. To examine appeals to hazard, article mentions of hazard were content analyzed. Appeals to hazard included anywhere in which the article mentioned the magnitude of the event, its probability, fear and its impact or potential impact on humans or the environment. Articles were also analyzed to determine if expressions of hazard were mentioned through the result of a quote from another source. Appeals to outrage were analyzed through the presence of articles mentions of anger, suspicion, distrust, contempt and directions concerning self-efficacy. Articles were also analyzed to determine if expressions of outrage were mentioned through the result of a quote from another source. Then at the article level of analysis the themes were added up and coded for presence and absence.

Table 13.1: Appeals and Mentions of Hazard and Outrage by Location

	Louisiana	Michigan
Appeals to Hazard mentioned in the article		
Magnitude	16 (100%)	17 (94%)
Probability	2 (12.5%)	1 (5.6%)
Mobility	7 (43.8%)	5 (27.8%)
Mortality	8 (50%)	2 (11%)
Expression of Hazard quoted by an individual		
Magnitude	16 (100%)	15 (83.3%)
Probability	2 (12.5%)	2 (11.1%)
Mobility	10 (62.5%)	9 (50%)
Mortality	6 (37.5%)	7 (38.9%)
Appeals to Outrage mentioned in the article		
Anger	5 (31.3%)	1 (5.6%0
Suspicion	9 (56.3%)	5 (27.7%)
Worry	7(43.8%)	3 (16.7%)
Distrust	2(12.5%)	4 (22.2%)
Efficacy	7 (43.8%)	2 (11.1%)
Expression of Outrage quoted by an individual		
Anger	7 (43.8%)	3 (16.7%)
Suspicion	13 (81.3%)	10 (55.6%)
Worry	15 (93.8%)	10 (55.6%)
Distrust	7 (43.8%)	7 (38.9%)
Efficacy	10 (62.5%)	3 (16.7%)

Table 13.2: Number of appeals for Image Restoration by Location

	Louisiana	Michigan
Denial	0	3
Shifting the Blame	2	2
Provocation	2	1
Defeasability	16	3
Good Intentions	1	1
Bolstering	37	19
Minimization	3	3
Differentiation	2	0
Transcendence	8	1
Corrective Action	29	10
Mortification	3	3

Finally, the articles were examined for the presence of Image Repair Theory tactics (Benoit, 1995). These include denial, shifting the blame, provocation, defeasibility, accident, good intentions, bolstering, minimizing, differentiation, transcendence, attack, compensation, corrective action and mortification.

DISCUSSION

The analysis reveals a number of interesting findings and discrepancies between local and regional coverage of the spills, in terms of appeals to hazard and outrage and in the presentation of image restoration tactics on the part of the parties responsible. In terms of copy appeals to hazard, both papers were equally likely to discuss the magnitude of the spills and equally unlikely to report on the probability of the event. This makes sense intuitively; give that we are looking at accounts of events that have already taken place; consideration of their relative probability seems irrelevant. However, it is noteworthy that the The Advocate was more likely to report on both morbidity and mortality. This seems to indicate that at least in this case, local reporting may shy away from the actual environmental and human harm and focus instead on the size of the event. This is problematic considering that those directly affected by an incident of this type are likely to need highly specific information concerning the likely damage associated with it, and may be more trusting of localized news sources for this type of information.

In terms of appeals to outrage, we see more in the way of anger, suspicion, and worry in The Advocate. More distrust is expressed in the local paper, which may make sense logically because the audience for a local paper may feel a stronger sense of relationship with the company responsible. At the same time, the local paper was less likely to provide information related to self-efficacy. Again, this is cause for concern. The audience of a regional level newspaper is less likely to have a need for information concerning self-efficacy, since they are less likely to be directly affected. Highly localized publics will turn to community media for specialized information concerning remedial steps that can be taken to neutralize any potential harm. In this instance, that appears to have been next to impossible.

In terms of quotes, there are not significant differences between the local and regional coverage in terms of the hazard categories. The frequency of quotes addressing outrage is somewhat different. At the local level, more quotes regarding worry over the ramifications of the incident were aired. Further, more quotes were printed regarding the need for self-efficacy. Again, perhaps these results are not surprising. As a localized crisis, it stands to reason that anyone providing a quote to a local paper would express more worry, as they are more likely to be affected directly by the spill. Further, given the abovementioned absence of consideration for self-efficacy in local media, it stand to reason that appeals to self-efficacy would be evident in quotes from individuals, as the question of "what do I do now?' is, ironically, disseminated alongside the information that fails to address it.

While there are no differences between the two papers in terms of the frequency of official statements, the subsequent image restoration tactics in these statements differ somewhat. At the regional level, there is greater use of defeasibility. At the local level, there is greater denial and mortification. In fact, the only documented incidents of mortification were found in the local paper.

This is consistent on a number of levels. First, defeasibility tactics are likely to work better at the non-local level. Those who have less direct contact, stand to lose less, and are in greater geographic proximity to the spill may be more easily swayed by defeasibility. Those living in the immediate path of the spill, who are privy to firsthand accounts of cleanup efforts, ecological damage, and physical harm, may be less likely to change their opinion of those responsible based on these appeals. Given this reluctance to be persuaded, two polar strategies—complete denial and mortification (or admission of wrongdoing and subsequent apology)—become the only viable strategies.

CONCLUSION

Although not the dormant medium today as it was in the past, newspapers still create and shape public opinion. They also have an information role which many count on. Moreover, older media such as news papers allowed for the interjection of editors, producers, executives, and others as information gatekeepers giving some the belief of more credibility to the medium. This, in some ways, may have fueled the rise of what Sundar (2008) calls the oldmedia heuristic, which suggests that aesthetic resemblance of newer social media to older media (i.e., newspapers) results in higher credibility judgments. Therefore these results help us in understanding how risk messages are communicated by the newspaper media and how they are used. If newspapers are to continue to be a respected media source, local papers such as the Kalamazoo Gazette need to make sure that issues of self efficacy are raised and elaborated in reporting. The ability to take some action during a crisis or in the presence of a risk can produce a sense of empowerment, thus creating an impression that the individual has some control in the situation (Seeger et al. 2003). Issues of self efficacy are important to all aspects of risk.

This chapter was an initial examining of the presence of hazard and outrage construct and Image Restoration Theory in newspapers. If further research on this area is undertaken, it would be advisable to examine difference between news reports and editorials and multioutlet markets.

REFERENCES

Becker, M. H. and Janz, N. K. (1987) Behavioral science perspectives on Health Hazard/Health Risk Appraisal. *Health Services Research, 22*, 537–51.

Benoit, W. L., and Hanczor, R. (1994). The Tonya Harding controversy: An analysis of image repair strategies. *Communication Quarterly, 42*, 41633.

Benoit, W. L. (1997). Image repair discourse and crisis communication. *Public Relations Review*, 23, 177–86.

Benoit, W. L. (1995). *Accounts, excuses, apologies: A theory of image restoration discourse*. Albany: State University of New York Press.

Brinson, S., and Benoit, W. L. (1996). Dow Corning's image repair strategies in the breast implant crisis. *Communication Quarterly, 44*, 2941.

Cohen, J. (1968). Weighted kappa: Nominal scale agreement with provision for scaled disagreement of partial credit. *Psychological Bulletin, 70*, 213220.

Coombs, W. T. (1999). *Ongoing Crisis Communication: Planning, Managing and Responding*. Thousand Oaks, CA: Sage.

Covello V. 1992. Risk communication, trust, and credibility. *Health and Environmental Digest* 6(1):14.

Lachlan, K. A., Spence, P. R., and Seeger, M. (2009). Terrorist attacks and uncertainty reduction: Media use after 9/11. *Behavioral Sciences of Terrorism and Political Aggression*, 1(2), 101–10.

Lachlan, K. A., and Spence, P. R. (2010). Communicating risks: Examining hazard and outrage in multiple contexts. *Risk Analysis*.

Lachlan, K. A., and Spence, P. R. (2009). Emergency communication: A framework for planning and targeting messages. *Journal of Emergency Management, 7* (4) 6972.

Lachlan, K. A., and Spence, P. R. (2007). Hazard and outrage: Developing a psychometric instrument in the aftermath of Katrina. *Journal of Applied Communication Research*.

Lombard, M., SnyderDuch, J, and Bracken, C. C. (2002). Content analysis in mass communication: Assessment and reporting intercoder reliability. *Human Communication Research, 28*, 587604.

Sandman P. M., Weinstein N.D., Hallman W. K. (1998). Communications to reduce risk underestimation and overestimation. *Risk Decision and Policy, 3*, 93–108.

Sandman, P. M. (2003, April). *Four kinds of risk communication*. The Synergist, pp. 26-27.

Scott, M. B., and Lyman, S. M. (1968). *Accounts, American Sociological Review*, 33, 46-62.

Seeger, M., Sellnow, T.L., and Ulmer, R.R. (2003). *Communication and organizational crisis*. Westport, CT: Praeger.

Sundar, S. S., and Nass, C. (2001). Conceptualizing sources in online news. *Journal of Communication, 51*, 5272.

Ware, B. L., and Linkugel, W. A. (1973). They spoke in defense of themselves: On the generic criticism of apologia, *Quarterly Journal of Speech, 59*, 273–83.

Wold, A., Davis, S., and Lemoine, D. (2010 May, 1). Cleanup effort impared. *The Advocate*. Retrieved from http://www.lexisnexis.com.

CHAPTER FOURTEEN
RESISTING CORPORATISM: CITIZENS FIGHT BACK AGAINST THE BP DEEPWATER HORIZON OIL SPILL DISASTER

STAN C. WEEBER

ABSTRACT

Long term stabilization of post-disaster recovery areas through "disaster mitigation" and "disaster reconstruction" tends to benefit the wealthy over the needy. Stabilization planning has been dominated by a neoliberal agenda in which the pursuit of free market competition and profits win out over victim interests and public safety. Thus, the inequalities exposed by disaster are reproduced and multiplied by the mitigation/reconstruction process. This chapter suggests that communities in the Gulf Coast resisted such corporatist policies during the British Petroleum Deepwater Horizon oil spill disaster of 2010. For instance, post-secondary and community education helped multiple publics to understand that the neoliberal agenda did not have to be accepted and could in fact be challenged at the local level. A related strategy of contestation and reform involved coordinated direct action to better the plight of oil spill victims. Here the community created its own rebuilding organizations that stressed community control of assets over absentee ownership in the spill zone. Such strategies were leveraged by public opinion and by organizational success.

INTRODUCTION

The aftermath of disaster, whether natural or manmade, unmasks wealth and income inequalities as well as abject poverty for the world to see. The receding flood waters in post-Katrina New Orleans exposed structural poverty and inequity deeply tied to race and class. In Haiti, desperately poor people with no place to go after the 2010 earthquake, including children separated from parents, huddled amidst the ruins of an abrupt reorganization of their lives that will last for years to come. Victims of the 2004 tsunami in Thailand wandered about aimlessly, grieving lost loved ones and not yet believing that their lives and livelihoods were washed away by a giant wave. Having lost all but their land, people sought out that land, bracing for the business/government coalition bent on rebuilding the Thai coastline on more business friendly terms.

The Deepwater Horizon oil disaster similarly removed the veil covering up fundamental evidence about the nature of social stratification and power in the Gulf of Mexico off of the Louisiana coast. Third and fourth generation fishing families found themselves out of work and precariously hanging on to their businesses. The U.S. government that is supposed to be looking out for their interests let British Petroleum call the shots in the first thirty days of the crisis, allowing BP to use their own resources or long familiar vendor resources to mitigate the crisis amidst substantial cost savings. More important, the government let BP control information and access to information, and allowed BP to construct its own narrative which for a time was able to limit the company's potential liability. In the meantime, Gulf Coast locals out of work and out of luck were not hired by BP in large numbers to help with the cleanup, and were for the most part not consulted about how the cleanup process was to proceed. The result was an unfortunate situation where the federal government, relying mostly upon BP, was supposed to "make things right" along the coastline at a time when locals knew that the crisis could mean the end of their jobs, the ruination of the coastal environment and possibly the destruction of their culture and ethnic heritage. The oil giant, when it leaves the scene, will have much of its resources intact. People indigenous to the disaster zone may find their lives rebuilt on a smaller scale, thus widening the resource gap between the oil company and an important portion of its purported clientele, the people of the Gulf Coast.

As the Deepwater Horizon case illustrates, the corporate rich that manipulate the long term stabilization of the disaster recovery areas through "disaster mitigation" or "disaster reconstruction" extend the advantages that the well off already possess over the needy. Stabilization planning, in at least three countries fairly recently, has been dominated by a neoliberal agenda in which the pursuit of free market competition and profits win out over victim interests and public safety – the existential interests of the unfortunate masses trapped in the disaster zone or scattered by evacuation. The most vulnerable victims suffer the most; some cannot return home; many find themselves worse off than before; and many find their way of life unworkable in the new normal that lies ahead. Thus, the inequalities exposed by disaster are reproduced and multiplied by the reconstruction process.

This chapter documents leveraged responses to such corporatist policies. For example, post-secondary and community education can help multiple publics to understand that the neoliberal agenda does not have to be accepted and can be challenged at the local level. A related strategy of contestation and reform involves coordinated direct action to better the plight of the disaster victims. Here the community with help from interested partners creates its own rebuilding organizations that stress community control of assets over absentee ownership or control in the disaster zone. These strategies are leveraged by public opinion and by organizational success.

THEORETICAL ORIENTATION: JOHN LOCKE AND THE SOCIAL CONTRACT

The beginning of sociology as an academic discipline coincided with the start of sociological theory's classical period. An overarching question to be answered by these classic figures concerned the problem of order, that is, given the giant social dislocations evident in the migrations of thousands of families from rural villages to new industrial cities, how would order be maintained? (Parsons 1937). The answer is complex because societies found themselves on new social terrain with new political regimes emerging. Because these regimes were more free and democratic than those of the feudal period, no one knew exactly how the problem of order would be resolved.

Social philosophers of the Enlightenment had already given thought to this problem, although they framed it differently. For these philosophers, it was not so much a problem of order as it was a matter of survival—as man/woman moved from a more primitive "state of nature" to a more sophisticated "civil society," certain adjustments were necessary. These writers asked us to consider that people formed a social compact or a social contract in order to survive in a new kind of society. The writers differed about why and how the contract is conceived, but all agreed in the utility of the idea of the social contract as a basic organizing principle for the new societies. Thomas Hobbes (1991) believed that the contract would be necessary to provide security and safety, or to save men from themselves. Rousseau (1954) argued that a general will or consensus emerges that directs the group on the path to fruition of the group's goals.

The writings of John Locke (1988) on the social contract appear to be especially germane to the kinds of social resistance found by citizens as they fought back against the BP oil spill. Locke believed that human beings had certain human rights, such as the right to life, liberty, and property, and that these were natural rights that could not be surrendered within the social contract. Thusly conceived, a social contract, though formed for the benefit of all, could be dissolved should it be determined that parties to the agreement are not fulfilling their portions of the contract. It was in this spirit, I believe, that citizens fought back against BP, feeling that their civil protectors in the form of governments at all levels had essentially abandoned their role as protectors of the people and had formed an alliance with BP. With their backs against the wall, self help became the strategy of choice.

RECONSTRUCTION VS. RECOVERY

Neo liberal policies on natural and/or manmade disaster management stress reconstruction over recovery. Such reconstruction policies are rooted in the free market philosophies of Chicago School economist Milton Friedman. Friedman, viewing the displacement of New Orleans school children in areas across the USA after Hurricane Katrina as unfortunate and tragic, also said that it opened the door for the radical reconstruction of the New Orleans schools by converting them into for profit charter schools. This was just the beginning of the social transformations that neoliberals had in mind for the city, as disasters in general are viewed as opportune moments to change the public sphere and to promote exciting market opportunities (Gunewardena and Schuller 2008; Saltman 2007). Only thirteen days after Katrina's landfall, the Heritage Foundation proposed thirtytwo free market solutions for New Orleans, several of which were subsequently implemented by the Bush Administration (Klein 2007). Reconstruction of critical infrastructure such as schools is needed after disasters, but it should not be pursued at the expense of individual families whose recoveries are uneven at best and nonexistent at worst.

Post-disaster reconstruction plans for New Orleans emphasized maximizing profits, the centerpiece of which were nobid contracts to Halliburton, Fluor, Shaw, and CH2M Hill totaling $3.4 billion (Klein 2007). Additionally, there was a lowering of the minimum wage; cheap labor was brought in by truckloads from the outside; and there were plans to completely redesign the city. Some residents rightly feared that these plans would destroy valuable aspects of New Orleans culture and bring about gentrification, making it impossible for them to return (Delgado 2010).

If reconstruction is incompetent at rebuilding or restoring homes and lives it could be attributable to the fact that rebuilding is not the primary purpose. Klein (2005: 2) stated that the purpose of reconstruction is to transform everything. She called this "a predatory form of disaster capitalism that uses the desperation and fear created by the catastrophe" to radically reinvent social and economic institutions. The reconstruction industry works with efficiency and dispatch so that privatizations and new land contracts get locked in before locals become aware of the consequences.

A template for action by the reconstruction industry was constructed following Hurricane Mitch's landfall in Honduras in 1998. Two months after Mitch, with the country still reeling from the devastation, the Honduran congress initiated "speed sell offs" which allowed the privatization of airports, seaports, and highways.

Additionally, accelerated plans were laid for privatization of the national electric company, the state telephone company, and portions of the water department. The World Bank and IMF placed a demand that the telecommunications company be privatized in return for releasing 4.4 billion in debt relief and 47 million in annual aid over three years. (Klein 2005; Stonich 2008).

As Hurricane Mitch demonstrated, with destruction there is opportunity for reconstruction, a chance to gain advantage from the unfortunate losses and redesign blighted areas with sophisticated cutting edge plans, as was the case in Aceh, Indonesia, ground zero of the 2004 tsunami (Klein 2005). In the disaster's aftermath, the World Bank was poised to help rebuild Thailand, Indonesia, and other areas affected by the giant waves. The affected countries were in need of immediate financial aid, and they had sought no long term debt relief. When the World Bank responded it was with loans and not grants. Fishing villages, about 80 percent of the victims needing relief, received minimal help. The bank instead pushed for corporate friendly expansion of industrial fish farms and tourism. Its recommendation to the government was to consider privatizing damaged public infrastructure such as roads and schools, arguing that rebuilding them could put undue strain upon public finances (Klein 2005; Gunewardena and Schuller 2008). Critics, with whom I heartily concur, refer to such reconstruction as a new form of colonialism. A researcher who works for the agency Focus on the Global South, commented: "We used to have vulgar colonialism. Now we have sophisticated colonialism, and they call it reconstruction" (quoted in Klein 2005: 2). A similar situation occurred in Haiti following the 2010 earthquake. The Heritage Foundation moved quickly, issuing a statement that the tragedy provided an opportunity to reshape Haiti's long dysfunctional government and economy and to improve the image of the United States in the region (Klein 2010).

A rebuilding effort in Haiti actually started before the earthquake, as both the United States and the World Bank were helping Haitians institute private development and privatization of governmental services. Following the exit of President JeanBertrand Aristide, the Bank ordered Haiti to adopt a publicprivate partnership form of governance in the health and education sectors; in exchange for that, the country received a 61 million loan. America for its part was encouraging Haiti's government to restructure and privatize selected public sector enterprises (Klein 2005; Schuller 2008). A similar plan in Sri Lanka prior to the tsunami offered loans from the International Monetary Fund and World Bank in exchange for agreements to open the economy to privatization and publicprivate partnerships (Klein 2007).

Thus conceived, reconstruction after the tsunami had little to do with replacing or recovering what individual families or groups of families had lost. Governments allowed hotels and industry to build on the coast as early as 2005 in Sri Lanka, Thailand, Indonesia and India. Meanwhile, laws and policies of the region's governments prohibited families from returning to their seaside land and rebuilding their homes. Indonesia's 12 volume reconstruction plan, for example, specified that hundreds of thousands living within two kilometers of the sea should be forcibly relocated inland to military style barracks in Aceh. As soon as this policy was announced, there was a rush by government insiders to buy land in the "new city" areas, and private sector contractors and commercial interests began jockeying for a share of the lucrative mega project contracts (ACHR 2005). Thus, the coast is not being restored to its pretsunami form. Instead, a governmentcorporate alliance together with foreign donors were teaming up to rebuild the areas as they had always wanted it to be developed: as a mecca for tourists, corporate fishing ventures, and privatized roads and airports. An citizen group called Thailand Tsunami Survivors and Supporters says that for businesspoliticians the tsunami was "the answer to their prayers, since it literally wiped these coastal areas clean of the communities which had previously stood in the way of their plans for resorts, hotels, casinos and shrimp farms" (Klein 2005: 1).

In the case of British Petroleum following the April 20 explosion and subsequent oil spill in the Gulf of Mexico, a number of differentiating circumstances emerged. This was a manmade disaster rather than a natural one. Second, there was no land area that needed to be reconstructed as such; only a well to be plugged and a large body of water to be cleaned along with marshes and coastlines. Third, there were not large numbers of people immediately dispersed or displaced from their homes, as is generally the case in natural disasters. Despite these factors that make the 2010 oil spill a unique circumstance, the way in which both BP and the U.S. government managed the crisis revealed that neoliberal policies were followed once again, repeating what had occurred in Haiti, New Orleans, and the Indian Ocean region after recent natural disasters.

The Obama administration initially downplayed the seriousness of the oil spill, playing into BP's effort to lowball the amount of oil gushing from the well and thereby reduce down potential liabilities. Early on, the president appeared to be conceding responsibility for the cleanup to BP. His initial behavior appeared to some to be feeble, indecisive, and somewhat aloof. No one person appeared to be clearly in charge of the situation, as General Russell Honore had been after Hurricane Katrina. Without clear leadership, the federal bureaucracy appeared, at least to some observers, to be in the way more than it was helping with the cleanup operation (Imhofe 2010).

For starters, the Obama administration failed to waive the Jones Act. This 1920s era law protects unionized labor from foreign competitors by prohibiting work in U.S. water by vessels that were either not manned by Americans or were built outside the U.S. Because of this, there was a forbidding of ships from numerous foreign countries that could have assisted in cleaning up the Gulf. By using familiar vendors and longstanding contacts, BP saved itself money by not having to pay whatever rates the foreigners might have demanded. Domestic ship owners were similarly rebuffed for different reasons, although the underlying economic goals remained the same. The Coast Guard, it appeared, was actively interfering with cleanup efforts in Louisiana. Oil sucking barges were stopped because the Coast Guard needed to confirm that the ships were equipped with fire extinguishers and life vests, and then there were further delays when it had trouble contacting the people who built the barges (Tennant 2010). Finally, for weeks the government stonewalled cleanup efforts by refusing to buy into Louisiana Governor Bobby Jindal's plan to erect sand berms near fragile ecosystems to prevent the oil from entering. After weeks of bitter rankling with various officials, Jindal's plan for about 45 miles of sand barriers was approved (Murphy 2010).

By controlling who participated in the cleanup, and when, BP was able to reduce down its costs. That control extended to the land based cleanup. Though it was responsible for paying the wages of an estimated 27,000 who signed up to work in the operation, the workers' pay was so low that the workers were classified as "volunteers" by BP. The workforce was made up mostly of employees of myriad government agencies and contractors (Webb 2010). BP also cut corners with respect to the lack of training it provided to cleanup workers. Evidence emerged that these workers were not receiving the proper training or equipment needed to safely handle the toxic mixture of oil and chemical dispersant contaminating the waters. Working conditions on the cleanup crew were so bad that some workers sought restraining orders against BP. Others, such as fishermen hired by BP to assist in the clean up, went out on strike in July to protest against the company's decision to force them to stay in quarter ships called "flotels." Workers, who were obliged to stay on the ships for periods of 18 days straight with on 3 days off, were not paid for their time on the flotels (Lee 2010).

A much larger structural issue is the determination of a $20 billion limit for damage payments, a limit that is wholly inadequate given the level of destruction for which the company is responsible. The BP spill released multiple times more oil than the Exxon Valdez spill in Alaska. If the fines from the Valdez accident were applied dollar for dollar to the BP disaster, the total would be several times larger than the 20 billion escrow account that BP set aside to settle claims (Kishore 2010).

After Valdez and subsequent federal legislation in 1990, governmentmandated readiness plans made oil spill response a lucrative business. Longtime friends or vendors of BP that are in the disaster response business cleaned up in more than one way from the Deepwater Horizon disaster. Nalco, the Illinois based water treatment firm sold more than $40 million of its oil dispersant Corexit. Clean Harbors, a Massachusetts disaster response company, expected second quarter sales to jump 15 to 20 percent or up to $70 million. Procter and Gamble shipped 7,000 bottles of Dawn dish soap to clean animals. There was also a significant presence by Seacor Holdings which has seven subsidiaries operating in the Gulf. Coast Guard personnel were ferried by Seacor's helicopters over the marshes; Seacor's 200 foot ships transported supplies. At subsidiary O'Brien's Response Management, consultants worked with BP at command posts (Cendrowski 2010).

Finally, there was a lack of access given to the press covering the event. Journalists trying to do their jobs in documenting the impacts of the oil rig explosion repeatedly found themselves turned away from public areas affected by the spill. They were denied access to information by BP and its contractors, and by local law enforcement, the Coast Guard and government officials (Kourlouris

2010). This enabled BP to build its own in house version of events and to announce this version as "fact" at press briefings (or alternatively, to not respond to legitimate questions asked). Without a valid countervailing viewpoint, BP has only to defend its own narrative when court proceedings against the company begin. The restricted press policy thus was part of an overall strategy of reducing down the liabilities of the company.

With emphasis on reconstruction, recovery becomes a sorely neglected agenda, in effect a public relations ploy to give off the impression that families are being helped when the grim reality is that many are falling through the cracks of the large, heavily bureaucratized operation. While the USA's FEMA issues reports that boast of activities accomplished and dollars spent, the harsh fact is that a substantial number of families end up getting little or no help at all (e.g., FEMA 2006; Sullivan 2005). Government reports fail to note these embarrassing statistics, and from a probusiness stance, too much help to families detracts from capital accumulation and profit. What evidence now exists to document the unwritten policy that the least help is the best help? Library shelves around the world are now full of volumes or documents that verify the lack of governmental response after Hurricane Katrina, the 2004 tsunami, the 2010 earthquake, and the BP oil spill (Bullard and Wright 2009; Brinkley 2006; Horne 2006; Dyson 2006 2007; Cooper and Block 2006; Troutt 2006; Hartman and Squires 2006; MontanaLeblanc 2008; Marable and Clark 2008; Potter 2007; Hidalgo and Barber 2007; Saltman 2007; Vollen and Ying 2008; Klein 2010; Van Auken 2010; Herz 2010; Katz 2010; Lariscy 2010; Freudenberg and Gramling 2011).

RESISTING CORPORATISM

Citizens after disasters have come to realize that they are deeply into the era of self help when it comes to recovery. Having discovered that the social contract binding them to government has been violated, self help becomes a realistic means of survival and selfpreservation. As Locke wrote, people had natural human rights that gave them all the authority they needed to resist or push back against government if they felt that the social contract was being not being carried out. This appeared to be the spirit with which communities fought back against corporatist policies after the BP oil spill. In response to the spill they created several worthy entities and agencies that were proactive in meeting citizens' needs given the lack of timely or sufficient government or corporate help. The following discussion emphasizes examples of such entities rather than a comprehensive analysis. Generally there are two lines of resistance, one in the field of higher education, the other in the field of community based organization, protestation, and reform.

EDUCATION

In southwestern Louisiana, a political sociologist teaches his students about a new research agenda fueled by Hurricane Katrina and the BP oil spill that includes an investigation into the controversy surrounding "reconstruction versus recovery" strategies and how citizens have fought back against the neoliberal machine (Hidalgo and Barber 2007; Weeber 2009). Geographically closer to the epicenter of the disaster zone, University of South Alabama sociologist Steven Picou works with coastal residents to help them cope with the oil spill by adopting uplifting community building attitudes toward their situation (Truong 2010). As he points out, one of the things that happens in a technological disasters such as the BP spill is that communities are characterized by conflict and resource loss, rather than an attempt to collectively build a consensus, come together, and rebuild the community better than it was before. Community conflict ensues or what he calls the "corrosive community." Individuals may show clinical levels of depression but the community is ill equipped to help it has been left dysfunctional and struggling by the disaster. Thus, Picou believes that outreach programs based upon principles of selfhelp are critical. One program he prefers is education on the Exxon Valdez spill and the lessons learned as well as related information resources available to disaster victims. A second program is called peerlistener training, which assists in changing a person's role from victim to active community helper by suggesting a number of recuperative coping strategies that help the person have a more positive outlook upon the negative situation that they find themselves in.

Elsewhere, teachins reminiscent of the Vietnam War teachins of the 1960s were conducted at California State UniversityFullerton and at Temple University. Temple's event during the week of October 8 was entitled, "The BP Disaster: Reckless Corporate Greed or Business as Usual?" Students for a Democratic Society, an apparent reincarnation of the 1960s group of the same name, held teaching symposiums in Asheville, North Carolina; Tallahassee, Florida; Houston, Texas (at Chevron); New Orleans and Grand Isle, Louisiana. The protest and disruption group Code Pink also used their innovative protests as teaching moments relative to the oil spill (Hong 2010).

At selected schools, universities sponsored presentations by off campus groups that were directed toward both the university and the larger nonacademic community. For example, at two public meetings on the Gulf Coast, the Socialist Equality Party taught people in academe as well as the greater community about the political implications of the BP spill, exploring the linkages between the spill and the capitalistic political system that produced it. These meetings took place in August at Tulane University in New Orleans and at the University of South Alabama in Mobile, respectively. At a similarly organized meeting directed toward education of both academic and non academic people was held June 8, an event called the Oil Spill Response Forum at the New School for Social Research in New York City. The purpose was to address the consequences of the oil leak, and speed up the transition to a better world. At the forum of 200 people, participants broke into discussion groups and brainstormed on how each attendee could impact the media and government, push for change in the business community, build public awareness, and offer direct assistance to Gulf communities.

Some teachins were community events not affiliated with universities, such as those at the Global Exchange in San Francisco and the Big Oil and Creative Nonviolent Action Mass Teach In at Oakland, California. A particularly significant event held on July 20 was sponsored by the Mobilization for Climate Justice West in San Francisco and involved hosting a discussion on the impacts of the BP oil spill on Gulf Coast communities, and enlisting positive solutions that people could use in the Bay Area to end dependence on fossil fuels and prevent future disasters.

While not a teachin as such, a fruitful cultural exchange took place between indigenous peoples of Ecaudor and Gulf Coast Native Tribes impacted by the Deepwater Horizon oil spill in July. The Ecaudoreans had been severely impacted for decades by Chevron's oil contamination in Ecaudor's rainforest. The Amazon leaders toured areas of the Bayou affected by the spill, engaged in community exchanges with the Houma and other Gulf Coast residents, and participated in a public meeting in the heart of the largest Houma community (Parker 2010).

COMMUNITY ORGANIZATION

The South Louisiana Vietnamese community utilized networks of self help and social support created after Hurricane Katrina to assist fishing families that were faced with loss of livelihood due to the oil spill. The fishermen could request replacement money from BP but they had to file a claim for up to $5,000 for losses or sign up for training to do oil cleanup work. In each case, they signed waivers agreeing to never hold the company liable for future losses or injury. Unfortunately, all the paperwork and all the training was in English, and most of the fishers could not read or speak English, and could not understand the legal fine print contained in the contracts. Further, fishermen complained about the proficiency and cultural competence of BP translators. Utilizing networks formed after Katrina, the Mary Queen of Vietnam Community Development Corporation trained some of its community members in oilspill jargon. Later, they served as translators to their people as they filled out BP's claims paperwork (Esclamado 2010).

Elsewhere, citizen groups determined that street protests, at least in their neighborhoods, would bring more publicity and provide a larger and more dramatic human statement of citizen dissatisfaction with the oil spill. On May 16, an ad hoc group of concerned activists picketed a BP gas station in Baltimore, Maryland. They expressed their outrage at BP for the massive oil spill, lax government regulations of the industry, and strongly opposed allowing any more offshore drilling permits (Hughes 2010). Also in May, the $500 million contract between the University of California at Berkeley and BP was the target of 150 demonstrators. The protest was held in front of the university's administration building, California Hall, where a "mock spill" event was held. Two students were arrested for dumping a molasseswater mixture on to the ground (Lord 2010).

In Alabama and Florida, there were no street protests but resistance coming from a most unlikely place—public officials. In both states elected officials believed that their jurisdictions had come to the end of their reasonable options for dealing with the spill, and chose to stand firm for the interests of the people they represent. In Magnolia Springs, Alabama Jamie Hinton, the local volunteer fire chief, led the campaign to save his town's waters from the approaching oil spill. He deployed nine spud barges flanked by containment boom, and by doing so, Hinton was faced with the possibility of being jailed for violating the federal and state chain of command at one time but persisted despite the obvious risks. Well aware that the federal unified command cared little about the interests of his small South Alabama town, and knowing that the local townsfolk knew the nearby waters better than anyone, he decided to stand down against the federal government. To everyone's surprise, Hinton's plan was approved and enacted (Busby 2010).

In Florida, the Okaloosa County Commission voted unanimously to allow their emergency managers to override and overrule federal emergency managers, thus taking on the responsibility of saving the white sand beaches in Fort Walton Beach that provide about 50 percent of the local economy. The Coast Guard told the commissioners that they would go to jail. The commissioners had tried but failed to get through to an 800 number for the unified federal command (Tennant 2010).

Perhaps the ultimate form of community organizing is the class action lawsuit, and several were formed in the weeks following the spill. Suits by hoteliers, property owners, fishermen and workers are likely to be combined into one mega suit.

LEVERAGE BY PUBLIC OPINION

The fact that so many in the Gulf Coast region had to help themselves during and after the oil spill exposed the neoliberal emphasis on reconstruction rather than recovery. Citizens in the affected areas as well as the public as a whole are beginning to gain a sense of the neoliberal agenda and to associate it with a lack of response to genuine needs of citizens. The media has been helpful in spreading the message and helping mobilize opposition. *Cable News Network* (*CNN*) for example sent young, idealistic journalists to the Gulf Coast where they spoke to disaffected citizens and travelled in boats to see damage first hand. The coverage clearly demonstrated how some individual families and their businesses were simply left to fend for themselves, which in turn fueled a public opinion reflective of moral outrage over the unfairness of the situation. A variety of opinion polling results demonstrated Americans' unhappiness with the way the government was responding to the needs of citizens during the oil spill crisis (Cohen 2010; Bigg 2010; Page 2010; *CBS* 2010; Chipman 2010; Muskal 2010). For example:

- 61 percent of Americans polled disapproved of Obama's actions in the aftermath of the spill
- 64 percent said BP should be charged with criminal counts by the federal government

- 62 percent gave a rating of "not so good" or "poor" to the federal government in its handling of the spill
- 81 percent said they disapproved of BP's response
- 53 percent believed Obama did a "poor" or "very poor" job of managing the crisis
- 73 percent said BP's response was "poor" or "very poor"
- 71 percent claimed that Obama was not tough enough in dealing with BP
- 75 percent said BP deserves a great deal of blame for the spill
- 72 percent disagreed with Obama's oil spill policy

The data show that a broad public dissatisfaction with how the crisis was handled that was spread fairly evenly between the government and BP, although BP's ratings were slightly more negative. This makes sense if BP was allowed to call the shots early on, and if neoliberal policies were followed which required a close alliance or collaboration between the two parties. The data also indicate a measure of the public's polarization and alienation from the adopted policies.

LEVERAGE BY ORGANIZATIONAL SUCCESS

A further leveraging tool is the success of organizations in fighting back against neoliberal mitigation and reconstruction. The success of the self help groups has provided templates for future groups to use as reference points in initiating action (Sierra Club 2010; Center for Biological Diversity 2010; MQVN 2010; Save Our Gulf 2010). The following discussion underlines exemplary activities, and is not intended to be a comprehensive or an all inclusive list of such accomplishments.

- The Oil Spill Response Forum in New York City adopted two direct action campaigns and three direct assistance projects related to the Gulf Coast oil spill
- The Center for Biological Diversity in Tucson, Arizona filed suit against the government for failing to assess the impacts of the spill on endangered whales and sea turtles
- The Center for Biological Diversity filed suit against the government for failing to make public its contacts with oil industry lobbyists since the spill began
- Sponsored by the Mobilization for Climate Justice West, over 200 people marched against Big Oil on August 29 in Oakland, California; 15 persons were arrested
- BP hired specially trained Vietnamese translators from Queen of Heaven church to assist Vietnamese fishermen in filing claims related paperwork
- Ninetyfive community activists from around the country drafted the Weeks Bay Principles for Gulf Recovery in October at Weeks Bay, Alabama
- Community cleanup efforts in Fort Walton Beach, Florida led to a declaration of "oil free beach" as of July 13.

CONCLUSION

Our post-modern world is driven by powerful media images. With the capping of the well underneath the Deepwater Horizon rig those images have stopped, and for that reason many now believe that the oil spill disaster ceases to exist. Numerous uncertainties remain, however, that will render the spill a continuing crisis for the Gulf Coast. If all the class action suits are joined into one mega suit, it may take years for the courts to hear the case and the compensation could be far less that what is anticipated today to be the total bill for cleaning up the oil spill and returning the Gulf Coast to its pre-spill condition. Uncertainties also remain with respect to environmental damage. No one knows for sure many oil plumes are staying under the water surface and what effect this has on the bottom dwelling plankton that larger marine species depend on. A top scientist has contradicted the government's claim that most of the oil that flooded into the Gulf from BP's busted well has degraded, insisting that more than half remains (Laney 2010).

This chapter brought some clarity to how the resource gap has grown between BP and an important part of its clientele living along the Gulf Coast. Neoliberal policies have been followed as has been the case ever since Hurricane Mitch during the Clinton Administration. In this era of self help, citizens fought back against BP through post-secondary and community education as well as community organization and action. Believing that their backs were against the wall and that all reasonable options had been exhausted, some of the actions bordered on extreme responses. Yet, Americans have defended their life, liberty and property before and they will do it again. Although most are probably unfamiliar with the writings John Locke, Locke's views helped inspire ideas about federalism that influenced the writers of the American Constitution.

The argument could be made that by encouraging self help for disaster victims, the neo liberal agenda basically wins the day by default and goes unchallenged. If citizens help themselves, businesses automatically spend less on recovery. This point is contested by the weight of the numerous success stories uncovered to date which in turn serve as templates for future action. This means that the post-disaster self help movement grows, and along with it, a growing perception that not enough is being done to support disaster victims. This is without question a public relations disaster for business. And, as the uncertainties of global climate change and deep water oil drilling play out into realities, there is a likelihood of even more disasters, and they may occur in parts of the world that have heretofore been relatively untouched or unaffected by the kinds of human tragedies considered in this paper. It is a near certainty that these newly affected groups will fight back against the neoliberal machine when their lives or livelihoods are on the line, and will look at past successes as a blueprint for how to proceed.

Therefore, it is important that disaster recovery problems and their links to neoliberal policy get attention at multiple events such as the World Social Forum and the U.S. Social Forum. Momentum towards building organized systems of advocacy for disaster victims could well grow as a result of the attention from these venues. In time, citizens must directly confront the source of the formulation and implementation of the policies, namely, the representatives of the executive and legislative branches of national governments. Reconstruction does not have to be abandoned, but a portion of its resources need to be reallocated for recovery.

This chapter outlined how citizens fought back against corporatist policies and why they fought. The last thought to consider is why fighting back is the right thing to do as a matter of principle. As a moral imperative, we must consider what a just and decent society would do in the best of all possible worlds to protect our most vulnerable people—the poor, the sick, the elderly, the disabled, the children and the ethnic minorities that are left behind by the disastrous event and then doubly victimized by the unfriendly, uncaring policies of neoliberalism (Quigley 2006). Native American tribes along Louisiana's coast close to the spill zone are now perilously close to ethnocide. Policies that leave such vulnerable victims poorer after disaster than they were before constitute an avoidable social harm that is morally untenable according to even the most basic principles of social justice. A new, higher standard must be established.

REFERENCES

ACHR. "Newsletter on the Tsunami Recovery in Asia." Asian Coalition of Housing Rights, August (2005): 12.

Bigg, Matthew. "Groups Sue BP for Harm to Endangered Gulf Wildlife." *Montreal Gazette*, October 20 (2010): 1.

Brinkley, Douglas. *The Great Deluge: Hurricane Katrina, New Orleans and the Mississippi Gulf Coast.* New York: Morrow, 2006.

Bullard, Robert and Beverly Wright. *Race, Place and Environmental Justice After Hurricane Katrina: Struggles to Reclaim, Rebuild, and Revitalize New Orleans and the Gulf Coast.* Boulder: Westview Press, 2009.

Busby, Guy. "Locals Step in to Save Weeks Bay from Oil Spill." *PressRegister* (Baldwin County), June 10 (2010): 1.

CBS. "Poll: 70% Say BP Handling Oil Spill Badly." May 25 (2010): 1.

Cendrowski, Scott. "The Cleanup: Who's Cleaning Up." *Fortune*, July 5 (2010): 16.

Center for Biological Diversity. Biologicaldiversity,org, accessed October 31, 2010.

Chipman, Kim. "Americans in 73% Majority Oppose Deepwater Drilling Ban." *Bloomberg*, July 14 (2010): 1.

Cohen, Joe. "Poll Shows Negative Ratings for BP, Federal Government." *USA Today*, June 7 (2010): 1.

Cooper, Christopher and Robert Block. *Disaster: Hurricane Katrina and the Failure of Homeland Security.* New York: Times Books, 2006.

Delgado, Sharon. "Responding to Hurricane Katrina." Earth Justice Ministries, June 5 (2010): 1.

Dyson, Michael. *Debating Race*. New York: Basic Civitas Books, 2007.
———. *Come Hell or High Water: Hurricane Katrina and the Color of Disaster*. New York: Basic Civitas Books, 2006.
Esclamado, Leo. "Vietnamese Americans Speak About BP Oil Drilling Disaster." Mary Queen of Vietnam, June 12 (2010): 1.
FEMA. "$4 Billion in the Hands of Disaster Victims." Federal Emergency Management Agency, January 4 (2006): 1.
Freudenburg, William and Robert Gramling. *Blowout in the Gulf: the BP Oil Spill Disaster and the Future of Energy in America*. Cambridge, MA: MIT Press, 2011.
Gunewardena, Nandini and Mark Schuller. *Capitalizing on Catastrophe: Neoliberal Strategies in Disaster Reconstruction*. Lanham: Rowman & Littlefield, 2008.
Hartman, Chester and Gregory Squires, *There is No Such Thing as Natural Disaster: Race, Class and Hurricane Katrina*. New York: Routledge, 2006.
Herz, Ansel. "U.N. Clash with Frustrated Students Spills into Camps." *Inter Press Service News Agency*, May 25 (2010): 1.
Hidalgo, Danielle and Kristen Barber. *Narrating the Storm: Sociological Stories of Hurricane Katrina*. Newcastle: Cambridge Scholars Press, 2007.
Hobbes, Thomas. *Leviathan*. Cambridge: Cambridge University Press, 1991.
Hong, Sharon. "Protesters Demand BP Be Shut Down Over Oil Spill." *Houston Chronicle*, May 24 (2010): 1.
Horne, Jed. *Breach of Faith: Hurricane Katrina and the Near Death of a Great American City*. New York: Random House, 2006.
Hughes, William. "Protesting BP in Baltimore, Maryland." *Vimeo*, May 16 (2010): 1.
Imhofe, James. "Bush's Response to Katrina Makes Obama Look Pitiful in Gulf." *Human Events*, September 6 (2010): 3.
Katz, Jonathan. "Protesters in Haiti Demand President Preval's Resignation, Accuse Him of Profiting from Quake. *Associated Press*, May 10, 2010.
Kishore, Joe. "Beyond BP." World Socialist Web Site, accessed October 28, 2010.
Klein, Naomi. "Haiti Disaster Capitalism Alert: Stop Them Before They Strike Us Again." Naomiklein.org, accessed June 5, 2010.
———. *The Shock Doctrine: The Rise of Disaster Capitalism*. New York: Metropolitan Books, 2007.
———. "The Rise of Disaster Capitalism." *The Nation*, April 14 (2005): 13.
Kourlouris, Bryan. "BP, Oil, Destruction and Profits." *Socialist World*, July 25 (2010): 1.
Laney, Kelsey. "BP Oil Spill Overview." *The Daily Titan*, August 30 (2010): 1.
Lariscy, Angel. "Black Oystermen Fight Discrimination in Gulf." *The Militant*, August 16 (2010): 1.
Lee, Hiram. "Fishermen on Strike Against BP." World Socialist Web Site, accessed October 28, 2010.
Locke, John. *Two Treatises of Government*. Cambridge: Cambridge University Press, 1988.

Lord, Vanessa. "Two Arrests in Protest Over Biofuels Deal." *The Daily Californian*, March 4 (2010): 1.

MQVN. Mary Queen of Vietnam Community Development Corporation, mqvncdc.org, accessed October 31, 2010.

Marable, Manning and Kristen Clark. *Seeking Higher Ground: the Hurricane Katrina Crisis, Race and Policy Reader*. New York: Palgrave Macmillan, 2008.

MontanaLeblanc, Phyllis. *Not Just the Levees Broke: My Story During and After Hurricane Katrina*. New York: Atria Books, 2008.

Morales, Lymari. "Americans Critical of Oil Spill Response; Keeping Close Tabs." *Gallup*, May 27 (2010): 1.

Murphy, Paul. "White House Approves Plan to Build Sand Berms." *WWL*, June 2 (2010): 1.

Muskal, Michael. "Poll: Bush Handled Katrina Better than Obama Dealt with Oil Spill." *Los Angeles Times*, August 22 (2010): 1.

Page, Susan. "Poll: Oil Spill's Economic Impact Worries Public." *USA Today*, June 6 (2010): 1.

Parker, Aubrey. "People From Ecuador to Louisiana Forge Alliances Against Global Oil Spills." *Circle of Blue*, July 8 (2010): 1.

Parsons, Talcott. *The Structure of Social Action*. New York: McGraw Hill, 1937.

Potter, Hillary. *Racing the Storm: Racial Implications and Lessons Learned from Hurricane Katrina*. Lanham: Lexington Books, 2007.

Quigley, Bill. "Six Months After Katrina: Who Was Left Behind Then and Who is Being Left Behind Now. Justice For New Orleans, February 21 (2006): 17.

Rousseau, Jean Jacques, *The Social Contract*. Chicago: H. Regnery Company, 1954.

Saltman, Kenneth. *Capitalizing on Disaster: Taking and Breaking Public Schools*. Boulder: Paradigm Publishers, 2007.

Save Our Gulf. Saveourgulf.org, accessed October 31, 2010.

Schuller, Mark. "'Haiti Is Finished!' Haiti's End of History Meets the Ends of Capitalism." In *Capitalizing on Catastrophe: Neoliberal Strategies in Disaster Reconstruction* edited by N. Gunewardena and M. Schuller. Lanham: Rowman & Littlefield, 2008, pp. 191-214.

Sierra Club. Sierraclub.org, accessed October 31, 2010.

Stonich, Susan. 2008. "International Tourism and Disaster Capitalism: The Case of Hurricane Mitch in Honduras." In *Capitalizing on Catastrophe*: *Neoliberal Strategies in Disaster Reconstruction* edited by N. Gunewardena and M. Schuller. Lanham: Rowman & Littlefield, 2008, pp. 47-68.

Sullivan, Bob. "Contacting FEMA: Is Anyone There?" *Red Tape Chronicles*, November 1 (2005): 1.

Tennant, Michael. "Gulf Oil Spills Greases the Skids for Federalism." *The New American*, June 21 (2010): 1.

Troutt, David. *After the Storm: Black Intellectuals Explore the Meaning of Hurricane Katrina*. New York: New Press, 2006.

Troung, Kelly. "A Sociologist Applies His Expertise to an OilWell Disaster." *Chronicle of Higher Education*, Volume 56, Issue 40, July 16 (2010): 6.

Van Auken, Bill. "Haiti: Hunger Sparks Growing Protests." World Socialist Web Site, accessed June 5, 2010.

Vollen, Lola and Chris Ying. *Voices from the Storm: The People of New Orleans on Hurricane Katrina and its Aftermath*. San Francisco: McSweeney's Books, 2008.

Webb, Tim. "BP Cleanup Contractors Told to Ignore Workers' Pay Claim 'Fraud.'" *The Guardian*, June 25 (2010): 1.

Weeber, Stan. *Post-Rita Reflections: A Sociological Journey*. Lanham: Hamilton Books, 2009.

CHAPTER FIFTEEN
DISASTER DISTRUST: RISK ASSESSMENT, CITIZEN SCIENCE AND TECHNOLEGAL DEBATES IN THE BP OIL SPILL[1]

SABRINA MCCORMICK

Burt's[2] boat gently floated up to the edge of the marsh. Silvery water ebbed at the roots of stiff, tilted grasses, leaving traces of metallic purple, blue and white. The oil was filtered through the muddy ground only a few feet from Burt's shrimp nets. Several scientists jumped out of the boat and began to take samples of the sediment. Another scientist reached into the water and filled a large clear bottle with a water sample. Burt watched, inexperienced with science. Most of his life has been spent shrimping in Southern Louisiana.

Burt is the third generation of a family of shrimpers living in the Bayou. A self-made man, he operates several major fishing operations. Although his livelihood and family traditions depend on it, he questions the safety of the shrimp they were meant to take to market once fishing waters were officially declared safe by the government. He and his workers had seen oil in the water and in the seafood they collected, and wondered if the food was safe. When a neighbor working with the Louisiana Bucket Brigade called to ask for help locating sites where samples of water, sediment and seafood could be found and tested for toxicity, he agreed, and

---

[1] This material is based upon work supported by the National Science Foundation under Grant No. 1051074.

[2] Names have been changed to protect the anonymity of the individuals in this chapter.

became a part of a new form of "citizen science." By collecting samples from the sites where he usually gathered shrimp and testing the samples for toxicity, both he and the scientists hoped to capture exposures otherwise missed in formal risk assessment.

Meanwhile, Jenna, a scientist from the National Oceanic Aerospace Administration (NOAA), traveled along the Gulf coast collecting oyster samples. Every morning, she took a small boat out into local oyster beds and gathered samples that were sent to the National Seafood Inspection Laboratory for analysis. Lab technicians tested seafood toxicity by first smelling specimens then performing a chemical examination. Jenna was one of many NOAA scientists that took sediment and seafood samples during the duration of the spill. Seafood was also collected dockside after fishermen pulled it in catches. NOAA used these data, and worked with the Food and Drug Administration to decide when waters should be opened and closed to fishermen.

Assessing risks of the oil spill, or disasters more generally, is a challenge complicated by the situated nature of knowledge-generation (Harding 2004) that results in differential perceptions and responses (Clarke and Short 1993; Neil et al.1994). These processes are critical in the case of the British Petroleum spill in the Gulf Coast since the identification of risks promises to have massive legal ramifications for multiple social actors. Exposures to communities, workers, and the environment have been neglected by experts in past oil spills, subsequently affecting the ability of these groups to gain desired legal outcomes (Ha 2008; Morita et al. 1999). A similar situation is developing with the Gulf oil spill as contention develops between government and industry assessors who claim the impacts are less than anticipated, while community groups continue to see risks in their daily lives. Clean up workers, local communities and fishermen are particularly vulnerable since they face daily exposures and lack an evidentiary base for their argument. Formal risk assessment processes do not necessarily capture the total range of risks since methods for measurement may be incongruent with the situational specificity of exposures that local people identify. In this way, the BP spill is similar to other instances in which discontinuities between lay and expert risk assessments have driven social contention, resulting in emergent lay assessments of risk (Sjöberg 1999). Yet, little is known regarding the outcomes of these disjunctures between lay and expert assessments of disaster risk, and how they can be bridged.

This chapter examines how differences between lay and expert assessments of risk lead to 'disaster distrust,' where an extreme event and the methods used to assess its risks foster tension and conflict across stakeholders. Research regarding the ways in which experts and lay audiences have understood these issues have often been categorized as risk perception and risk assessment, respectively. I reframe the understanding of risks on the part of lay people and experts as analogous processes based on diverse evidentiary bases and methods. I investigate how human health impacts of the British Petroleum spill are identified through local citizen science endeavors and formalized risk assessments. The first is largely embodied and experiential, although increasingly codified through scientific testing. The second is assessed through abstract scientific methods. Each has its limitations, although, together, they may complement or inform one another.

I argue that the gap between these two approaches during risk assessment and legal decision-making, results in disaster distrust. This distrust is particularly important considering that these types of approaches and their differential levels of credibility is particularly important to technolegal decisions regarding spill clean up and recovery. While lay knowledge is largely discredited in legal settings, expert assessments are the basis of compensation for spill impacts. Therefore, the intersection of lay and expert approaches to risk assessment represents the site in which legal disputes and disaster recovery are negotiated. The potential economic, social, and political ramifications of these differences raise the stakes regarding codification and legitimacy of perceptions, science and data collection. These differential approaches to understanding what the impacts of the spill are on health and livelihoods result in legal battles between stakeholders that hold diverse perspectives on such risks. Most commonly, while plaintiffs argue that BP is disregarding the impacts on their lives, BP argues that there is insufficient evidence to substantiate these claims. Government risk assessors and independent scientists stand in the middle of these two stakeholders as the arbiters of defining which evidence is credible and therefore usable in these long-term struggles. Decisions regarding the credibility of evidence promise to have critical impacts on the lives of communities in the states affected by the spill, as well as a ripple effect across the nation as the products and services provided by these communities either flourish or fail.

Issues of evidence, scientific methods, and localized understandings of risk hold particular significance during extreme events and disasters. During these moments existing social processes and norms that otherwise go unnoticed are highlighted revealing social relations. Spill events, such as this one, are important moments for social change (Kurtz 2004). However, they are affected by historically contentious relationship between communities and the oil industry. Such systemic problems have exacerbated crisis moments, and are critical elements to consider in risk management (Slovic 1993). These dimensions shape risk perception and assessment (Beamish 2001). Disasters are also particularly charged because their rapidity of onset and need for address may demand new resolution of previously vague and undefined social relations. In this way, technolegal aspects of risk assessment processes affect disaster recovery and long-term response.

The most basic aspect that determines whether or not there is sufficient and appropriate evidence to establish a claim is sampling. Sampling of seafood and air are the basic measures upon which such assessments are made. As such, it is rife with contention and therefore a central factor in the development of disaster distrust. Therefore, I focus on sampling methods used to understanding exposures as the central focus of disaster distrust. Air and food-borne exposures are some of the most important in the case of the BP oil spill. Both government agencies and non-governmental organizations have mapped these samples into a spatial framework to assess risk to diverse communities and workers for both government risk assessments and community-based sampling.

New approaches to mapping risk by social movements during the BP spill have resulted in tactical innovations that respond to local community concerns. These new forms of citizen science have aimed to overcome historical limitations to expert assessments of oil spills by reflecting embodied risk perception of community members and workers (Palinkas et al. 1993). Although there has been little study of such technological advances in disasters, they are similar to the rapidly expanding online social networking and information sharing that is intimately affecting other social processes and arenas (Castells et al. 2009). Online information sharing is a growing tool for many types of social actors, and in a range of social processes (DiMaggio et al. 2001). In the case of the BP spill, these new networks of knowledge have the potential to shape social relations and risk perceptions by engendering the collectivization of embodied knowledge in affected communities through a more detailed and grounded method of mapping (Corburn 2005).

It remains to be seen whether these efforts affect the outcomes of legal struggles, yet it is important to study lay-driven risk mapping that has already affected local organizing, and may also shape future studies of risks to local communities or workers. This focus takes an alternative approach to that traditionally adopted in surveillance studies where expert codifications of risk control the behaviors and outcomes of those being surveyed (Hacking 1999). By seeing these innovations in movements, this research also seeks to contribute to an understanding of how technology is used in social movements (Hess 2005), and the role of technology in governance (Jasanoff 2004).

In order to investigate how this disaster distrust develops, I collected qualitative interviews, ethnographic observations, and video data with local social movement organizations, grassroots groups, spill workers, fisherman, local residents, scientists and government representatives within five months of the spill event. Interviewees were identified through a snowball sample that began with local social movement organizations and government officials identified based on documents. Documents were also collected from government websites and reports. This sample resulted in thirty-one interviews with fishermen, oil cleanup workers, government officials, independent scientists, and community residents. Relevant government websites were also analyzed for data regarding risks, and relevant reports were gathered and included in this process.

HEALTH IMPACTS AS THE CRUX OF DISASTER DISTRUST

One of the most central concerns of legal struggles regarding the impacts of the spill is that regarding its effects on human health. The long-term human health impacts of the Gulf Oil Spill are wide-ranging and, as yet, largely unidentified. Few studies have focused on this topic, while there has been more research tracking impacts on the ecological context that may have impacts on human health in the later term. Government agencies have studied exposures to a greater or lesser degree, depending on the site and the type of exposure. However, dedicated health studies are few. Therefore, ephemeral data regarding short-term exposures to community members are being lost.

Most human health assessments have found little threat to human populations, including workers, local communities, or those who consume seafood. For instance, air monitoring by the Environmental Protection Agency during the period of the active spill found only a handful of days where airborne exposures were above safe levels. In August, the National Institute for Occupational Safety and Health (NIOSH) released an interim report that emphasized the risks to oil workers due to heat stress, while minimizing that 46 percent of respondents also had a rash and other neurological symptoms (NIOSH 2010). Although these official monitoring processes and reports demonstrate minimal concern regarding human health risks, local groups have continued to be concerned about threats to human health.

A range of exposures have been neglected by experts in past oil spills, including those to communities, workers, and the environment (Ha 2008; Morita et al. 1999). Cleanup workers who face immediate and long-term risks are particularly vulnerable to such oversight. Exposures in the workplace are often some of the most potent environmental health risks. This ignorance of worker concerns has occurred as a part of avoiding legal ramifications for lax worker protections and by inadequate measurement of exposures (Rosner and Markowitz 2003). Therefore, the BP spill represents an opportunity for improved accountability regarding risks to communities and workers of oil spills.

THE DEVELOPMENT OF DISASTER DISTRUST

The average person is going to go up to the table [at the EPA exhibit] saying 'I can eat the seafood,' [after they see the exhibit] because even if it's in the seafood, it's fine. . . It's like they're trying to pull the wool over our eyes. Their credibility is gone and I don't feel that I can trust what they're saying. (Resident of Montegut, Louisiana)

My initial research questions were focused on how lay people assessed risk of the BP spill. However, as I expanded my research, I realized that the predominant reaction amongst interviewees and in ethnographic observations was suspicion, as reflected in the quote above. Much of the distrust seemed to center around the role of evidence that either supported or refuted risks. Residents like this one, and across multiple communities in Southern Louisiana, felt that their trust in the government had been shattered. The question became why and how this distrust developed. As they talked about their experiences as workers on the cleanup, and their engagement with government or BP officials, several themes emerged. First, interviewees felt that health and ecological impacts that affected their day-to-day lives were going unrecognized. One fisherman indicated this point well when he said:

> The oil is still out here and you can see it. I never seen nobody clean it up, you know. You go out there and all you see is people shrimpin' with the oil still out there in the marsh. Me, personally, I don't want to sell no seafood to no public. This is not something I want to do. You got scientists. I went to meetings and they said they have fifty scientists telling you that the seafood's safe. I'm going to tell you what. BP done paid these scientists off. I wouldn't, to eat seafood that came out of an oil tank, I don't know. It ain't something we was taught. We don't want no food stamps, no handouts, just give us back our waters so we can work.

He felt, as did many others, that the science was misrepresenting the impacts. The root of suspicion laid in the differences between localized experiences with risks and the expert assessments that were being communicated via local town hall meetings, the media and other fora.

Discrepancies between scientific sampling and the informal, daily identification of exposures by lay populations in context are the first, most fundamental problem in the generation of disaster distrust. For non-experts, experiencing risks day-to-day means there is a sort of constant sampling across types of exposures and places in which risks might occur. This can be measured for specificity through lay science or citizen science approaches where non-experts are involved in the identification of risks, collection of data, and/or analysis of results (Gibbs 1995). Expert sampling is conducted through particular sites represents areas where risks are suspected, and the data collected there reflect whether or not this is the case. However, the infrastructure for sampling and the locating of such sites may be constrained by a lack of resources necessary to gather sufficient information, by past scientific assessments of risk that do not relate to exposures specific to a particular event, or by the lack of expert knowledge of localized risks. Samples taken by lay people, on the other hand, may be regarded as ad hoc, inaccurate, and biased. Both sampling approaches are limited in scope, but provide some basis for assessment.

EXPERT AND GOVERNMENT SAMPLING PROCESSES

During and after the spill, the Environmental Protection Agency (EPA), the Occupational Safety and Health Administration (OSHA), the National Oceanic Aerospace Administration (NOAA), and BP have been taking airborne samples to assess exposures through five main methods—stationary air monitors, a Trace Atmospheric Gas Analyzers mobile air monitoring van, an ASPECT (Airborne Spectral Photometric Environmental Collection Technology) airplane, and handheld air monitors on cleanup boats. Stationary equipment and its siting were partially coordinated in cooperation with state and city agencies that had such infrastructure prior to the spill. Monitoring by EPA was largely on shore, and so the most comparable to that conducted by community members, which was also mainly on shore. Therefore, this section is most concerned with EPA monitoring as a corollary to lay perceptions.

Air monitoring conducted by the EPA gathered data for several chemicals, including ozone, volatile organic compounds (VOCs), hydrogen sulfide, polyaromatic cyclic hydrocarbons (PAHs), PM 2.5, and PM 10. PAHs were the greatest concern as chemicals released in the catalyzation of oil, although PAH data were not always available for public consumption. Samples of these chemicals were taken at stationary devices in Mississippi, Louisiana, Alabama and Florida. There were thirteen air sampling stations in southern Louisiana where the majority of the oil reached the shore. EPA made these data available on its oil spill website (Environmental Protection Agency 2010).

By selecting particular types of exposures and dates of sampling, the data are available and downloadable. The data were organized such that they could be transformed into a statistical analysis framework, not as identifiable points that could be related to a monitoring site. Therefore, the listing of exposure data was difficult to read for a lay person, and almost impossible to relate to a particular residence. These expert exposure assessments are different from lay assessments in a variety of ways, which indicate why local community perceptions of risk are in direct opposition to those conceptualized by federal regulators.

CITIZEN SCIENCE DATA COLLECTION

> Now, it's, you know, like, what kind of health impacts it's going to have on their lives? I know from what I experienced there's some mean, bad stuff out there in the air. Everyday I am overwhelmed and stressed because, like I say, I can't even feel comfortable sitting outside on the swing with my grand-baby. It's not right. It's not right to have to feel that way, and know, knowing the chemicals that are in the air. We've got to find out how we're being affected.(Bayou Resident)

Interviewees like the one above were concerned about how to collect samples that would better measure their exposures. This resulted into engagement with a new, online, lay mapping system. Lay mapping has often been used in instances where communities face unmonitored, and often unjust, environmental exposures that go otherwise undetected or dismissed by experts and government officials (Allen 2003; McCormick 2009a). Citizen science efforts to collect data and map oil spill exposures were conducted by a range of organizations, from the Louisiana Shrimpers Association to SkyTruth, an environmental advocacy organization that maps ecological resources. Many of these samples were not made public, although potentially relevant in lawsuits and for individuals.

The largest and most central repository for citizen science data collection was Ushahidi, or "testimony" in Swahili. It is an online mapping system was used during the Haitian earthquake by the State Department and others, and has been called the future of disaster response. The BP spill is its first application in the United States, and to an environmental crisis. The organization that created Ushahidi worked with the Louisiana Bucket Brigade to create "The Oil Spill Crisis Map," as a centralized, accessible information database reflecting community experiences and risk perception.

The Spill Crisis Map is the only interface that allows the public to upload exposure data through cell phonebased text messages and webbased submissions. Such 'crowdsourcing' that allows the public to drive data aggregation is a new form of citizen science, where lay people engage in research design, data collection, and analysis. This new approach to risk and exposure has the potential to address past research gaps and public controversy regarding disasters (Backstrand 2003; Edwards and von Winterfeldt 2006). The information gathered in this crowd-sourcing tool, including lay perceptions of damaged ecosystems, health affects, and airborne exposures, reflects experiences otherwise unaddressed in expert risk assessment methods, as is often true when lay people collect information about risks in their own lives (Brown 1992). The central organization behind the Map, the Louisiana Bucket Brigade, has historically conducted air monitoring in communities living fenceline to neighboring oil refineries, and generated shifts in science and industry (Ottinger 2010). The organization is a historical and contemporary collaborator with other environmental justice groups in the area and plays a key role in framing the debate about local impacts of the oil spill (Rolfes 2010; Weber 2010).

The site has been increasingly populated since its inception two weeks after the spill occurred. It has response categories for a wide variety of spill impacts, including: oil in water, oil on shore, wildlife, odor, health effects, smoke, birds, marine wildlife, livelihood threatened, property damaged by oil, solutions and ideas, community meetings and organizing, needs, cultural loss, tainted seafood, dispersant in use, and sampling results. These factors are quite different than those in governmental mapped risk assessments, such as that created by NOAA that measures marine transport, shipping fairways, and critical habitat areas. Traditionally, there has been limited citizen science that challenges expert knowledge of disaster risks. This situation has changed in the case of the BP spill with the implementation of Ushahidi.

By December 1, 2010, reports had been received from Texas, Louisiana, Mississippi, Alabama and Florida, as well as from Mexico, Cuba, and the Cayman Islands. The majority of them were in Southern Louisiana, in Terrebonne and Lafourche Parishes. An analysis of the response map conducted by the Brigade found that by October 10, there were 2,628 reports to the map (Louisiana Bucket Brigade 2010). The majority of reports were about oil on water, followed by oil on shore, odor, and health effects. In the report, EPA's air monitoring sites were overlaid with reports regarding odors and health complaints. This showed how there were many reports in places where no air monitors existed. Many of the reports included photos, media reports and video—all types of lay data not captured by expert assessments. This aggregated data was reported on by multiple news sources across the country, and resulted in the Director of the organization, Anne Rolfes, testifying before Congress on two occasions. While the impact of this source of information on policymaking and disaster responses is as yet unclear, it serves as the central repository for localized risk assessments from the date of the spill onward.

The second central facet driving the development of disaster distrust was the lack of opportunity to resolve differences in risk perception and assessment. There were few pathways through which local residents and workers could effectively express their concerns and grievances, and gain satisfactory answers. A series of hearings, town halls, and dockside chats were held in states where the spill was felt. However, contention often emerged during these events, and subsequently went unresolved. As one resident said at a town hall in Grand Isle, Louisiana, one of the areas worst hit by oil on shore:

> I got in my gulf cart yesterday and rode down a block, went to the water and there's oil. But yesterday they said there's no oil. How can that be?...I'm scared because I already have problems with my health...I started getting sick at the end of May. I went to the doctor. I thought it was something else. . . EPA has just finally got someone to come out and say that they're lying to us. Why are they doing that? Why are we being sold out?

She was not satisfied with the answers to her questions, and few other opportunities existed for her to engage in dialogue. This lack of engagement between risk assessors and those facing risks played an important role in fostering contention. This was particularly true regarding risks to human health and livelihoods. This lack of collaboration was reflected in the scientific approaches to data collection and analysis that resulted in different conclusions being drawn on the part of the public and formalized assessors of risk.

RESOLVING DISASTER DISTRUST NOW AND IN THE FUTURE

Disaster distrust develops between stakeholders either impacted by and/or involved in recovery and response to an event. Its roots are in the different perceptions of risks or impacts across those involved in assessing risks and providing compensation for impacts. Because differences in expert and lay processes of understanding risk are difficult to avoid, resolving disaster distrust depends on the creation of participatory institutions through which risks can be jointly identified and addressed. This approach would also offer a more comprehensive picture of impacts, therefore improving outcomes for affected communities. In this case, as well as many others, such efforts necessarily implemented at the local level would also serve the greater national good. The best management of resources and community recovery in the areas affected by the spill will improve the possibility or preserving the social and ecological resources in the region upon which the rest of the nation draws.

Bridging the gap across stakeholders is a critical step to reducing disaster distrust and improving such outcomes. In this case, that requires the address of community concerns regarding exposures to toxins in the air and water, and improved engagement with decisionmakers. Expanded sampling to assess potential exposures is a critical route to achieving both of these ends. The National Institutes of Health has planned a $10 million budget for a Gulf Worker Study, and an additional $10 million from BP to supplement this project. A portion of the funding should be used to support communitybased participatory research (CBPR) that can include local perspectives on risk (Minkler 2000). Using this approach to research can result in power sharing on the part of the researchers with local people, including who defines the research problem, and then analyzes, owns and acts on the information (Cornwall 1995). It can also build on strengths of the community such as skills, networks, and social capital (Israel et al. 1998). It allows community members to take part in data collection (Northridge et al. 1999), which fosters mutual trust and cooperation (Schell and Tarbell 1998). As such, communitybased participatory research has improved relevance of research to community groups, improved adherence to findings, resulted in innovative research, and offered a new ethics to research (McCormick 2009b). Thus, CBPR should be included in the NIH support for research on spill effects. In the future, more participatory forms of risk assessment implemented by government agencies in disaster scenarios is likely to reduce disaster distrust, foster democracy, and result in outcomes that improve the lives of those whose lives are affected.

REFERENCES

Allen, Barbara. 2003. *Uneasy Alchemy: Citizens and Experts in Louisiana's Chemical Corridor Disputes*. Cambridge, MA: MIT Press.

Backstrand, Karin. 2002. Civic Science for Sustainability: Reframing the Role of Experts, PolicyMakers and Citizens in Environmental Governance. *Global Environmental Governance* 3(4): 24-41.

Beamish, Thomas D. 2001. Environmental Hazard and Institutional Betrayal: LayPublic Perceptions of Risk in the San Luis Obispo County Oil Spill. *Organization & Environment* 14 (1): 5-33.

Brown, Phil. 1992. Popular Epidemiology and Toxic Waste Contamination: Lay and Professional Ways of Knowing. Journal of Health and Social Behavior 33: 267-81.

Brown, Phil, Zavestoski, Stephen, Mayer, Brian, McCormick, Sabrina and Webster, Pamela. 2002. Policy Outcomes of Environmental Health Disputes. *Annals of the American Academy of Political and Social Science* 584(1): 175-202.

Castells, Manuel, Fernandez-Ardevol, Mireia, Linchuan Qiu, Jack and Sey, Araba. 2009. *Mobile Communication and Society: A Global Perspective*. Cambridge, MA: MIT Press.

Clarke, Lee and Short, Jr., James F. 1993. Social Organization and Risk: Some Current Controversies. *Annual Review of Sociology* 19: 375-99.

Collier, John. 1995. Photography and Visual Anthropology. In *Principles of Visual Anthropology*, John Collier, ed. Germany: Walter de Gruyter.

Cornwall, Andrea. 1995. "What Is Participatory Research?" *Social Science and Medicine* 41(12): 1667-76.

Coughlin, S.S. 1990. Recall Bias in Epidemiologic Studies. *Journal of Clinical Epidemiology* 43(1): 8791.

DiMaggio, Paul, Hargittai, Eszter, Neuman, W. Russell, and Robinson, John P. 2001. Social Implications of the Internet. *Annual Review of Sociology* 27: 307-36.

Edwards, Ward and von Winterfeldt, Detlof. 2003. Public Values in Risk Debates. *Risk Analysis* 7(2): 141-58

Environmental Protection Agency. 2010. Air sampling site reference.

Feinstein, A. R. 1985. Experimental requirements and scientific principles in case-control studies. *Journal of Chronic Disease* 38: 127-33.

Hacking, I. 1999. Making Up People. In M. Biagioli (Ed.), *The Science Studies Reader*. New York: Routledge, pp. 161-71.

Hess, David J. 2005. Technology and Product-Oriented Movements: Approximating Social Movement Studies and Science and Technology Studies. *Science, Technology & Human Values* 30 (4): 515-35.

Israel, B. A., Schulz, Amy J., Parker, Edith A., and Becker, Adam B. 1998. "Review of Community-Based Research: Assessing Partnership Approaches to Improve Public Health." *Annual Review of Public Health* 19:173–202.

Jasanoff, Sheila. 2004. States of Knowledge: The CoProduction of Science and Social Order. New York, NY: Routledge.

Killian, Lewis K. 2002. An Introduction of Methodological Problems in Field Studies in Disasters. In *Methods of Disaster Research*. United States: International Research Committee on Disasters.

Kurtz, Rick S. 2004. Coastal Oil Pollution: Spills, Crisis, and Policy Change. *Review of Policy Research* 21(2): 201-19.

Kusaka, Akemi Moritukinori, Deguchi, Yoji, Moriuchi, Akio, Nakanaga, Yuko, Iki, Masayuki, Miyazaki, Shigekazu and Kawahar, Kazuo. 1999. Nakhodka Oil Spill. *Environmental Research* 81(3):185-94

Markowitz, Gerald and Rosner, David. 2003. *Deceit and Denial: The Deadly Politics of Industrial Pollution*. Berkeley, CA: University of California Press.

McCormick, Sabrina. 2009a. Democratizing Science Movements: A New Framework for Contestation. *Social Studies of Science* 37: 609-23.

———. 2009b. *Mobilizing Science: Movements, Participation and the ReMaking of Knowledge*. Philadelphia, PA: Temple University Press.

Ha, Mina, Lee, Won Jin, Lee, Seungmin and Cheong, HaeKwan. 2008. A Literature Review on Health Effects of Exposure to Oil Spill *Journal of Preventive Medicine and Public Health* 41(5):345-55.

Louisiana Bucket Brigade. 2010. *Oil Spill Crisis Map: Initial Analysis and Review*. New Orleans, Louisiana.

Minkler, Meredith. 2000. Using Participatory Action Research to build Healthy Communities. *Public Health Reports* 115(23): 191-97.

Neil, N., Malmfors, T., and Slovic, P. 1994. Intuitive toxicology: expert and lay judgments of chemical risks. *Toxiological Pathology* 22(2): 198-201.

Northridge, Mary E., Yankura, Joanne, Kinney, Patrick L., Santella, Regina M., Shepard, Peggy, Riojas, Ynolde, Aggarwal, Maneesha, Strickland, Paul and the Earth Crew. 1999. "Diesel Exhaust among Adolescents in Harlem: A CommunityDriven Study." *American Journal of Public Health* 89:998–1002.

Ottinger, Gwen. 2010. Buckets of Resistance: Standards and the Effectiveness of Citizen Science. *Science, Technology & Human Values* 35 (2): 244-70.

Palinkas, Lawrence A., Downs, Michael A., Petterson, John S. and Russell, John. 1993. Social, Cultural, and Psychological Impacts of the Exxon Valdez Oil Spill. *Human Organization* 52(1): 113.

Peterson, Charles H., Rice, Stanley D., Short, Jeffrey W., Esler, Daniel, Bodkin, James L., Ballachey, Brenda E., Irons., David B. 2003. LongTerm Ecosystem Response to the Exxon Valdez Oil Spill. *Science* 302 (5653): 2082-86.

Rolfes, Ann. 2010. Congressional testimony. House Energy Committee. Washington, D.C.

Salam, Muhammad Towhid, Li, YuFen, Langholz, Bryan, and Gilliland, Frank Davis. 2004. Early-Life Environmental Risk Factors for Asthma: Findings from the Children's Health Study. *Environmental Health Perspectives* 112: 760-65.

Schell, L.M., and Tarbell, A.M. 1998. "A Partnership Study of PCBs and the Health of Mohawk Youth: Lessons from Our Past and Guidelines for Our Future." *Environmental Health Perspectives* 106:833–40.

Sjöberg, Lennart. 1999. A Dilemma in Risk Management. *Human Ecology Review 6 (2): 19.*

Slovic, Paul. 1993. Perceived Risk, Trust, and Democracy. *Risk Analysis* 13 (6): 675-82.

Tierney, Kathleen. 1999. Toward a Critical Sociology of Risk. *Sociological Forum* 14 (2): 215-42.

Vaughn, Diane. 1996. *The Challenger Launch Decision: Risky Technology, Culture, and Deviance at NASA*. Chicago: University of Chicago Press.

Weber, Sam. 2010. Crowd-sourcing the Gulf oil spill. Public Television. http://www.pbs.org/wnet/needtoknow/environment/crowdsourcingthegulfoilspill/459/

Zota, Ami R., Ettinger, Adrienne S., Bouchard, Maryse, Amarasiriwardena, Chitra J., Schwartz, Joel, Hu, Howard, and Wright, Robert O. 2009. Maternal Blood Manganese Levels and Infant Birth Weight. *Epidemiology* 20 (3): 67-373.

CHAPTER SIXTEEN
THE PRESIDENT, THE NEWS, AND THE OIL SPILL: AN EXAMINATION OF NATIONAL AND STATE NEWSPAPERS' FRAMING OF OBAMA AND HIS ADMINISTRATION'S RESPONSE TO THE DEEPWATER HORIZON GULF OIL SPILL

JAS M. SULLIVAN AND MEGHAN S. SANDERS

INTRODUCTION

On April 20, 2010 a massive explosion occurred in the Gulf Coast causing the Deepwater Horizon, a rig owned by BP, to sink. At the time of explosion, 126 people were on board the football-sized rig—eleven have been presumed dead, after United States Coast Guard spent nearly two days in a substantial rescue effort. Initial reports indicated that nearly 210,000 gallons was leaking from the oil well per day, which having tremendous ramifications for the environment, wildlife, Gulf Coast tourism, and jobs (especially for fisherman). The cost of the clean up (and compensation to Louisiana citizens) may exceed $20 billion dollars—a tab BP has set aside for this sole purpose.

The reaction to the incident from federal, state, and local officials was immediate, with many calling for an investigation. President Barack Obama stated, "We have an obligation to investigate what went wrong and to determine what reforms are needed so that we never have to experience a crisis like this again" (Rowland and Seper, 2010). Some legislators in the US Congress expressed their intention to hold hearings on safety regulations. For example, Dan McLaughlin, an aid to Senator Nelson (DFL), "asked why the government had not done more to ensure that the oil companies could shut down a well if a catastrophic failure happened" (Mufson and Shear, April 30, 2010). Others, Governor Bobby Jindal (Louisiana) positioned his affected state to receive funding, not only from BP, but also from the Federal government. Jindal said:

> We need Congress to work quickly to not only invest these BP fines into restoring our coast, but also to help fund the 18 coastal restoration projects already authorized by Congress and to accelerate revenue sharing from Outer Continental Shelf leases, which will bring in $200 million to the state starting in 2017, so we can invest these funds into coastal restoration projects immediately. All of these components will be integral parts of our comprehensive effort to make coastal Louisiana whole again (Sep 28, 2010).

While political leaders were focusing their attention on solving the problems confronted by their constituents, shifting blame, calling for hearings, and securing financial compensation, environmentalists and those who were directly affected, like fisherman, were squarely focused on their own priorities. Environmentalists argued that they "hoped this would serve as another wakeup call on this issue that there is no such thing as safe oil drilling. Once the oil starts leaking in the in the ocean, that damage is irreversible" (Wan, April 24, 2010). Other than just propagating a position on the dangers of oil exploration off the coast of U.S., many environmentalists have played an important role in fighting the negative effects of the Gulf oil spill by directly aiding in cleanup efforts involving both the wildlife and the marshes off the coast.

The effects of the oil spill have especially been devastating both psychologically and economically for those whose livelihood is dependent on fishing. The Gulf Coast is fertile ground for fish, crab, oysters, and crawfish, and for the most part, all fishing activities were shut down after the oil spill. Research has shown traumatic experiences such as this lead often times to suicide and depression. A few months after the spill, 55-year old Allen Kruse, a fish boat captain, committed suicide. In addition, domestic violence calls have increased greatly around the communities affected by the spill, and many psychologists agree that, "suicides and other psychological issues such as depression remind Gulf Coast of the mental toll of the BP oil spill," and that "something people thought was practically impossible has happened. There will be severe and long-lasting impacts, and this will probably not be the last suicide" (Picou, Jun 29, 2010).

The major worry among those affected has been on how to sustain their livelihood—from Louisiana fishers, food processors in Mississippi, business owners and tourism workers around the Gulf Coast. One business owner in the Gulf region indicated, "I'm dealing with anxiety. It affects your quality of life, your property values. There's just so many ways this affected us but nobody is talking about that" (Skoloff, September 19, 2010). A Louisiana fisherman indicated, "When you have people who rely on you for their existence, it puts some serious pressure on you. And who's going to hire a 56-year old man to do something else? I got diabetes and high blood pressure. I can't even go back to driving trucks" (Skoloff, September 19, 2010). The seriousness of the crisis and the urgency for help can be heard from the comments of another fisherman. He said, "All I want is what I lost, then I'm whole. Everything I'm making now is going to pay off debt from this summer. I'm behind on my child support payments, house payments. I need money"(Skoloff, September 19, 2010).

As illustrated in the above paragraphs, the effects of the oil spill are far-reaching and devastating. The spill has impacted the environment, wildlife, and people's way of living. In many ways, it has also taken a political twist—with many criticizing the response of President Obama and his administration. A May Pew Research survey found that 36 percent of respondents disapproved of Obama's handling of oil spill, while 47 percent indicated he could have done more in responding to the oil spill. A June NBC News/Wall Street Journal Survey poll found that 50 percent of those polled disapproved of the way Obama was handling the aftermath of the Gulf oil spill. Another poll conducted by the Public Policy Polling surveyed 403 likely voters in Louisiana (between August 21 and 22nd), and their findings showed that 61 percent of those polled disapproved of how Obama handled the aftermath of the oil spill, while 70 percent approve of how Jindal handled the aftermath of the oil spill.

Many received their information regarding the crisis through news coverage provided by both sate-based and national news outlets, rather than through personal experience. This being the case, news coverage could have been a strong influencing factor in the perceptions people of the federal government and its commander-in-chief. Consequently, the purpose of this paper is to explore the way in which the media framed both President Obama and the Obama administration's response to the Deep Horizon spill. Our goal in this paper isn't to show a direct causation, but it is our intention to explore the various frames, responsibilities and evaluations provided by news.

THEORIES OF FRAMING EFFECTS

Framing has been defined in the following ways:

"essence of public opinion formation" (Chong, 1993: pg. 870);
"the process by which a source (a newspaper, news story, or perhaps a single individual) defines the essential problem underlying particular social or political issues, and outlines a set of considerations purportedly relevant to that issue" (Nelson et al., 1997: pg. 222); and

"subtle alterations in the statement or presentation of judgment and choice problems" (Iyengar, 1991: pg. 11).

Even with a "lack of conceptual clarity and consistency about what exactly framing is" (Nelson et al., 1997b: pg. 22), a great deal of research has evaluated the effects of framing.

A study conducted by Nelson and Oxley (1999) examined just how framing causes different attitudes and opinions. In their experiment, they framed a proposal for a new construction of a hotel and convention complex from either an economic or environmental perspective to investigate the influence these frames would have on support for the project. Results showed "the participants in the economic framing condition expressed more favorable attitudes toward the proposal hotel and convention complex than did participants in the environmental framing condition" (Nelson and Oxley, 1999: pg. 1047).

Another study conducted by Nelson, Clawson, and Oxley (1997) showed the effects of framing a Ku Klux Klan rally. Using news coverage, one frame of the situation showed the rally as an exercise in free speech, while the other frame showed the rally as a disruption of public order. Their study found that free speech participants were more tolerant of the situation. The authors found that the "choices journalists make about how to cover a story—from the words, phrases, and images they convey to the broader angel they take on controversy can result in substantially different portrayals of the very same event and the broader controversy it represents" (Nelson et al, 1997: pg. 576). "These alternative portrayals, or frames, can exert appreciable influence on citizens' perceptions of the issue and, ultimately, the opinions they express" (Nelson et al., 1997: pg. 576).

The effects of frames on political matters have also been studied by Patterson and Jamieson (1996). The scholars examined the effects of frames found in the media during ad campaigns and news coverage, as well as during the health care reform debate of 1994. Their conclusions are parallel with other conclusions made about the effects of frames and framing. For example, Kellstedt (2000) found that there is a correlation between racial policy preference and frames in the media. According to Kellstedt, frames in the media influence either egalitarianism or individualism values. Kellstedt explained that individualism promotes individual prosperity with an ownership of a hard work ethic, while egalitarianism promotes equal opportunities for prosperity and success. Thus, in egalitarianism, the government plays a role in ensuring equal opportunities and assistance for disadvantage people. When a frame is focused on egalitarianism values, individuals are more likely to favor government policies on race; however, when a frame is focused on individualism values, individuals are more likely to favor conservative government policies on race (Kellstedt, 2000).

The media's roles in forming voters' national economic evaluations were studied by Hetherington (1996). Hetherington specifically focused on the 1992 presidential election. Hetherington found that negative frames about the economy in the media influenced voter's perception about the economy and therefore, influenced voters' preferences toward Clinton instead of Bush. This is an example of thematic framing. "Thematic framing of the 'deteriorating' economy would encourage media consumers to attribute its condition to George Bush" (Hetherington, 1996: pg. 375). The way that the media framed the economy had a major impact in the election polls. "There are good reasons to be concerned about the effects of the way the news frames politicians, policy, campaigns, and governance, not only for the quality of democracy but also for perceptions about the press" (Capella and Jamieson, 1996: pg. 75).

Terkildson and Schnell (1997) examined the effects of framing toward women's rights and gender roles. After examining national news weeklies, they found that frames relevant to women's rights and gender roles took five forms: a traditional frame, an antifeminism frame, and economic rights/workplace frame, a political roles frame, and a feminism or gender equality frame. They concluded that the type of frames shaped people's feelings and opinions about the subject matter. Negative frames toward the women's movement produced negative attitudes, while positive frames toward the women's movement produced positive attitudes.

Iyengar (1990) studied the effects on how news stories framed poverty. Iyengar found that the frames on poverty were either thematic or episodic. For example, an episodic frame would depict individuals responsible for poverty; however, a thematic frame would hold society responsible for poverty (Iyenger: 1990). Thus, the use of thematic or episodic frames can produce drastic differences in opinion about societal issues and shift of responsibility. Iyenger (1990) found that when a thematic frame was employed in news coverage on poverty, people believed that society was responsible for poverty; however, when an episodic frame was employed, people believed that people in poverty were responsible for their own destitute.

> The effects of these frames demonstrate clearly that belief about who or what is responsible for poverty vary considerably, depending on how poverty is framed. In the case of television news coverage, the thematic frame engenders a stronger sense of societal responsibility, whereas the episodic frame engenders a stronger sense of individual responsibility" (Iyenger, 1990: pg. 35).

Scholarly toil has not only studied the effects of frames, but researchers have also sought to determine which individuals are the most susceptible to the effects of frames and framing. Vulnerability to framing effects has important implications on how framing may target populations and influence preferences based on these considerations. The idea that certain populations may be more susceptible to the effects of frames has been discussed in the literature with some controversy. Some research has shown that people with higher levels of knowledge are more likely to be affected by frames. Yet, other studies have shown the contrary.

One study conducted by Druckman and Nelson (2003) showed that individuals with more knowledge about a particular situation or topic are more susceptible to the effects of frames than individuals who are less knowledgeable about the topic.

Specifically, they wanted to know if conversations can eliminate elite framing. One group of participants were exposed to a free speech framed article, while another group of participants were exposed to a specialinterests framed article about campaign finance reform. Then the researchers divided the participants into groups to discuss the articles so the authors could study the influence of interpersonal relationships, combined with elite frames. There were three groups: a no discussion group, a mixed discussion group combing participants of both articles, and an unmixed discussion group including individuals with the same articles. Specifically, the authors found that "Conversations that include only common perspectives have no effect on elite framing, but conversations that include conflicting perspectives eliminate elite framing effects" (Druckman and Nelson, 2003: pg. 729). Thus, elite framing had the greatest effect on the participants who were not included in a discussion group, and more knowledgeable. The findings suggest, "although elites may attempt to use framing to manipulate an unknowing and naïve populace, it is knowledgeable people who are relatively more apt to use the frames (pg. 740).

Oxley and Clawson (1997) also found that people with higher levels of knowledge are more susceptible to frames in the media. They argue that these populations of people are more susceptible because they are more likely to draw connections between the frame and their own opinions. Therefore, the more individuals are aware of the situation being framed; the more prone they are to the effects of those frames.

A study conducted by Nelson et al. (1997) also concluded that people with higher levels of knowledge are more likely to be affected by frames. In their study they used the topic of the welfare debate. Those who were more knowledgeable about the welfare debate were more prone to framing effects on the topic. Not only did they conclude that knowledgeable people are influenced by frames the most, they found that people respond to frames differently than to messages that try to change their beliefs. Frames appear to activate existing beliefs and cognitions, rather than adding something new to the individual's beliefs about the issue" (pg. 235236). By extension, their results suggest, "that mass media, and other institutions of mass political communication, can profoundly influence public opinion even without any overt attempt at persuasion or manipulation" (Nelson et al., 1997: pg. 236).

On the contrary, the literature that supports the claim that less knowledgeable people are more susceptible to frames is extensive, as well. One study that supports this claim was conducted by Kinder and Sanders (1999: pg. 90). These authors argue that less knowledgeable people possess fewer strongly held prior opinions (and frames) and thus exhibit increased susceptibility. Eagly and Chaiken (1993); McGuire (1985); Zaller (1992) all found that greater knowledge or sophistication about an issue inhibits attitude change. The more sophisticated are less likely to yield to or be persuaded, because they are likely to be already familiar with the argument; the additional information, even if novel, will be trivial compared to their existing mental stockpile of information; and greater knowledge enables the recipient to more easily dismantle and dismiss counter-attitudinal arguments).

From the research literature on framing, what we do know is that framing occurs and it has an important role in creating attitudes and opinions regarding public policy and political candidates. Thus, our goal in this paper is to explore the ways in which the media has framed both President Obama and the administration—specifically, regarding the oil spill response. In addition, we examined any differences in framing that may exist as a function of the location of the news (state versus national) and the political ideology of the outlet (conservative versus liberal).

METHOD

A quantitative content analysis was conducted using a census of newspaper stories published April 20, 2010 to August 20, 2010 in the following newspapers: *The Advocate* (Baton Rouge), *The TimesPicayune* (New Orleans), *The Washington Post* (a liberalleaning newspaper), the *Washington Times* (a conservative leaning newspaper), and the New York Times (a liberalleaning newspaper). State newspapers did not have an ideological viewpoint. Articles were gathered using LexisNexus Academic database. All editorials, letters to editors, and other opinionbased pieces were excluded, so the unit of analysis was news article. This resulted in the analysis of 662 articles. The majority of the articles (61.5 percent) came from national newspapers and were written by journalists working for those papers (96.5 percent). For the state news articles, more were published by the *Picayune* (33.9 percent) as opposed to the *Advocate* (4.6 percent). In regards to the articles published by national newspapers, 56.5 percent came from the more liberalleaning newspapers.

To assess the presence of frames in the news coverage we assessed how often President Obama was mentioned in the news stories. The mention could include instances of his name either in the headline or text. An administration mention included a mention of the Obama administration, the U.S. administration, mentions of specific federal agencies (i.e. U.S. Department of Interior, U. S. Fish and Wildlife Service, Office of Emergency Management, etc.), the White House, or the general phrase "a/the federal agency." An evaluation of either President Obama or the administration was assessed by whether the article provided some kind of assessment, either by the journalist's own words or through quotes of others, of the president's or the administration's performance regarding the oil spill. Performance assessment included reactions to the event, evaluation of response time, evaluation of the response plan, evaluation of leadership skills and qualities (i.e. was he/the administration capable of handling the job), appropriateness and timeliness of response/reaction, assessment of visits to affected regions, assessments of outcomes of visits to affected regions, adequacy of amount of attention/priority devoted to the event. Coders also recorded who was responsible for making the evaluation (i.e. a state official, federal official, affected resident, the writer or other). Each evaluation was also noted as either being subjectively positive or negative (i.e. blame attribution).Coders also made note of the origin of each article (i.e. an independent original report, reprint, wire story), as well as generally if the article placed blame on any person or entity (i.e. previous federal administration, BP, etc.).

Judgments about each article were made by two undergraduate student coders. Reliability was assessed using Cohen's kappa. No variable was below 83 percent agreement.

RESULTS

According to a descriptives analysis, 45.5 percent of the coded articles mentioned President Obama, while 77.8 percent made some reference to the Obama administration (see Table 16.1 for full summary). A series of chisquare analyses were conducted to determine whether the various groups of newspapers presented different frames of the Obama administration, and of the president (see Table 16.2 for full summary).

STATE VS. NATIONAL NEWSPAPER COVERAGE

According to the results, state newspaper articles were equally likely to mention (55.3 percent) as they were to not mention President Obama (44.75 percent). However, national newspapers were much more likely to not mention the president in their coverage of the spill (60.5 percent). The difference in these trends was statistically significant, c^2 (1)=15.81, $p<.001$, Cramer's V=.16. Evaluations of the President's handling of the oil spill crisis were significantly less like to be present in both types of papers (State: 16.5 percent; National: 10.6 percent), c^2 (1)=4.87, $p<.05$, Cramer's V=.09. When evaluations were present, they were most likely to come from a state official in statebased newspaper articles (44.4 percent) and from the journalist's perspective in the national newspaper articles (58.1 percent). Evaluations were also more likely to be negative in nature for both newspaper types (State: 60.5 percent; National: 65.0 percent) than they were to be subjectively positive or praising, c (1)=.18, $p>.05$, Cramer's V=.05.

Unlike the references to the president, both newspaper types were significantly more likely to mention the administration (State: 83.9%; National: 73.8%) than to not make a reference, c^2 (1)=9.16, $p<.01$, Cramer's V=.12. However, the majority of these mentions did not come in the form of an evaluation of the administration's handling of the situation, c^2 (1)=7.22, $p<.01$, Cramer's V=.11. State newspaper articles only provided an evaluation 30 percent of the time, while national newspaper articles made an evaluation 20.8 percent of the time. As with evaluations of the president, when evaluations were present in state-based articles, they were most often provided by state officials (59.2 percent). National newspaper evaluations of the administration were provided based on the journalist's personal perspective (36.9 percent) and other

Table 16.1: Summary of Oil Spill News Content

	N	Percent
Newspaper		
Advocate	30	4.6%
Picayune	223	33.9%
New York Times	212	32.2%
Washington Post	160	24.3%
Washington Times	33	5.0%
Location		
State	254	38.5%
National	405	61.5%
Story Origin		
Wire Service	21	3.2%
Original Article	634	96.5%
Reprint	2	.3%
Obama Mention		
Yes	300	45.5%
No	359	54.5%
Obama Administration Mention		
Yes	514	77.8%
No	147	22.2%
Blame to Prior Administration	11	1.7%
Blame to Obama Administration	55	8.3%
Blame to Jindal Administration	1	.2%
Blame to BP	256	38.7%
Blame to Other	11	1.7%
Obama Evaluated		
Yes	85	12.8%
No	577	87.2%
Obama Administration Evaluated		
Yes	160	24.2%
No	500	75.8%

Table 16.2: Summary of Variables of Interest by Newspaper

	Advocate		Picayune		NY Times		Wash. Post		Wash. Times	
	N	%	N	%	N	%	N	%	N	%
Obama Mention**										
Yes	24	82.8	114	51.1	79	37.3	48	30.4	33	100
No	5	17.2	109	48.9	133	62.7	110	69.6	0	0
Obama Administration Mention*										
Yes	25	83.3	189	84.8	152	71.7	114	71.7	31	93.9
No	5	16.7	34	15.2	60	28.3	45	28.3	2	6.1
Obama Evaluated**										
Yes	7	23.3	35	15.7	26	12.3	6	3.8	11	33.3
No	23	76.7	188	84.3	186	87.7	154	96.3	22	66.7
Obama Evaluation Source**										
State Official	2	28.6	18	47.4	0	0	1	16.7	0	0
Federal Official	0	O	0	0	1	3.8	1	16.7	0	0
Affected Resident	0	O	3	7.9	1	3.8	0	0	1	9.1
Journalist/Writer	3	42.9	4	10.5	22	84.6	1	16.7	2	18.2
Other	2	28.6	13	34/2	2	7.7	3	50.0	8	72.7
Obama Evaluation										
Positive	1	16.7	16	43.2	11	42.3	2	50.0	1	10.0
Negative	5	83.3	21	56.8	15	57.7	2	50.0	9	90.0

	Advocate		Picayune		NY Times		Wash. Post		Wash. Times	
	N	**%**	**N**	**%**	**N**	**%**	**N**	**%**	**N**	**%**
Obama Administration Evaluated**										
Yes	14	46.7	61	27.5	47	22.2	23	14.5	15	45.5
No	16	53.3	161	72.5	165	77.8	136	85.5	18	54.5
Obama Administration Evaluation Source**										
State Official	6	42.9	39	63.9	2	4.3	4	17.4	1	6.3
Federal Official	1	7.1	5	8.2	11	23.0	4	17.4	4	25.0
Affected Resident	0	0	0	0	1	2.2	1	4.3	0	0
Journalist/Writer	5	35.7	4	6.6	25	54.3	3	13.0	4	25.0
Other	2	14.3	13	21.3	7	15.2	11	47.8	7	43.8
Obama Administration Evaluation*										
Positive	2	22.2	4	6.3	14	30.4	0	0	1	6.3
Negative	7	77.8	59	93.7	32	60.6	22	100	15	93.8

Note: $*p<.01$, $**p<.001$

sources (29.8 percent). These differences were statistically significant, c^2 (4)=49.51, $p<.001$, Cramer's V=.56. In regard to what kind of evaluation existed, in both levels of news the evaluations were primarily negative (State: 91.8 percent; National: 81.9 percent), although national news articles were positive slightly more often (18.1 percent; State: 8.2 percent).

When examining where news articles tended to place blame or responsibility for the spill, the results suggest that the news did not place the responsibility on neither the previous federal administration (98.0 percent) nor the Obama administration (94.1 percent). Rather, articles tended to place blame more often on BP (State: 46.1 percent; National: 33.8 percent), although blame overall was not discussed the majority of the time.

NEWSPAPER IDEOLOGY

Conservative-leaning newspapers mentioned President Obama (100 percent) and the administration (93.9 percent) significantly more often compared to more liberal-leaning papers (Obama: 34.1 percent; Administration: 71.6 percent) and statebased newspapers (Obama: 55.2 percent; Administration: 84.6 percent), Obama c^2 (2)=68.37, $p<.001$, Cramer's V=.32; Administration: c (2)=20.00, $p<.001$, Cramer's V=.17. While the majority of the newspapers did not provide an evaluation of the President, when an evaluation was present it was significantly more likely to appear in a conservative article (33.3 percent) compared to liberal (8.6 percent) and state-based articles (16.6 percent), c (2)=21.66, $p<.001$, Cramer's V=.18. Liberal newspaper articles were just as likely to provide a positive evaluation (43.3 percent) as they were a negative one (56.7 percent). However, conservative articles overwhelmingly provided negative evaluations (90 percent), while state-based articles were negative to a lesser degree (60.5 percent). When evaluations about the president's performance were present in Conservative articles, they came most often from other sources (43.8 percent), followed by federal officials (25 percent), and evaluations implied by the journalists' word choices (25 percent). Liberal-based articles' evaluations were most often implied by the journalists (40.6 percent), but for state-based news articles the evaluations were provided primarily by state officials.

When it came to evaluating the Obama administration, the results painted a different picture, as illustrated in Figure 16.1. Overall, more articles provided an evaluation of the administration. Again, conservative newspaper articles were more likely to provide an evaluation (45.5 percent), followed by liberal (29.8 percent) and statebased articles (24.3 percent),c (2)=18.32, $p<.001$, Cramer's V=.17. However, unlike with evaluations of President Obama, all newspapers were more likely to provide negative evaluations of the administration than positive ones (see Figure 16.1). For conservative articles, evaluations were most likely to come from other sources (43.8 percent), while liberal articles' evaluations came from journalists' interpretations (40.6 percent). State sources were most often the reference (60 percent) for state-based articles.

Figure 16.1: Praise and Blame Frames of Obama and Obama Administration by Newspaper Ideology

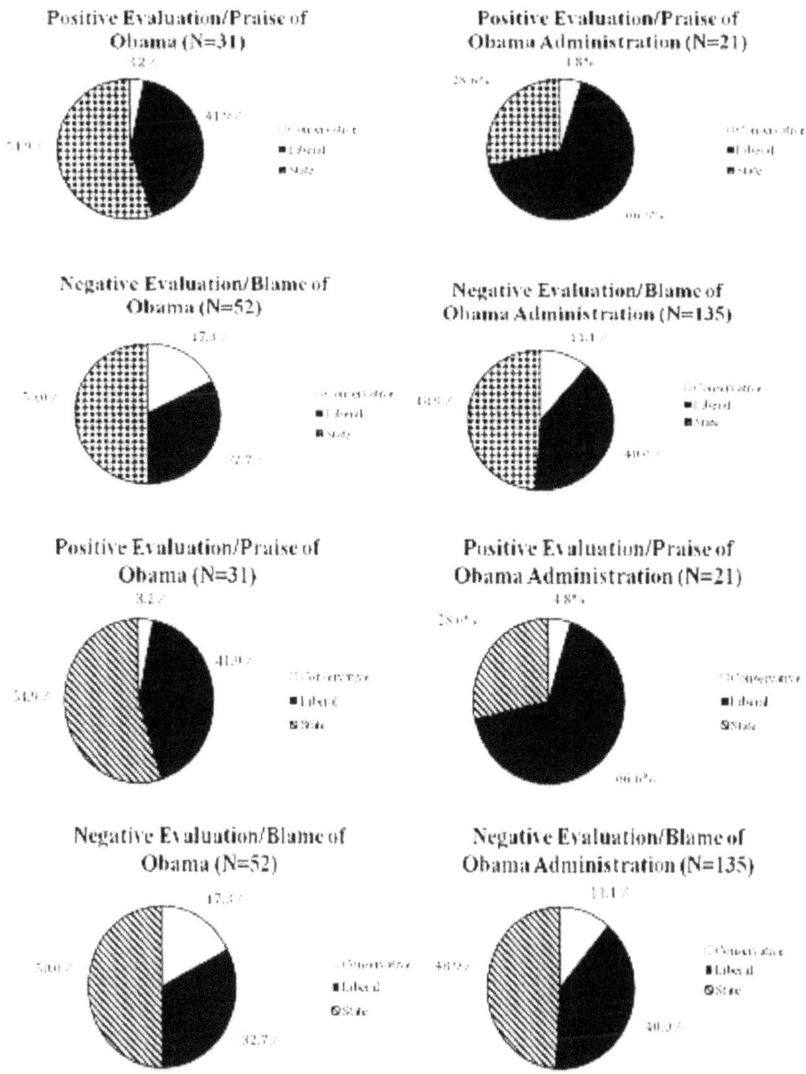

When it came to discussions of blame, there were no significant differences when the conversation of blame involved the previous administration, C (2)=78, $p>.05$, Cramer's V=.03. Conservative (97 percent), Liberal (98.7 percent) and statebased articles (98 percent) did not contain information that placed responsibility for the spill on actions and decisions made by the Bush administration. While the results approached significance for attribution of blame to the Obama administration, C (2)=5.29, p=.07, Cramer's V=.09, the newspaper articles did not overtly place blame for the spill on the current administration (Conservative: 97 percent; Liberal: 89.6 percent; State: 94.1 percent). Again, blame was not often discussed overall in regards to BP but when it was, statebased papers were more likely to do so (46.2 percent) than were liberal news articles (33.7 percent), and conservative articles (39.4 percent).

DISCUSSION

While the effects of the oil spill have been devastating and farreaching, it has also taken a political twist—with many criticizing the response of President Obama and his administration. As evidence, the surveys taken immediately following the spill has shows dismal numbers. For example, between 50 to 61 percent of the people surveyed (in various polls) disapproved Obama's response to the spill. What we do know is that many received their information regarding the oil spill crisis through news coverage, rather than through state and national news. With this being said, news coverage could have been influence in the perceptions of President Obama and his administration. Consequently, the purpose of this paper was to explore the ways in which the media framed both President Obama and his administration.

A quantitative content analysis of both national and state newspapers revealed the following: both state and national newspapers were equally likely to mention President Obama, when Obama was mentioned, the evaluations were more likely to be negative in both types of papers; regarding the evaluations of Obama's administration, newspapers were more likely to be negative; with respect to the ideology of the newspapers, conservative newspapers were more likely to mention Obama and his administration than liberal newspapers; when evaluations were present, it was significantly more likely to appear in conservative articles, and they were more likely to provide an overwhelmingly negative evaluations; and all newspapers were more likely to provide negative evaluations of the administration.

What are the implications of our findings? Newspapers often frame a story and leaders in a specific manner, and research in both communication studies and political science has shown that the manner in which a newspaper frames a story often leave an impression in the public's mind, especially, in the absence of direct experience. In the case of newspaper coverage of the oil spill, we argue that it is very plausible that the public's negative view of Obama and his administrations are due to the way state and national newspapers have covered the story. In other words, if all an individual reads were negative stories about the failures of Obama and his administration's response to the oil spill, it would seem that the public perception would also be negative. Where would readers get disconfirming or alternative information—if one exists? The only way to really get an accurate assessment of Obama and his administration's response would be to experience it directly—which for most people is an impossibility—or read more than one type newspaper (i.e., state, national, liberal, and conservative). In the case of the oil spill, both conservative and liberal newspapers were equally likely to mention Obama and his administration, but conservative papers were more likely to give a negative evaluation. Consequently, in a state like Louisiana (which is politically conservative and thus, the high probability that majority of the electorate get their news from conservative media), it's not surprising that *Public Policy Polling* found that 61 percent of those polled in Louisiana disapproved of how Obama handled the aftermath of the oil spill. While we can't directly attribute Louisiana and rest of the nation's perception of Obama and his administration response to the oil spill, what we have shown in this study is that they have received overwhelming amount of negative coverage in both state and national newspapers, which could have impacted the public's perception.

REFERENCES

Cappella, Joseph N., and Kathleen Hall Jamieson. 1996. "News Frames, Political Cynicism, and Media Cynicism." *Annals of the American Academy of Political and Social Science* 546: 71-84.

Chong, Dennis. 1993. "How People Think, Reason, and Feel about Rights and Liberties." *American Journal of Political Science* 37(3): 867-99.

Druckman, James N., and Kjersten R. Nelson. 2003. "Framing and Deliberation: How Citizens' Conversations Limit Elite Influence." *American Journal of Political Science* 47(4): 729-45.

Eagly, Alice H., and Shelly Chaiken. 1993. *The Psychology of Attitudes*. Fort Worth, TX: Harcourt Brace Jovanovich.

Hetherington, Marc J. 1996. "The Media's Role in Forming Voters' National Economic Evaluations in 1992." *American Journal of Political Science* 40(2): 372-95.

Iyengar, Shanto. 1990. "Framing Responsibility for Political Issues: The Case of Poverty." *Political Behavior* 12(1): 19-40.

Iyengar, Shanto. 1991. Is Anyone Responsible? How Television Frames Political Issues. Chicago: University of Chicago Press.

Jindal, Bobby. Sep 28, 2010. "Governor Jindal Statement on Report Released from Secretary Mabus on Coastal Restoration Efforts Following Oil Spill." *The Office of Governor's Website*.

Kellstedt, Paul M. 2000. "Media Framing and the Dynamics of Racial Policy Preferences." *American Journal of Political Science* 44(2): 245-60.

Kinder, Donald R., and Lynn M. Sanders. 1990. "Mimicking Political Debate with Survey Questions: The Case of White Opinion on Affirmative Action for Blacks." *Social Cognition* 8(1): 73-103.

McGuire, William J. 1985. "Attitudes and Attitude Change." In Gardner Lindzey and Elliot Aronson (eds.), *The Handbook of Social Psychology*. fourth edition. Hillsdale, NJ: Erlbaum.
Mufson, Steven and Shear, Michael. April 30, 2010. "Gulf of Mexico Oil Spill Reaches Louisiana Coast; Drilling Plans Criticized Damage Could be Worse than Exxon Valdez." *Washington Post.*
Nelson, Thomas E., Zoe M. Oxley, and Rosalee A. Clawson. 1997. "Toward a Psychology of Framing Effects." *Political Behavior* 19(3): 221-46.
Nelson, Thomas E., Rosalee A. Clawson, and Zoe M. Oxley. 1997. "Media Framing of a Civil Liberties Conflict and Its Effect on Tolerance." *The American Political Science Review* 91(3): 567-83.
Nelson, Thomas E., and Zoe M. Oxley. 1999. "Issue Framing Effects on Belief Importance and Opinion." *The Journal of Politics* 61(4): 1040-67.
Picou, J. Steven. Jun 29, 2010. "The Gulf Oil Spill and Its Psychological Effects." http://www.suite101.com/.
Rowland, Kara and Seper, Jerry. June 2, 2010. "Feds Open Criminal Probe of Oil Spill; Obama, Holder Vow Legal Justice, Regulatory Reform." *Washington Times.*
Skoloff, Brian. Sunday, September 19, 2010. "Final oil well sealing is small comfort to suffering, anxious Gulf residents." *Associated Press.*
Terkildsen, Nayda, and Frauke Schnell. 1997. "How Media Frames Move Public Opinion: An Analysis of the Women's Movement." *Political Research Quarterly* 50(4): 879-900.
Terkildsen, Nayda, and David F. Damore. 1999. "The Dynamics of Racialized Media Coverage in Congressional Elections." *Journal of Politics* 61(3): 680-99.
Wan, Sara. April 24, 2010. "Search Discontinued for 11 Missing After Oil Rig Explosion in the Gulf of Mexico." *The Washington Post.*
Zaller, John R. 1992. *The Nature and Origins of Mass Opinion*. New York, NY: Cambridge University Press.

EDITORS

Ashraf M. Esmail is an Assistant Professor of Criminal Justice at Southern University in New Orleans. His research interests include urban/multicultural/peace education, family, cultural diversity, political sociology, criminology, social problems, and deviance. He is currently working on two books on Racial Identity and Multicultural Education. He currently serves as the senior editor of the Journal of Education and Social Justice. He served as the conference chair for the National Association of Multicultural Education 18th Annual International Conference in New Orleans in 2008. He served as Program Chair for the 20th Annual Conference in Las Vegas in 2010 and will serve as chair for the 2011 conference in Chicago. He also was elected as President of the National Association for Peace Education in 2010.

Lisa A. Eargle is a Professor of Sociology at Francis Marion University in Florence, SC. Her research focuses on a variety of issues, including development, environment, social justice and education. Recent publications include "Disaster Impacts on Education: Hurricane Katrina and the Adaptation and Recovery of New Orleans Area Colleges and Universities," a chapter for the second edition of *The Sociology of Katrina: Perspectives on a Modern Catastrophe* (published by Rowman & Littlefield), and "Economic Reforms, Values, and Politics as Determinants of CrossNational Sex Trafficking Network in Developing Countries: A Theoretical Overture Using Global Commodity Chain Approach" (published by the Journal of *Race, Class And Gender*). She is currently working on two books, Environmental Sociology and Disasters in Society textbooks. Courses taught by her include Urban Sociology, Environment, Population, Criminal Organizations, Criminology, Social Problems, and Quantitative Research Methods. Dr. Eargle has also conducted workshops at major conferences which addressed the use of Geographic Information Systems in teaching and research, and how to integrate Environmental Issues into the Sociology Curriculum.

CONTRIBUTORS

A.E. Luloff is Professor of Rural Sociology, Department of Agricultural Economics and Rural Sociology, Penn State. Luloff teaches, conducts research, and writes about the impacts of rapid social change, as a result of demographic shifts, on the natural and human resource bases of community. Changes in land cover and use, particularly in areas at the ruralurban fringe, and the impact of rural development policy on small and rural communities have been central to his work. He is cofounder and current Executive Director of the International Association for Society and Natural Resources (IASNR) and former editor of *The Journal of the Community Development Society of America*. Some of his major publications include: *American Rural Communities: Trends and Prospects* (coedited with L. Swanson); *Persistence and Change in Rural Communities A Fifty Year Followup to Six Classic Studies* (co-edited with R. Krannich); and *People, Places and Landscapes: Social Change in High Amenity Rural Areas* (forthcoming and coauthored with R. Krannich and D. Field).

Amanda K. Goddard is a graduate student at Western Michigan University in Kalamazoo, where she graduated received a B.A. with a major in organizational communication. Her primary research interests are risk and crisis communication, organizational crisis, and organizational learning.

Brandon E. Fleming is a Research Assistant in the Liberal Arts and Sciences institute for Research and Community Service. He has worked for mentoring agencies and afterschool programs to help "at risk" children's pathways to success.

Daina Cheyenne Harvey is a doctoral candidate in sociology at Rutgers University. He specializes in the areas of culture, cognition, marginality, and racial inequality. His dissertation focuses on how people move on from large scale instances of social disruption. In this work he asks what do people do with disruption and how do disruptions effect the way we see and understand society and social relations. In particular, he is looking at how residents of the Lower Ninth ward in New Orleans are dealing with the secondary disruptions stemming from Hurricane Katrina, especially the cultural and cognitive practices that help them move on from the largest disruption in recent US history.

Dee Wood Harper is Professor of Sociology and Criminal Justice at Loyola University New Orleans. He earned a BA from Vanderbilt University and the MA and PhD from Louisiana State University. He was a National Institute of Mental Health postdoctoral fellow in social psychiatry at Tulane University School of Medicine before joining the faculty at Loyola. His scholarly research has been wide ranging but currently focuses on tourism and crime, violence and the criminology of disaster. His research has appeared in the *American Journal of Sociology, American Journal of Economics and Sociology, Sociological Symposium, Annals of Tourism Research, International Journal of Law and Information Technology, Artificial Intelligence and Applications, Proceedings: Artificial Intelligence Applications Conference, Criminal Justice Review, Homicide Studies, Deviant Behavior* and *Police Practice and Research: An International Journal.* In 2010, Dr. Harper co-edited and authored several chapters in the first book on the subject of disasters and crime, *Crime and Criminal Justice in Disaster*, published by Carolina Academic Press.

DeMond Shondell Miller is a Professor of Sociology and Environmental Studies and Director of the Liberal Arts and Sciences Institute for Research and Community Service at Rowan University (Glassboro, New Jersey). He has worked as principal investigator to facilitate research projects involving natural and technological disasters, environmental issues, and community satisfaction. His primary area of specialization is environmental sociology (disaster studies and the study of the social construction of place), community development and community organizing, and social impact assessment. Dr. Miller has presented and published several professional papers; recent examples of such work can be found in *Space and Culture: An International Journal of Social Spaces, Journal of Black Studies, Journal of Public Management and Social Policy, Sociological Spectrum,* and *The International Journal of Culture, Tourism and Hospitality Research, Environmental Justice*. Recently, he has contributed to several edited volumes including *Through the Eye of Katrina: Social Justice in the United States* and *The Sociology of Katrina: Perspectives on a Modern Catastrophe*; he is the coauthor of *Hurricane Katrina and the Redefinition of Landscape* with Jason D. Rivera and co-editor of *African American and Community Engagement in Higher Education: Community Service, Service Learning and Community-Based Research* with S. Evans, C. Taylor, and M. Dunlap, *How Ethnically Marginalized Americans Cope with Catastrophic Disaster: Studies in Suffering and Resiliency* with Jason D. Rivera and *Community Disaster Recovery and Resiliency: Exploring Global Opportunities and Challenges* with Jason D. Rivera.

James Hawdon is a Professor of Sociology and Director of Graduate Studies at Virginia Tech. His research interests center on issues of community, including how communities try to control crime and how communities respond to mass tragedies, disasters and critical incidents. He has published in the areas of criminology, deviant behavior, the sociology of drugs, the sociology of policing, and in disaster research / critical incident analysis. He is a member of the Academy of Critical Incident Analysis at John Jay College of Criminal Justice, and he serves on the editorial board of the *Journal of Critical Incident Analysis*.

Jas M. Sullivan is an Assistant Professor of Political Science and African and African American Studies at Louisiana State University. His research and teaching focus on African-American politics, political psychology, racial identity and stereotypes, and education policy.

Jason David Rivera is a Research Associate in the William J. Hughes Center for Public Policy at The Richard Stockton College of New Jersey. His research focuses on social vulnerability to natural and manmade disasters with an emphasis on minority experiences. Additionally, his research highlights institutional structures that have historically perpetuated social vulnerability within minority and low-income communities. His research findings have been incorporated into policy recommendations that make mitigation, response, and recovery more effective. Examples of his work can be found in the *Journal of Black Studies*, *Journal of Applied Security Research: Prevention and Response in Asset Protection, Terrorism and Violence*, *Journal of Public Management and Social Policy*, *Sociological Spectrum*, *The Journal for the Study of Radicalism*, *Space and Culture*, *The Sociology of Katrina: Perspectives on a Modern Catastrophe*, *Through the Eye of the Storm: Social Justice in the United States*, and *Dangers in the Incommensurability of Globalization: SocioPolitical Volatilities*. He is coauthor of *Hurricane Katrina and the Redefinition of Landscape*, with DeMond S. Miller and coeditor of *How Ethnically Marginalized Americans Cope with Catastrophic Disaster: Studies in Suffering and Resiliency* with DeMond S. Miller and *Community Disaster Recovery and Resiliency: Exploring Global Opportunities and Challenges* with DeMond S. Miller.

Jason S. Gordon is an Assistant Professor of Extension in the Department of Forestry at Mississippi State University. With a background in rural sociology and forest resources, his extension programming focuses on assisting rural and urban communities improve their well-being through sustainable natural resource management. Gordon's research concerns community and natural resources in the contexts of community organization and development, environmental risk perceptions, and under-served stakeholders. All of his work seeks to explore the intersection between nature and community agency. His research has addressed land use and conservation, hazards and risk, forest recreation, bio-fuels, private forest landowners, and under-served stakeholders. His work has been published in *Society and Natural Resources* and *Rural Sociology*.

Jeffrey R. Wozniak is an ecosystem ecologist who has utilized wetlands as model systems to address research questions regarding the bio-geochemical cycling of nutrients, ecosystemlevel responses to natural and anthropogenic drivers, and the importance of habitat management in restoring and sustaining ecosystem goods and services. Dr. Wozniak has conducted aquatic research in a wide range of systems include subalpine wetlands in Colorado, lake systems in Pennsylvania, freshwater marshes of the Florida Everglades and salt marshes along the Texas Gulf Coast. He is currently a Research Fellow at the Texas Research Institute for Environmental Studies at Sam Houston State University in Huntsville, Texas.

John Barnshaw is a PostDoctoral Fellow in the Department of Sociology at the University of South Florida where his research focuses on the cultures and consequences of risk in a variety of contexts ranging from infectious disease, to disasters, to financial crises. Dr. Barnshaw received his Ph.D. in Sociology from the University of Delaware, where he was Projects Coordinator at the Disaster Research Center (2006-2009) and a University Dissertation Fellow (2009-2010). Currently, his research blends behavioral finance, public policy and social problems theorizing to explore how financial risk taking led to the recent economic crisis. Dr. Barnshaw also has ongoing research in understanding how disasters exacerbate social inequality over time at the individual, organizational and societal levels. His research has been recently published in *Health Sociology Review, Social Forces, Annals of the American Academy of Political and Social Science,* and *Contemporary Sociology.*

John Ryan is Professor and Chair of Sociology at Virginia Tech where he teaches the Sociology of Law. His research interests include the study of culture production and consumption, critical incident analysis, and violence and crime control within communities. One line of research has been to examine the relative effectiveness of violence intervention programs that focus on structural issues (for example, enforcing housing codes) versus those that focus on cultural issues (for example, parenting or self-esteem classes). He has also researched dimensions of community solidarity, the effectiveness of community policing, and the relationship among grassroots neighborhood anti-crime organizations, community policing, social disorganization and crime. He is former Chair of the Social Science Panel of the Ford Foundation Minority Fellowship Program.

Jude Clemente is an energy security analyst and technical writer in the Homeland Security Department at San Diego State University. He holds a B.A. in Political Science from Penn State University and an M.S. in Homeland Security from SDSU. He also holds certificates in infrastructure protection and emergency preparedness from the Federal Emergency Management Agency, the American Red Cross, and the U.S. Department of Homeland Security. His research specialization is energy security at the national and international level. His most recent work has appeared in Oil and Gas Journal, Pipeline and Gas Journal, Capture Carbon Journal, and The Journal of Energy Security. He is a member of the National Defense Transportation Association and energy adviser to B and S Corporation and Penn State's Research Project on North America Energy Supply.

Kelly Frailing earned her BA in psychology from Grinnell College and her master's in criminal justice from Loyola University New Orleans. She recently received her PhD from the University of Cambridge Institute of Criminology where she studied mental health courts. She has published several articles on mental health courts, including in the *International Journal of Law and Psychiatry* and the *Cambridge Student Law Review*. She is also interested in disasters and crime and has published numerous articles and book chapters on her research in this area, including in the *Natural Hazards Observer*, the *American Journal of Economics and Sociology* and in both editions of *The Sociology of Katrina: Perspectives on a Modern Catastrophe*. Kelly also co-edited and contributed a number of chapters to the first book on this topic, entitled *Crime and Criminal Justice in Disaster*, published by Carolina Academic Press.

Kenneth Lachlan (PhD, Michigan State University, 2003) is Associate Professor and Director of the Communication Studies program at the University of Massachusetts Boston, where he is also a research affiliate with the Center for Rebuilding Sustainable Communities After Disasters. He also serves as an investigator for the National Center for Food Protection and Defense. His research focuses on psychological responses to mediated information, with a particular emphasis on crisis and risk communication campaigns. Past and ongoing projects include evaluating audience perceptions of emergency messages, evaluation of audience informational needs, media dependencies during crises and emergencies, and the use of media as a coping mechanism in the aftermath of a major event. Other projects have examined the use and function of varying media amongst different sub-populations, with a particular focus on deviations between mainstream and historically under-served groups in terms of how well their needs are met by crisis communication campaigns. At present, he is also field testing a persuasion model for use under conditions of extreme psychological duress that may be able to help emergency communication efforts manage public response. He has published his research in journals such as *Journal of Broadcasting and Electronic Media, Media Psychology, Journal of Applied Communication Research, Journal of Emergency Management, Behavioral Sciences of Terrorism and Political Aggression,* and *American Journal of Public Health*. He is also the coauthor of Straight Talk about Communication Research Methods (2010, Kendall-Hunt) and serves as a consultant for businesses and nonprofit organizations in the Boston area.

Lori Peek is an assistant professor of Sociology and co-director of the Center for Disaster and Risk Analysis at Colorado State University. Dr. Peek also serves as the Associate Chair for the Social Science Research Council Task Force on Hurricane Katrina and Rebuilding the Gulf Coast. Her research focuses on the intersection of social inequality, vulnerability, and disasters. She is the author of *Behind the Backlash: Muslim Americans after 9/11* (Temple University Press, 2010) and coeditor of *Displaced: Life in the Katrina Diaspora* (University of Texas Press, 2012). In 2009, the American Sociological Association Section on Children and Youth honored her with the Early Career Award for Outstanding Scholarship. Dr. Peek teaches classes on race and ethnic relations, the sociology of disaster, and qualitative research methods. Since arriving at Colorado State, she has been named the Greek Life Professor of the Year and has received the Alumni Association Best Teacher Award, the College of Liberal Arts Excellence in Teaching Award, and the Waterpik Excellence in Education Award.

Lynn Letukas is a Doctoral student in the Department of Sociology at the University of Delaware. Her current research focuses broadly on social problems construction in areas of political discourse, public policy, globalization and risk. Letukas has been involved in a variety of research projects ranging from the evacuation of lower Manhattan following the September 11, 2001 attacks on the World Trade Center to supply chain and convergence in New Orleans following Hurricane Katrina, as well as the handling mass causalities following the 2004 Indian Ocean tsunami. Letukas has presented at numerous national and regional conferences across the United States and has given lectures in sociology, geography and education.

Meghan S. Sanders is an Assistant Professor at the Manship School of Mass Communication at Louisiana State University. Since 2001 she has conducted experimental and survey research specializing in the impact of media stereotypes, and cognitive processing of and emotional responses to entertainment media. She received her Ph.D. in mass communication from The Pennsylvania State University.

Michelle Meyer Lueck is a Ph.D. candidate in Sociology at Colorado State University focusing on the sociology of disasters and environmental sociology. Her research interests center on the interplay between environmental conditions and social vulnerability, including how to foster positive environmental behavior and increase disaster resilience and mitigation. Michelle is a research assistant with the Center for Disaster and Risk Analysis at CSU, where she is involved in a National Science Foundation and National Oceanic and Atmospheric Administration funded project on hurricane risk perception. Her dissertation research focuses on the intersection of individual and community hurricane resilience in terms of social networks and collective efficacy along the U.S. Gulf and Atlantic Coasts. Michelle has worked on various research projects including government disaster response planning, organizational energy conservation, nonprofit program evaluation, segregation and environmental justice issues, undergraduate career expectations, social network analysis, and environmental attitudes and behaviors. Her teaching interests include social research methods, social statistics, stratification, environmental sociology, and sociology of disasters.

Patric R. Spence is Assistant Professor in the School of Communication at Western Michigan University. He is also a research affiliate with the Center for Rebuilding Sustainable Communities After Disasters and also serves as an investigator for the National Center for Food Protection and Defense. His research focuses crisis and risk communication, examining audience perceptions of risk and emergency messages produced by emergency management, organizations, government and news agencies. Specifically, looking at how these messages motivate the public to take action in light of perceived threats during the life-cycle of a crisis. Other research examines how the physical and psychological needs of under-served populations are handled in the context of public health events and disasters; industry response, and the role of new media in disaster preparation, response and recovery. He has written widely on the issues of race and class surrounding Hurricane Katrina and issues of gender and information seeking in disasters. His research has recently been cited in the National Consensus Statement on Integrating Racially and Ethnically Diverse Communities into Public Health Emergency Preparedness, released by the Office of Minority Health, in the U.S. Department of Health and Human Services. Recent work has been published in *New Directions for Evaluation*, communication Research Reports, *Journal of Applied Statistical Methods*, the *Journal of Emergency Management*, *Journal of Health Care for the Poor and Underserved*, and *Risk analysis*. Recent book chapters can be found in *Through the Eye of Katrina: Social Justice in the United States* (Carolina Academic Press), *Minority Resiliency and the Legacy of Disaster* (The Edwin Mellen Press) and *Interracial Communication: Contexts, Communities and Choices* (Kendal/Hunt).

Shyamal Das earned his PhD from Southern Illinois University at Carbondale. Currently he is an Associate Professor and the Coordinator of the Sociology Program at Elizabeth City State University in North Carolina. Dr. Das has published numerous articles on economic development, women's empowerment and sex work, sex trafficking, biosocial explanations of criminality, men's promiscuity, inertia of culture, disaster, etc. His articles appear in various journals, including the *Journals of Social Studies, Personality and Individual Differences, Sexual Archives, Sociological Bulletin, Natural Sciences*, and *Race, Class, and Gender*. Currently, he is involved in research on development, Islamism, and corruption

Stan C. Weeber is an Associate Professor of Sociology and Criminal Justice at McNeese State University in Lake Charles, Louisiana. His interests in sociology include political sociology, social movements, sociological theory, and crime/deviance. The author or editor of eighteen books, his work has appeared in

The American Sociologist, *The Sociological Quarterly*, the *Journal of Public Management and Social Policy*, *International Review of Modern Sociology*, the *Canadian Review of Sociology and Anthropology*, the *Journal of Popular Culture*, the *Journal of Global Analysis*, and several other journals. In addition, Dr. Weeber serves on the editorial boards of four international sociology journals. In 2010, he participated in the Oxford Roundtable on Social Justice, a social issues think tank at Oxford University.

Stephen Zavestoski is an Associate Professor of Sociology and Environmental Studies at the University of San Francisco, is at work on a book entitled "Growing up Toxic in the Age of Globalization: A Social Movement's Struggle for Justice Following the Bhopal Chemical Disaster," which draws on his research examining the mobilization of transnational social movement networks in response to the increased risk of industrial hazards brought on by globalization. He is also the co-editor, with Phil Brown, of *Social Movements in Health* (2005, by Blackwell). His research areas include environmental sociology, social movements, and sociology of health and illness. Dr. Zavestoski's previous research has also covered topics such as ecological identity, consumerism, and the effects of the Internet on public participation in environmental regulatory rule-making processes.

Timothy J. Haney is an Assistant Professor of Sociology at Mount Royal University in Calgary, Alberta. Before finishing his Ph.D. at the University of Oregon, he was a graduate student at Tulane University, and from 2003 to 2006, he was a resident of New Orleans. Experiencing Hurricane Katrina sparked his interests in the sociology of disaster and environmental justice, particularly applied to the Gulf Coast region. His current research involves neighborhood differences in response to and recovery from Hurricane Katrina, the long-term recovery of coastal communities following the oil spill, as well as the intersections of gender, space, and employment in post-welfare reform U.S. cities.

Tomás Mac Sheoin is an independent scholar who writes on the chemical industry and social movements. He has edited a special issue of *Global Social Policy* (vol. 9, no. 3, December 2009) to mark the twenty-fifth anniversary of the Bhopal disaster, written on the common factors between Bhopal and the BP Texas City refinery accident (*Monthly Review* Vol.62, no. 4, pp.21-33, August 2010) and written a book on the Asian chemical industry, *Asphyxiating Asia* (Goa: Other India Press, 2003).

www.ingramcontent.com/pod-product-compliance
Ingram Content Group UK Ltd.
Pitfield, Milton Keynes, MK11 3LW, UK
UKHW020127290926
15471UKWH00003B/328

9 780761 859789